Palgrave Studies in the History of Science and Technology

Designed to bridge the gap between the history of science and the history of technology, this series publishes the best new work by promising and accomplished authors in both areas. In particular, it offers historical perspectives on issues of current and ongoing concern, provides international and global perspectives on scientific issues, and encourages productive communication between historians and practicing scientists.

Charles J. Gross

Technological Innovation and the Rise of Aviation, 1903–1941

Wings for the American Military

palgrave
macmillan

Charles J. Gross
U.S. Air Force
Granville, OH, USA

ISSN 2730-972X ISSN 2730-9738 (electronic)
Palgrave Studies in the History of Science and Technology
ISBN 978-3-031-71107-7 ISBN 978-3-031-71108-4 (eBook)
https://doi.org/10.1007/978-3-031-71108-4

Cover credit: Naval History and Heritage Command

This Palgrave Macmillan imprint is published by the registered company Springer Nature Switzerland AG
The registered company address is: Gewerbestrasse 11, 6330 Cham, Switzerland

Acknowledgements

I am deeply indebted to my colleagues and friends who have contributed their expertise and valuable time to this project, especially Dr. Roger D. Launuis, Dr. John D. Sherwood, Dr. Daniel Harrington and Professor Allan R. Millett. The assistance of archivists and librarians (and their web sites) at many organizations was also invaluable. They included the: National Archives; Manuscript Division of the Library of Congress; Herbert Hoover Presidential Library and Museum; Franklin D. Roosevelt Presidential Library and Museum; Harry S. Truman Presidential Library and Museum; Nimitz Library at the U.S. Naval Academy; U.S. Air Force Academy Library; Pentagon Library; Ann Arundel County, Maryland Public Library; Granville, Ohio Public Library; Naval History and Heritage Command; Air Force Historical Research Agency; National Museum of the United States Air Force; Air Force Materiel Command Office of History; Air Force Historical Support Division; Smithsonian National Air and Space Museum; U.S. Army Military History Institute; Hathi Trust Digital Library; Internet Archive; Aerospace Industries Association of America web site; University of North Texas Libraries; Defense Technical Information Center; and Wright State University Libraries.

Finally, I am indebted to my most loyal supporter, my wife Barbara. Her patient encouragement over many years aided me immeasurably in completing this work.

Granville, Ohio — Charles J. Gross

Praise for *Technological Innovation and the Rise of Aviation, 1903–1941*

"This book skilfully combines information from existing studies with considerable primary source research to fill a longstanding gap in our understanding of the institutional, political, and technological history of American military aviation."

—Alan D. Meyer, *Associate Professor, Department of History, Auburn University*

"Charles J. Gross is an accomplished American air power historian, having authored the definitive one-volume history of US airpower entitled American Military Aviation: The Indispensable Arm. In this new book, he examines the pre-World War II history of US airpower, looking closely at the innovative research, development and procurement programs conducted by the nation's armed forces and the National Advisory Committee for Aeronautics in conjunction with industry during the interwar period. This book reveals how American aviation achieved global pre-eminence by leveraging innovation and continuous improvement. This is a seminal work and should be required reading for current airpower leaders and policymakers grappling with similar issues today."

—John Darrell Sherwood, *Naval Historian*

Contents

List of Figures

List of Tables

CHAPTER 1

Introduction

Just as the Japanese surrender ceremony was ending aboard the battleship USS *Missouri* on 2 September 1945, the cloudy skies above Tokyo Bay opened up to a massive display of America's air power. As planned by allied commander General of the Army Douglas MacArthur, some 465 Army Air Forces B-29 *Superfortress* bombers had flown from their bases in the Mariana Islands followed 450 carrier-based Navy fighter aircraft embarked on Task Force 38 carriers steaming offshore roared over the armada of 258 allied warships gathered in that harbor. Inspired by the ancient Roman practice of making defeated enemies "pass under the yoke" and his own public relations instincts, MacArthur had choreographed an enormous pageant compared to which "... a royal wedding would have looked like a middle-school play."[1] Writing in *Air & Space Magazine*, author Cory Graff observed that "MacArthur wanted as many airplanes as possible, a demonstration of overwhelming force that would be an admonition designed to quell any slight hope of resistance. It also served as a warning to all other nations that the U.S. was now the world's sole superpower."[2] The massive flyover demonstrated the enormous technological and quantitative progress American military aviation had made after the Germans overran Poland in September 1939.

American aviation, whose scientific and technological innovations were largely financed by the U.S. armed forces and the National Advisory

C. J. Gross, *Technological Innovation and the Rise of Aviation, 1903–1941*, Palgrave Studies in the History of Science and Technology,
https://doi.org/10.1007/978-3-031-71108-4_1

Committee for Aeronautics (NACA) during the interwar period, achieved a global preeminence during war which it has retained to the present day. How and why did that happen? This volume addresses how the American forces, working with NACA and selected firms in the aircraft industry promoted scientific and technological innovation in aviation to advance their military objectives albeit sometimes haltingly. They were aided immensely in that effort by foreign technologists and technologies. Their sometimes hit or miss efforts created a world-class technological/industrial base in aviation which, along massive federal spending and strong political support, enabled the U.S. to dominate the skies during World War II and beyond.[3]

The results of that process were never predetermined or inevitable. There was no master plan that promoted the final outcome. Some projects, notably the Army Air Service's *Barling* bomber in the 1920s, were expensive failures. Pure chance played a role sometimes. The P-51 *Mustang* achieved its iconic status largely because an English test pilot suggested replacing its original power plant with the British Merlin Rolls-Royce engine. Lighter-than-air vehicles proved to be very expensive mixed bags in the long run. While dirigibles performed valuable service during World War II, large rigid airships like USS *Shenandoah* were costly technological dead ends and operational failures. Those vehicles are beyond the scope of this book.

Although there is growing wealth of published scholarship about U.S. military aviation before World War II, most of it focuses on operational matters, individual services, specific campaigns, industry, famous individuals including Brig Gen William L."Billy" Mitchell or significant episodes like World War I. While this work covers some of the same ground as Irving B. Holley, Jr.'s highly regarded 1964 volume, *Buying Aircraft: Materiel Procurement for the Army Air Forces. Special Studies. U.S. Army in World War II,* the latter focuses on the Army Air Forces' aircraft materiel procurement policies and practices during World War II. Holley's other major pioneering work, *Ideas and Weapons: Exploitation of the Aerial Weapon by the United States in World War I: A Study in the Relationship of Technological Advance, Military Doctrine, and the Development of Weapons*, explored how America mishandled the development of Army aviation during World War I. This work concentrates on the critical role that the Army and Navy air arms, working with the aviation industry, NACA and engine makers, played in developing aviation technology as well as effective procurement policies and practices from the Wright brothers first flight to the Japanese attack on Pearl Harbor.

Although early progress in American aeronautics has been associated in the popular mind with skilled individuals like the Wright brothers, solitary individuals had a limited impact. Historian Roger D. Launius emphasizes that "The kind of lone wolves that make up the folklore, reinforced by the reality of a few bona fide geniuses are rare indeed. In the twentieth-century aeronautical engineering the increasing depth of information in the individual disciplines ensures that no one person can now master the multifarious skills necessary in the research, design, development, and building of a piece of aerospace hardware."[4]

The underlying goal of those military aviation programs was to field ever more reliable aircraft that flew faster, higher and further while carrying heavier ordnance loads. The innovative design activities undertaken by those entities, aided by foreign technologies as well as highly skilled immigrant scientists and engineers, were primarily responsible for fostering an aeronautical revolution during the interwar period. To a lesser extent, the advent of viable commercial airliners in the U.S. during the 1930s aided that process.

During World War II, the U.S. produced more military aircraft than any of the other major combatants. It also built progressively larger, more complex, technically advanced and increasingly capable aircraft as the conflict roared on. Having been well behind other major powers when the war began in Europe, America fielded the war's most powerful global air armada and supported those of its allies. With the exception of jets, the quality of its aircraft equaled or surpassed those of other major combatants in virtually every other major category. How had the U.S. come so far so fast? Conventional thinking has emphasized the importance of national political and economic mobilization during the war years including close cooperation between political, military, technical and business elites. Those factors were critical to the success of the massive American wartime aviation program but weren't sufficient to fully explain it.

During World War I, the U.S. had mounted a huge and costly military aviation program almost from scratch. It had fallen well short of the federal government's ambitious objectives because the nation lacked adequate technical, managerial and industrial base for the rapidly innovating world of European combat aviation during the conflict. But, that base and the innovative culture which sustained it was firmly in place when the U.S. began preparing for a possible global war in the late 1930s. Those early years of American military aviation suggest that the armed forces were much more important in the history of American aviation than has generally been recognized. Some senior officers in the Army and Navy as well as Congress and officials in the executive branch were much more supportive of aviation than either advocates of an independent air force or a separate naval flying corps understood or were willing to admit. Numerous high-profile government boards and commissions tackled the issue of its place in the nation's military posture after the November 1918 Armistice. In an era marked by disenchantment with the results of the Great War, major cuts in military spending and a unilateralist foreign policy, the armed forces of the U.S. devoted substantial resources to their fledgling air arms and the development of the technological and industrial base needed to sustain them. Military aviation was neither ignored nor under-appreciated during the interwar period. The core issue was how it was going to be controlled for what purposes and how much was to be spent on it out of constrained federal military budgets.

This history examines the sometimes reluctant efforts of the military services to promote the technical and industrial foundations of American aviation prior to the Japanese attack on Pearl Harbor. Although it discusses hardware, this work focuses upon the innovative ideas, individuals, institutions, technological advances and politics that shaped the development of aviation in the American armed forces. While their approach to innovation was incremental in the short run, it produced revolutionary improvements in aircraft in the long term.

World War I had underscored the multiple roles aviation could play in conjunction with conventional ground and naval surface forces as well as the importance of technological superiority to the victor, especially in air-to-air combat. That lesson was not lost on a handful of American Army and Navy flying officers. Along with their civilian allies in government, industry and the scientific community, they worked actively to develop aviation as a vital combat arm in their respective services while vigorously promoting technological innovation and a strong industrial base for American aviation. During the 1920s, the U.S. fielded some of the most advanced military aircraft in the world gauged by the international performance records they held. However, that situation changed after Japan and the major European powers launched major rearmament programs.

The scientific, technological and industrial base which had made this possible had been deliberately promoted as a matter of policy by the Army and Navy in cooperation with their civilian allies. Military funding plus the close cooperation that had existed between the armed forces and a loose coalition which included the NACA, universities and the aviation industry as well as key political figures were critical to the massive surge in U.S. military aviation program which helped destroy the Axis war machine and establish the global preeminence of the nation's commercial aviation. More often than it is generally understood aviation and aeronautical innovation had been strongly supported by senior non-flying officers including Rear Adm David W. Taylor, the head of the Navy's Bureau of Construction and Repair (BuC&R) which ran the service's shipbuilding and aircraft construction programs during World War I.

World War I established the modern American aviation industry based on science and engineering. A massive aviation program was touted as the quickest way for the U.S. to win the war. It was part of an ongoing pattern of overselling the potential impact of emerging technologies. The war saw the creation of the Army's Aviation Engineering Division at McCook Field in Dayton, Ohio and the Naval Aircraft Factory (NAF) in Philadelphia, Pennsylvania. Those organizations played key roles in the development of aviation technology and its application to a variety of military needs during the interwar period. The conflict convinced a dedicated cadre of military officers, and some key civilian officials that the airplane was destined to play a key role in future wars.

Following the 1918 Armistice, the U.S. quickly demobilized the majority of its armed forces, including most of its fledgling air forces. But, the years that immediately followed saw the passage of critical legislation that put military aviation on a solid footing. They also featured the emergence of federal military and civilian organizations which served as major catalysts for aeronautical progress in the U.S. However, during that period, public opinion and official policy rejected both a large peacetime military establishment and the radical idea that the nation's defense should be based primarily on air power. Unable to secure a separate air force and denied enough funds to build a formidable standing air arm, Maj Gen Mason Patrick, Chief of the Air Service (later Air Corps), concentrated instead on developing the technical, industrial and commercial foundations of American aviation. His naval counterpart, Rear Adm William A. Moffett, the first head of the Navy's Bureau of Aeronautics (BuAer) formed in 1921, skillfully dealt with major challenges posed by battleship-oriented admirals, naval fliers who wanted a separate aviation corps akin to the Marines and the Air Service's "Stormy Petrel," Brig Gen Mitchell, who led the unsuccessful charge for a separate air force.

Moffett, like Patrick, was a skilled military insider and bureaucratic political operator. He was instrumental in developing a strong naval aviation program well integrated into the fleet. Both senior officers worked closely with the NACA, their respective service technical organizations, industry and Congress as well as senior Army and Navy leaders to advance their respective aviation programs. Stimulated primarily by government-sponsored research and development programs, American aeronautical technology advanced rapidly.

The late 1920s featured the beginning of a crucial transition for American (and international) aviation. That "Golden Age" of innovation witnessed the origins of the airframe revolution that produced modern (pre-jet) civilian and military aircraft. Slow, box-like biplanes made primarily of wood, cloth and wire were replaced by sleek, all-metal monoplanes with much more powerful and reliable engines. While the armed services continued to actively promote aeronautical research and development during this period, more emphasis was placed on procurement of small numbers of more advanced aircraft for operational use with the fleet and ground forces. That change was driven primarily by the increased availability of funds due to aircraft modernization and expansion programs approved by Congress in 1926. In concert with a loose coalition of scientists and engineers, businessmen and politicians, senior aviators in the Army and Navy evolved materiel organizations and acquisition practices designed to minimize the impact of idiosyncratic senior officers like Mitchell and outdated federal procurement regulations while fielding some of the most capable warplanes in the world.

During the 1930s, Army and Navy fliers continued struggling to overcome the challenges imposed by public isolationism and pacifism, conservative military doctrines as well as ineffective government procurement practices and initial federal spending cuts in response to the Great Depression to lay the foundations of America's massive World War II aviation program. With the development of Boeing's B-17, airmen like Brig Gen Henry H. "Hap" Arnold believed they at last had acquired the weapon that could do the job although their Army superiors continued to insist that substantial resources had to be poured into aircraft designed to support the ground forces. Like their naval counterparts, senior airmen learned to work within the Army bureaucracy and the American political system to achieve incremental progress toward their ultimate goal of a separate air force by demonstrating the military, civil support and diplomatic utility of aviation.

The Army and Navy remained the largest and most important customers of the American aviation industry and the NACA during the interwar period. They continued to be aided by aeronautical technology imported from overseas and the significant contributions of immigrant scientists and engineers. Much of the technology developed for the armed services during that period was incorporated in a revolutionary generation of new American commercial airliners epitomized by the iconic, all-metal streamlined Douglas DC-3 airliner. In turn, some of the technological innovations featured in modern, pre-jet commercial airliners often were incorporated in military aircraft. However, little of that technological progress could have happened nearly as rapidly as it did during the interwar period without the large infusion of military funding that enabled a few aircraft and engine firms to maintain large and expensive technical design staffs.

Despite the deteriorating security situation in Asia and Europe, the U.S. was unwilling to embark on a massive rearmament program until the late 1930s. Consequently, the performance of its warplanes which had been among the best in the world early in that decade, had fallen well behind by September 1939 when Germany ignited World War II in Europe. The aviation industry, along with the rest of the massive American economy, was mobilized to conduct a global air war against the Axis. Until that period, military aviation procurement policies and practices as well as tight funding had drastically limited the number of firms that could effectively compete for those contracts.

Army and Navy aviation modernization programs during the interwar period played a critical role in developing the technological and industrial foundations of American air power in World War II despite the fact that the nation's standing air strength had been well behind those of the other major belligerents when President Roosevelt asked Congress for a declaration of war against Japan. Before the conflict ended, the U.S. outproduced all of those other nations. With the exception of jet fighters, its most advanced military aircraft equaled or surpassed those of other powers in virtually every category.

Although a politically unified nation, massive federal government spending and industrial mobilization on an unprecedented scale explained much of that success, those critical factors were not enough to explain the wartime dominance of American aviation. Unlike World War I, the U.S. had several years to begin building up its air power before the attack on Pearl Harbor. Foreign military aircraft orders played a key role in accelerating aircraft production in America. Moreover, without the innovative technological and aviation industry foundations established during the interwar period with the critical support of the armed forces, it is likely that the massive American aviation program wouldn't have been able to play a critical role in allied victory in World War II. That innovative process continued through the war as ever more capable versions of pre-war aircraft and designs entered the inventories of the nation's armed forces. It set the stage for the postwar international dominance of American military and commercial aviation that continues to the present day.

Given the economic, political and strategic circumstances of the interwar period, American military aviation programs, aided by an infusion of foreign talent and ideas, performed extremely well. Criticizing them as inadequate because their leaders failed to foresee the full implications of revolutionary technologies including radar and jet propulsion is unwarranted. Moreover, complaints about inadequate aviation budgets as well as the hostility (or ignorance) of the Navy's battleship-oriented "gun club" and the Army's ground-focused General Staff miss the mark. In many respects the period between World War I and World War II was a "Golden Age" of American aeronautical innovation.

Notes

1. Cory Graff, "Final Mission: Staging Japan's Surrender," *Air & Space Magazine*, September 2020, smithsonianmag.com; "Formal Surrender of Japan, 2 September 1945—Aircraft Flyover as the Ceremonies Conclude," Naval Historical Center, links revised 7 March 1999, ibiblio.org/hyperwar/OnlineLibrary.
2. Graff, "Final Mission: Staging Japan's Surrender," *Air & Space Magazine*, September 2020, smithsonianmag.com.
3. Ibid.
4. Roger D. Launius, ed., *Innovation and the Development of Flight* (College Station, TX: Texas A&M University Press, 1999), 5.

CHAPTER 2

The Origins of American Military Aeronautics, 1903–1917

"You can't convince me that thing will fly" declared Congressman Joseph G. "Uncle Joe" Cannon, the powerful Speaker of the House of Representatives. He was observing preparations for the acceptance trials of the Wright brothers' first military aircraft during July 1909 at Fort Myer, Virginia.[1] His skepticism was widespread. Who could believe that an obscure pair of bicycle mechanics and tireless tinkerers from Dayton, Ohio had mastered a complex set of technical problems whose solutions had eluded trained scientists and engineers for years. The secretiveness of the Wrights—they had kept their invention under wraps for several years after Orville's historic first flight in December 1903 while trying to perfect it and secure a patent—contributed to widespread doubts about their claims.[2]

Reluctant Patrons: The Armed Forces

Like other technological pioneers, the Wrights believed their invention would usher in a new era of international peace and prosperity. But, failing to find private funders for promoting peaceful commercial applications of aviation, they turned to the military. Utopian dreams were out, a pragmatic search for cash was in. However, the U.S. armed forces proved to be reluctant patrons.[3]

C. J. Gross, *Technological Innovation and the Rise of Aviation, 1903–1941*, Palgrave Studies in the History of Science and Technology, https://doi.org/10.1007/978-3-031-71108-4_2

Meanwhile, by 1913 European technological aviation innovations had quickly outstripped American developments. They included the monocoque[4] fuselage, a completely enclosed airframe constructed from a stamped sheet metal developed by the French. American designs still featured wood, wire, and cloth open fuselages, exposing pilots and passengers to the weather. French manufacturer, Eugene Bleriot developed an aircraft operating system that remained standard for decades; foot pedal controls operating the airplane's rear rudder, and a central "joystick" for manipulating the ailerons that replaced the Wrights' wing-warping device in Europe as the standard control method.[5]

The Army had initially shown little interest in the Wrights' innovation. It was a conservative institution cautious about exploring radical new technologies. The service had been seriously embarrassed in the autumn of 1903 when Professor Samuel Pierpont Langley's experimental "Aerodrome," had twice burrowed into the muddy bottom of the Potomac River while attempting to fly.[6] The Army had invested $50,000 (approximately $1.5 million in 2020 dollars) in the project. Congress and the press complained loudly that the government had wasted public money on hair-brained schemes. The controversy convinced some senior officers that it should avoid risking its reputation and limited funds on such highly speculative ventures.[7]

The Navy, which had been dealing with revolutionary technological innovations since before the American Civil War, had an organizational culture which found it easier to promote technological innovations and incorporate them in its operational schemes than the Army. But, it had rebuffed a proposal backed by then Assistant Secretary Theodore Roosevelt and Navy Secretary John D. Long prior to the Spanish-American War to sponsor Professor Langley's research. The Navy's General Board[8] concluded that aviation should be the Army's business. During 1909, the Navy rejected a proposal by its own Bureau of Equipment to purchase two experimental aircraft on the grounds that the technology was still too primitive to be of any value to the fleet.[9]

Following their historic flight, the Wrights devoted two more years to conducting painstaking experiments to perfect their design.[10] Twice in 1905, they tried to persuade the federal government to buy their invention sight unseen. They were virtually ignored by the Army. Their unwillingness to reveal details about their invention before they had secured clear patent rights to it and the Army's continuing embarrassment over the Langley fiasco contributed to its reluctance to give them a full hearing.[11]

Army Airplane No. 1

The Army was finally compelled to take a serious look at the Wrights' invention by a civilian special interest group, the Aero Club of America, and pressures from within its own Signal Corps. Organized in 1905, the club consisted of wealthy individuals who were determined to promote aeronautics. Although hopeful about the commercial potential of flying, they believed that only the federal government would be willing to risk the money needed to initially develop the technology of flight. In 1907, the club's president, millionaire Corlandt Bishop, asked his brother-in-law, Congressman Herbert Parsons to intercede with the War Department on behalf of the Wrights. Instead, Parsons took the issue directly to President Theodore Roosevelt. The latter asked Secretary of War William Howard Taft to have the Army reconsider the Wrights' invention. Meanwhile, Brig Gen James Allen, the Army's Chief Signal Officer (CSO), had directed his assistant, Maj George Owen Squier, to perform a detailed study of aeronautics and its potential military applications. Squier, an 1887 United States Military Academy (USMA) graduate and holder of a doctorate in physics from Johns Hopkins University, soon convinced his boss to create a new organization in his office to promote aviation. On 1 August 1907 the Aeronautical Division of the CSO's was established with responsibility for all things concerning aircraft, balloons and related matters in the Army. Allen saw aviation an extension of the Signal Corps' communications mission and a guarantor of its organizational survival.

The combination of external and internal pressures worked. Following a personal appeal by Wilbur Wright in November 1907, Allen issued specifications for the world's first military aircraft and advertised for bids the following month. But, initial War Department efforts to convince Congress to appropriate funds for military aeronautics proved fruitless.[12]

Squier drew upon his extensive studies of aeronautics to write draft Specification No. 486 for the Army's first aircraft. Less than a page long, it was developed with the help of several scientifically trained individuals. Years later that practice, vastly expanded in scope and institutionalized, became a major force in weapons modernization by the nation's armed forces.[13]

Specification No. 486 illustrated the purely experimental character of early Army aviation. It gave no military mission for the flying machine. The document merely delineated desired performance characteristics. The aircraft had to achieve a speed of 40 miles per hour (mph) while carrying two persons whose combined weight was 350 pounds and fuel for a non-stop flight of 125 miles. The machine also had to be capable of being assembled in about an hour and be transportable in Army wagons. A premium of 10% of the contract price would be paid for each additional mph speed over 40, while a similar penalty would be levied for each one below that target.[14]

At that time, federal government contracting was regulated by an 1860 act of Congress. That legislation provided that the government would normally obtain goods and non-personal services only through public advertising of competitive bids that responded to detailed specifications. Contract awards would be made to the lowest responsible bidders. A 1901 revision of the legislation liberalized the exception to advertised competitions to include emergencies or cases where competitions were impractical. The system of open, competitive bidding later became a major barrier to progress in the eyes of those responsible for the development of aviation. It allowed firms that had spent substantial funds designing, developing and testing successful aircraft to lose separate competitions for the more lucrative production contracts.[15]

The Army received 41 bids in response to Specification No. 486. One was from an inmate in a federal penitentiary. Cost estimates in those 41 responses ranged from $500 to $10,000,000. Only three of the proposals actually met Army specifications. The Wrights' submission was one of them. Although, the Signal Corps preferred to deal with them, their $25,000 bid, cut by 50% from the original offer, was the highest of the three. This placed them last on the list of potential contractors under existing government procurement practices. The Army's only other viable alternative was to buy from all three successful bidders. However, it lacked enough money to accept all of them. The problem was solved when additional money was obtained from a special fund maintained by President Roosevelt. That money turned out to be unnecessary when only the Wrights were able to fulfill the terms of their contract.[16]

Orville Wright delivered an aircraft to the Army on 20 August 1908, just prior to the contract deadline. Acceptance tests were scheduled at Fort Myer under the watchful eyes of a special aeronautical board appointed by Allen. Squier headed it. President Roosevelt and his cabinet along with members of Congress observed. The Navy sent two official representatives. Army 1st Lt Thomas Selfridge volunteered to serve as an observer at the acceptance trials and ride as a passenger with Orville. He was killed when the aircraft crashed on September 17th. Orville was seriously injured and his machine badly damaged. The Army delayed the final tests and delivery date until the following summer (Illustration 2.1).[17]

After a series of delays and accidents, Orville completed the first of the Army's requirements for the aircraft on 27 July 1909 witnessed by President William Howard Taft. Accompanied by 1st Lt Frank P. Lahm, he stayed aloft for 1 hour, 12 minutes and 40 seconds. Orville completed

Illustration 2.1 1st Lt Selfridge and Orville Wright (*Photo* 1st Lt Selfridge seated in the Wright airplane with Orville at Fort Myer on 17 September 1908, the day that Army officer was killed when the plane crashed. The Lewis E. Goodier Jr. Collection, Early Air Service History: Vol. II, photo by VIRIN: 090713-F-1405A-099.JPG, Air Force Historical Research Agency (AFHRA), afhra.af.mil)

the official trials with a speed run of 42 mph on July 30th over a 10-mile measured course.

Airplane No. 1 had fulfilled all the endurance and speed requirements of the Army contract. Under the terms of its fixed-price, incentive fee contract, its performance won the Wrights a $5000 bonus in addition to the basic $25,000 contract award. The airplane was formally accepted by the Army on 2 August 1909. The remaining provisions of the contract were fulfilled later that year when Wilbur trained two Army officers as pilots at a temporary flying field established near College Park, Maryland.[18] Women achieved a historic milestone during that period when the Wrights hired Ida Holdgreve, a Dayton, Ohio seamstress.In about 1910, she had answered a local ad that read, "Plain Sewing Wanted." The Wrights were hiring a seamstress to work at their Dayton plant. She got the job. Ida became first woman known to work in the American aircraft industry on a factory floor. She sewed the fabric coverings for wings and stabilizers on Wright airplanes.[19]

Army flying was suspended at College Park in November 1909 after Airplane No. 1 was badly damaged in a crash. Operations did not resume there until June 1911. During the following month, the Signal Corps designated that location as the Signal Corps Aviation School. Subsequently, Army flyers conducted experimental cross-country and high altitude flights, plus experimentation in radio-telegraphy, aerial photography and signaling from that location. During May 1912, they began testing new aircraft built by the Wrights and Glenn Curtiss. The Army terminated its flying operations at College Park in November 1912 because it concluded that the weather was better in the southern section of the nation (Illustration 2.2).[20]

Rough Seas: Early Naval Aviation

Naval aviation had a more difficult start than its Army counterpart. Historian Laurence M. Burke II observes that the "... Signal Corps proactively sought out jurisdiction over all aircraft and faced no real opposition from other branches of the Army. There was no similar history of aviation in the Navy making it far from clear just which part of the Navy's organization would, or should, be responsible for aviation. This led to an early competition between several of the Navy's semiautonomous bureaus for control of the new technology."[21]

Illustration 2.2 Ida Holdgreve, Seamstress (Photo of Ida Holgreve courtesy of Wright State University Libraries' Special Collections & Archives)

Navy Secretary George von L. Meyer received conflicting advice about aviation from senior officers. Several bureau chiefs opposed spending money on it. Others supported it. The General Board, headed by Admiral of the Navy George Dewey, believed that the service should explore the potential uses of aircraft. He recommended that the Navy examine problems associated with developing new designs for airplanes that could be embarked on its scouting warships. Meyer ignored Dewey's advice.[22]

Meyer had also received a flood of letters from civilian aeronautical enthusiasts. A note from influential millionaire publisher John B. Ryan may have caught his attention. Ryan headed a private civilian organization known as the "United States Aeronautical Reserve (U.S.A.R.)." Its goal was to foster the development of aviation in the U.S. It claimed some 3,200 members including President Taft, Wilbur Wright, bicycle racer and pioneer aviation inventor Glenn Curtiss and retired Lt Gen Nelson A, Miles, the Army's former commanding general. Ryan asked Meyer to designate a naval officer to contact regarding aviation matters.

Given the U.S.A.R.'s influential membership, it was not surprising that Meyer soon designated a talented officer to review aviation-related correspondence. CAPT Washington Irving Chambers, an 1876 graduate of the United States Naval Academy (USNA), was an accomplished engineer who had had devoted his career to developing better large caliber guns, armor, torpedoes, mines and battleship designs for the Navy through the application of scientific methods. During 1910, Rear Adm William Swift, Secretary Meyer's aide for materiel, convinced Chambers to relinquish his prized battleship command to return to Washington, D.C. to handle aviation mail in addition to his regular duties. Chambers immersed himself in the available literature on aeronautics, discussed it with experts, read naval attaché reports, attended aeronautical exhibitions and became an aviation enthusiast. He insisted that airplanes must fit current naval needs and doctrine. Assistant Secretary of the Navy Beekman Winthrop supported Chambers and ordered the Bureau of Steam Engineering (BuEng) and the Bureau of Construction and Repair (BuC&R) to coordinate with (not work for) him on aviation matters. Historian Stephen K. Stein observes that "Virtually all of the important decisions concerning naval aviation during its first years were approved by Winthrop while Meyer was out of the office."[23]

But, Chambers lacked formal authority over aviation policy and budget matters. Muddying the bureaucratic waters even further, Meyer directed the Navy's BuC&R to monitor aviation developments. The confusion worsened because the responsibility for naval aviation was split between Chambers and three different bureaus. He exercised tenuous influence over personnel, policymaking and general control of the program. The Bureau of Navigation (BuNav) handled personnel and training. BuC&R was responsible for aircraft while BuEng had charge of engines.[24]

During April 1911, Chambers arranged to be reassigned to BuNav because Congress had placed the Navy's aviation funding in that organization's budget. However, Chambers was never even given a clerk to assist him and was almost completely ignored by other officers in that organization. Despite the fact that he was BuNav's second most senior officer, Rear Adm Reginald F. Nicolson, the bureau's chief, suggested that Chambers "work at home." Nevertheless, Chambers found space in a dank corner of a basement room in the War, State, Navy building. He quipped that it was "...a good place to catch a cold."[25]

Despite those obstacles, Chambers proved to be an effective advocate for naval aviation. He closely monitored aeronautical developments at home and abroad while lobbying senior officers, government officials and prominent civilians to support the development of naval aviation. Chambers also befriended Glenn Curtiss, a mechanic and former motorcycle racer, self-taught aeronautical engineer and pioneer aircraft designer. Curtiss, was much more approachable and open-minded about working with the sea service than the Wrights.[26]

Curtiss was a member of the Aerial Experiment Association (AEA) established by Scottish-born American inventor Alexander Graham Bell and others in 1907. Bell, who was working to improve aircraft performance, learned of Curtiss' reputation for mechanical skill and developing superior lightweight engines. The latter designed and piloted the fourth successful aircraft developed by the AEA. Known as the "June Bug," its AEA-developed innovations included ailerons[27] instead of wing warping and wheeled landing gear rather than skids.

After the association was dissolved in 1909, Curtiss obtained the commercial rights to its designs and patents. He decided to begin manufacturing and selling airplanes. The Wrights sued him for patent infringement in 1909, a legal battle that lasted for years and hindered the development of American aviation.[28]

The scope of the Wright patent's was so broad that basically anyone who built an airplane with any control surfaces violated their patent. It had a chilling effect on technological progress. The Wrights engaged in patent lawsuits on both sides of the Atlantic. But, European powers ignored reciprocal patent protection agreements with the U.S. during World War I in the name of military expediency.

The patent legal battle had not deterred CAPT Chambers' efforts to bring aviation to the Navy. Understanding that he had to do something substantial to demonstrate that planes could operate with the fleet, he tried to convince Meyer to approve an experimental launching from a warship, only to have the latter dismiss the airplane as a "carnival toy." Unwilling to give up, Chambers approached Assistant Secretary Winthrop when Meyer was out of the office. Winthrop approved the experiment.

Chambers and Curtiss arranged to have the latter's top pilot, Eugene Ely, launch a Curtiss aircraft from a warship. Curtiss traveled to Hampton Roads, Virginia where he personally supervised the construction of a launching platform over the bow of the Scout Cruiser No. 2 USS *Birmingham*. Ely successfully launched a Curtiss biplane from the ship on 14 November 1910 and flew to the nearest land. It was a near disaster because a propeller was damaged on takeoff when his plane hit the water.

Water spots on his goggles and low visibility caused him to lose his sense of direction forcing him to seek out the nearest land after initially heading out to sea where he probably would have died; he couldn't swim.

A Navy board consisting of Chambers, Lt N. H. Wright and Naval Constructor William McIntyre witnessed the experiment and wrote an official report. Meyer remained unimpressed. Essentially, he wanted a seaplane, a project that Curtiss was already working on. The Navy had no money available (or was unwilling to allocate it) for such aeronautical experiments.

Efforts by San Francisco businessmen to organize an aviation meet in their city gave Chambers and Curtiss an opportunity to achieve another historic milestone. Curtiss saw the proposed event as another opportunity to help convince conservative senior Navy leaders to buy his aircraft. Subsequently, Ely became the first aviator to land on a warship. He touched down on an improvised platform on the stern of the armored cruiser USS *Pennsylvania* on 18 January 1911. Later that day, Ely took off from the cruiser and flew back to where he had begun his journey. Despite that impressive demonstration, it didn't generate significant support for aviation among senior military leaders or Congress.[29]

On 17 February 1911, Curtiss demonstrated the potential of his new hydroaeroplane (floatplane) in San Diego, California's harbor. He landed next to Armored Cruiser No. 4 USS *Pennsylvania* and was hoisted aboard by one of the ship's cranes. After a brief visit onboard, he and his plane were returned to the water before talking off. Chambers believed that such planes could be useful to the fleet.[30]

In March 1911, Chambers was primarily responsible for securing the first naval aviation funds. Congress provided a $25,000 allocation for experimental purposes in the Naval Appropriation Act for Fiscal Year 1912. After newspapers announced the allocation, Chambers received a flood of offers to build airplanes, many of them from cranks and charlatans. He refused to purchase anything but complete, functioning products that had been well tested by reputable firms. During the winter of 1911, the Navy had committed itself informally to purchase a Curtiss "hydroaeroplane" and one of his land planes. It also decided to buy a Wright land plane.

On 8 May 1911, Chambers prepared specifications to purchase two Curtiss biplanes. That date is officially designated as the birthday of naval aviation. Like the original specifications for the first Army aircraft, those requirements were experimental and did not indicate a specific naval mission although Chambers expected that the aircraft would be used primarily to train pilots. He proposed that the airplanes should be equipped to operate from both land or water, attain a speed of at least 45 mph as well as carry a pilot and one passenger. They would also be provided with dual controls enabling the pilot or passenger to fly the machines. Although the specifications never were incorporated in the official purchase contract, they indicated which types of aircraft Chambers thought the Navy should buy. Subsequently, his specifications served as the basis for building the Navy's first aircraft, designated the A-1 "Triad" floatplane that could operate on land, water and air because of its retractable landing gear and a pontoon float. Powered by a single Curtiss V-8 engine that generated 75 horsepower (hp), its maximum speed was 60 mph. Designed and built by Curtiss, it cost the Navy $5, 500. Curtiss later designed a mechanical system for the aircraft, consisting of a cable mounted on a downward slope from the boat deck to the bow, which enabled the "Triad" to be launched from a warship.

During May 1911, the Navy issued requisitions with broad specifications for the characteristics of its first airplanes. They featured: a Curtiss slow-flying land plane for training; a Curtiss hydroplane equipped with a pontoon; and a Wright Model B land plane to be capable of flying no less than 40 mph. They were delivered in July 1911. The A-1 was a wheeled aircraft while the second, the "Triad" was a seaplane with retractable landing gear that enabled it to operate from both land and water. The contract required the manufacturer to train one pilot and one mechanic for each plane (Illustration 2.3).

Curtiss personally conducted the first flights of the naval aircraft on 1 July 1911 near his Hammondsport, New York plant. He made four flights that day. On the first, he flew alone for five minutes at an altitude of 25 feet. On the second, he flew with Navy Lt Theodore G. Ellyson, as a passenger. Operating alone, Ellyson piloted the remaining two flights. In another of his multifaceted efforts to promote military aviation and his own economic interests, Curtiss had volunteered in November 1910 to train a naval officer free of charge to become a pilot. Chambers had selected Ellyson. Eventually, Curtiss also trained Lt John H. Towers.[31]

Towers, a 1906 Naval Academy graduate, was arguably the most influential of those pioneer naval aviators. The Rome, Georgia native was intelligent, studious, inquisitive, reserved, self-reliant, very quiet and modest. Upon graduating from Annapolis, he gained a reputation as a steady and very conventional officer. As his career progressed, Towers

Illustration 2.3 A-1 TRIAD Naval Aircraft (*Photo* A-1 TRIAD, the Navy's first aircraft, at Hammondsport, New York home of the aircraft's designer and manufacturer Glenn Curtiss. CAPT Chambers, Lt Theodore G. Ellyson, and Lt John H. Towers were also pictured in this undated photo. National Naval Aviation Museum, Naval History and Heritage Command (NHHC), undated, www.history.navy.mil)

became increasingly convinced that aviation would play a critical role in the Navy's future. Initially, he saw flying as a way to spot the fall of shells from the new high-powered big guns of battleships. Unlike some of his Army counterparts, Towers didn't believe that the airplane would completely revolutionize warfare and render existing forms of combat and conventional military organizations nearly obsolete. Instead, he worked quietly during the interwar period within the existing system to convince the Navy, Congress, senior federal officials and the public to support a strong naval aviation program.[32]

Greenbury Point

CAPT Chambers discovered that the Navy had land available across the Severn River from the Naval Academy known as Greenbury Point. He secured it as a site for aviation experiments and training. It became the Navy's first aviation base.

The young Navy flyers didn't receive a warm welcome at their alma mater. Officers assigned to the Academy had little interest in aviation. The school also lacked the administrative processes to provide parts and other material for aeronautics. The fledgling naval aviators chipped in with their own money to help purchase needed items. Chambers turned a blind eye while his pilots and mechanics stole badly needed oil, parts and tools from the nearby naval station. They also wrote private firms to get free gas and oil on the ostensible grounds that they wanted to test samples of their products. Few mechanics were available in the early days of naval aviation. As late as May 1912, only six enlisted men were attached to Greenbury Point. Major overhauls were primarily accomplished by the officers. The naval aviators shared their experiences and often socialized with Army pilots training at College Park, only 30 miles away.

Greenbury Point was the site of significant developments during its three years of operation. Flights from that location tested early airborne wireless radio transmissions, though with limited success because of electrical interference from engines and vibration during flight. Its aviators also conducted experiments in spotting submerged submarines in the Chesapeake Bay. On the ground, they and their enlisted mechanics tested various types of wire to determine which ones would be more structurally sound for use in aircraft construction.[33]

Meanwhile, Maj Gen William P. Biddle, the Marine Corps' Commandant, had become interested in the Navy's aviation experiments. Since 1900, the Corps and the Navy had been working to form an "advanced base force" of Leatherneck infantry and artillery to occupy and defend forward naval bases for the fleet. Biddle thought the Corps might benefit from the addition of trained aviators. He ordered two officers, 1st Lt Alfred A. Cunningham and 1st Lt Bernard L. Smith, to Annapolis for flight training.

On 22 May 1912, Cunningham became the first Marine officer to report for flight training at Greenbury Point. During 1911, while assigned to the Marine Barracks in Philadelphia, Cunningham had become an aviation enthusiast. He had started experimenting with an airplane which he rented for $25 a month from a civilian aviator. His suggestion that aviation would add an important reconnaissance capability to the Marines had led to Maj Gen Biddle's orders sending him to Annapolis.[34]

Despite funding, organizational and attitudinal challenges within the Navy, Chambers continued to push for aeronautical progress in his service. The first attempt to launch an airplane from a ship by a catapult designed by him failed at Annapolis on 31 July 1912. Naval Constructor (i.e., naval architect and engineer) H.C. Richardson redesigned the catapult that was successfully tested during the following November at the Washington Navy Yard. Subsequently, an improved version was installed on the Armored Cruiser No. 12 USS *North Carolina*. A Curtiss AB-3 flying boat conducted the first successful test of catapult launching from a warship underway on 12 July 1916. That November, Rear Adm David W. Taylor, the Navy's Chief Constructor and head of its BuC&R, requested suitable designs from manufacturers for high-speed seaplanes that could be catapulted from warships, then land alongside and be hoisted back aboard by crane.

The sea service continued to develop requirements for improved seaplanes. Meanwhile, under the direction of Naval Constructor Richardson, the service tested models of flying boats at its model basin in the Washington Navy Yard. Those tests generated an enormous amount of useful information about a variety of factors including the influence of form at high speeds and the most efficient hulls. Based on those tests, Richardson developed the ideas which shaped the designs of U.S. Navy seaplane hulls and floatplane floats.

Rear Adm Taylor was one of the sea service's top technological innovators. He had been sent to the Royal Navy College where he won top scholastic honors. Before he was assigned to head the BuC&R in December 1914, Taylor had built the Navy's first towing tank for ship model testing in 1898. During 1914, he had supervised the design of the Navy's first wind tunnel. Taylor was an acknowledged international expert in the relationship between a ship's hull design and engine power. Subsequently, he also became heavily involved in aeronautics.

Based on his department's early experiences with airplanes, the Acting Navy Secretary Philip Andrews, published the sea service's general specifications for naval aircraft on 25 July 1912. CAPT Chambers and his team developed a prototype steam catapult that could launch aircraft from battleships and cruisers. Chambers was partially successful in his battle to obtain adequate funding for naval aviation. After the three Navy bureaus involved in aviation slashed his budget estimates, Congress eventually appropriated a total of $65,000 well below the $150,000 he requested.

Chambers deserved credit for recruiting the initial cadre of talented young officers into naval aviation including Lt Towers and Lt Cdr Henry C. Mustin. He advocated incremental improvements in aeronautics that would enable aircraft to gradually fit into existing naval doctrine and organization. Despite support from the Navy's General Board, he struggled against the opposition of many of the service's battleship-oriented senior officers and Meyers. Like Squier, Chambers insisted that aviation was destined to play a crucial role in warfare. He understood that, without a strong research and development program, it would never realize its full potential.[35]

Improvising Army Aviation

The Army's meager aviation program also continued to struggle during the years before the U.S. entered World War I. In the autumn of 1910, it ran out of money. Its resources consisted of 1st Lt Benjamin D. Foulois and a single Wright aircraft conducting flying experiments in the San Antonio, Texas area. That February, the Signal Corps had moved its flying program to Fort Sam Houston near "The Alamo City" because of the cold, windy weather at College Park. Fearing the return of Foulois and the aircraft to Washington, D.C. because the service had run out of aviation funds, San Antonio citizens raised $100,000 ($2.7 million in 2020 dollars) and donated it to the War Department so the Army could continue aerial experiments in their city.

That initative reflected what historian Joseph Corn identifies as "the winged gospel." But that enthusiasm was not reflected by the federal government at the time. Although the War and Navy Departments, attracted by aviation's potential and increasingly concerned by foreign developments, had each requested $500,000 appropriations for aeronautics earlier that year, Congress had failed to act on their submission. Instead, American security seemed assured by wide oceans and a powerful Navy. Under those conditions, there was little congressional interest in spending significant funds on something as nebulous and unproven as aviation.[36]

Nevertheless, there had been progress on the doctrinal front after the Army acquired its Wright Flyer in 1909. Army historian Edgar F. Raines, Jr. writes that "... a number of officers had given considerable thought to aerial observation. Field Service Regulations, 1910, published within months of the Flyer's successful acceptance tests at Fort Myer, Virginia, specified the formation of an aerial company in each corps-size unit upon mobilization but left the question of mission open until the Signal Corps had acquired experience with the new machine."[37]

Meanwhile, North Island near San Diego, California had become the epicenter of U.S. military aviation during 1912. Responding to an invitation from Curtiss, Chambers had notified the former in December 1911 that he would send three aviators and their aircraft there after 1 January 1912 for additional training and experimentation. The Signal Corps had also decided to concentrate its flying training and experimental activities at North Island. Pilots favored its good year-round flying weather. The location also permitted training on both land-based aircraft and hydroplanes. The initial detachment of naval aviation personnel arrived on 4 November 1912. They were followed by the first Army aircraft the next month. North Island was officially designated as the "Signal Corps Aviation School" in December 1912.[38]

The first year at North Island was grim. Many airplanes crashed and four pilots were killed. The local press complained about the "niggardly" War Department policy of using obsolete aircraft or experimental aircraft to train military flyers. By the spring of 1914, eight qualified Army aviators had been killed. Lt Col Samuel Reber, who headed the Signal Corps' Aeronautical Division, was determined to fix those problems. He was convinced that it was critical to secure technically trained individuals to improve the airplanes the Army was flying.

Reber hired Grover C. Loening during July 1914 to fix the terrible problems at North Island. Loening was the holder of the first ever American master's degree in aviation and aeronautics. It had been granted by Columbia University in 1910. He had served as the general manager of the Wright Company in Dayton during 1913–1914. Loening had corresponded frequently with a number of young military aviators including Army 1st Lt Henry H. Arnold as well as CAPT Chambers concerning potential career opportunities in aviation. Eventually, he secured a job with the Signal Corps. Loening and another civilian, Oscar Brindley, were given a free hand to reorganize training, engineering, maintenance, and procurement activities at North Island (Illustration 2.4).

During 1914, Loening and Brindley opened an "Experimental and Repair Depot" with the former in charge. It was responsible for all Signal Corps aeronautical experimental work as well as the development and

Illustration 2.4 North Island, California 1912 (*Photo* San Diego School, North Island, 1912. Catalog #: USN 902875. Copyright Owner: National Archives. NHHC, history.navy.mil)

construction of new aircraft structural features. In addition, it performed all major aircraft repairs.[39]

While early efforts by the Army to produce its own aircraft designs at North Island were unsuccessful, several important aeronautical advances had emerged from private designers by 1914. Open frame, pusher-type aircraft like those first built by the Wrights and Curtiss were being replaced by more efficient tractor designs with enclosed fuselages. Pushers, with the engines located behind the wings and pilot, were unstable and extremely dangerous. Frequent crashes and numerous deaths had caused Army pilots to lose faith in them. The main danger was that in a crash, the engine—located behind the pilot—would crush him. Tractor engines and propellers were placed in front of the aircraft. In 1914, an Army board condemned all pushers in the service. The Navy banned pushers in the summer of 1916.

Curtiss continued to develop more capable seaplanes. He responded to a contest sponsored by an English newspaper offering a $50,000 prize for the first aircraft to fly across the Atlantic in less than 72 hours. Led by English engineer Benjamin D. Thomas, his firm designed a series of innovative planes which eventually resulted in its "Model H-2." Funded by Philadelphia department store magnate Lewis Rodman Wanamaker, the giant aircraft featured a hull designed for at sea operations, biplane wings, engines in a "puller configuration" and twin-bladed propellers. The prototype first flew in June 1914. Named the "America," it was taken to Great Britain to attempt a transatlantic flight.

After the Great War erupted in Europe that summer, the competition was cancelled. However, the Royal Navy decided to use the Curtiss design. It purchased the 2 existing airframes and ordered 12 more. Later, the British engineered their own version of the Curtiss seaplane. As the war ground on, the Curtiss firm continued to refine the design culminating in its "Model H-16." Although purchased by both the U.S. and Royal navies, only the latter's actually saw combat in European waters. American versions focused primarily on anti-submarine patrols along U.S. coasts.[40]

Meanwhile, the big aeronautical controversy at North Island in 1914 had been the dangerously intense rivalry between advocates of aircraft built by the Wrights and those constructed by Curtiss. Aviators could not fly aircraft built by both firms because their control systems were so different and they had not been taught to fly both. Arguments between the pilots over the relative merits of both firms' machines were undermining the efficiency of the school and encouraging dangerous flying competitions. To eliminate that friction, a board of officers was appointed to write specifications for a standard aircraft that would eliminate the distinction between "Wright" and "Curtiss" pilots.[41]

The work of the board was accepted by the CSO, Brig Gen George P. Scriven. In July 1914, he announced a formal competition to select a standard aircraft for field service with the Army that would replace the variety of experimental and training types then in use. A purse of $30,000 was to be distributed among the three best designs. The Signal Corps wanted to encourage aircraft manufacturers to develop advanced designs and test the newer types of aircraft that were already being built. Each entry would be graded on a point system that stressed such factors as construction, maneuverability, and stability. A board, consisting of five officers and Loening was designated to test the entries. The Glenn L. Martin and Curtiss firms dominated the contest. The resulting purchases included 8 Curtiss JN-2s and 17 Martin TT trainers, each included the innovative standardized landing gear of two wheels, an axle and a tail skid developed by Loening's organization.[42]

In July 1914, Congress gave aviation its first definite legal status within the Army. Prodded by the Aero Club of America and growing international tensions, it had established the Aviation Section as part of the Signal Corps in law on 18 July 1914.[43] The law also limited the new aeronautical organization to no more than 60 officers detailed from other Army combat arms (infantry, artillery and cavalry) and 260 enlisted men. Although strengthening the legal position of Army aviation, Congress had decided, in effect, that it would remain an experimental program.[44]

Meanwhile, some members of the Army's combat arms showed an interest or least an awareness of the new technology's military possibilities and limitations. For example, the *Journal of the United States Cavalry Association* (JUSA) initially focused on aviation's shortcomings. Historian Lori A. Henning stresses that "They included technological flaws due to weather, mechanical limitations and unreliability, and unfavorable comparisons to cavalry capabilities. American cavalrymen were reassured that the unreliability of airplanes meant that they could not be depended on as the sole means of reconnaissance."[45] According to Henning, the JUSCA "... also reported on aviation's positive aspects.... unlike the army's present 'eyes and ears' (the cavalry), the future 'eyes and ears' (airplanes) would 'have an uninterrupted view, not only of the enemy's front, but of his flans, center and rear'."[46]

Money remained the key problem for American military aviation. Other nations were spending far more on aeronautics than the U.S. which had devoted only $435,000 of the estimated $85,000,000 ploughed into military aviation around the world from 1908 through 1913.[47] Historian Herbert A. Johnson concludes that "Before 1914, army aviation had barely survived a parsimonious congressional attitude toward flying in the military. As European wars and mobilizations became known to the American people and their legislative representatives in Washington, there were more and larger surges of concern for an adequate military program."[48]

Integrating Aviation With the Navy

Naval aviation navigated a different course than its Army counterpart. Despite the efforts of some officers to establish a separate Navy flying corps akin to the Marines, naval aviation gradually won itself a critical role within the mainstream of the Navy. Its flyers remained naval officers first and pilots second. Although facing an uphill battle like their Army counterparts, naval aviators early on found more support from some senior members of the service's hierarchy than is generally realized. The real argument was over the pace of development and what form it would take including seaplanes, float planes, land-based patrol aircraft and aircraft carriers.[49]

Naval aviation made its experimental operational debut with the fleet in 1913. The General Board had asked Chambers to prepare his small group of aviators and aircraft to participate in the annual fleet exercise. In January 1913, four aircraft plus their pilots and mechanics under Towers' command were dispatched to Guantanamo Bay, Cuba to participate in that training. It constituted the Navy's entire aviation element at the time.[50]

Assistant Secretary of the Navy Franklin D. Roosevelt appointed a board in October 1913 headed by Chambers to develop a "comprehensive plan for the organization of a Naval Aeronautic Service."[51] Roosevelt had taken plane rides that summer and become convinced of aviation's importance. Like his cousin Theodore, he had waited until the Secretary of the Navy was out of the office to act on controversial policy initiatives he strongly believed in.[52]

The board's first action was to relocate the Greenbury Point detachment to Pensacola, Florida. On 20 January 1914, the first naval air station (NAS) was activated there with the arrival of the detachment's aircraft. The site's location adjacent to Pensacola Bay was attractive because naval aviation was focused on seaplanes and the sea service owned a recently closed Navy yard there. Its temperate climate allowed year-round flying.[53]

During 1914 the Marine contingent at Greenbury Point was transferred to Pensacola. Early that year, 2 Marine officer pilots and 10 of their enlisted mechanics plus 2 seaplanes provided by the Navy were dispatched to Culebra Island, Puerto Rico. They participated in the Corps' first full-scale advanced base exercise. It was a new concept in which Leathernecks would occupy undefended harbors and secure them for use by the fleet in offensive campaigns across the Caribbean and Pacific. The Marine aviation contingent's role in that first test was to provide reconnaissance and scouting support of a Marine brigade defending the island against an enemy counterthrust.

During 1915, both Marine and Navy flyers began training with land-based planes at the Signal Corps' new school at San Diego. Army historian Graham A. Cosmos observes that "Secretary of the Navy Josephus Daniels had arranged with the Army for this training in the belief that defense of advance[d] bases and, in the case of the Marines, possible joint operations with the Army, required an aviation force able to operate from land as well as water."[54]

The Mexican Revolution

Events abroad played a critical role in the early development of U.S. military aviation. The first of those came south of the nation's border when a civil war erupted in Mexico after General Victoriano Huerta seized power in February 1913. Refusing to recognize his regime, President Woodrow Wilson concentrated Army troops in Texas along the border, including a hastily thrown together provisional Army air unit designated the First Aero Squadron. The unit had eight pilots and nine airplanes. Only three of the aviators were considered fully qualified for flight duties. Fortunately, armed hostilities were avoided. However, the Army flyers became extremely angry because of frequent accidents and poor leadership. They wrote a letter to Scriven demanding major changes, most of which were implemented.

After 1913, the air unit spent most of the following two years training and testing aircraft at San Diego. That September, Scriven's office had issued a blunt public statement describing the status of military aviation in the U.S. as critically deficient and appealed for a much stronger effort to strengthen that program. Congress appropriated $125,000 for Army aviation in 1913 and doubled that amount the following year. However, War Department officials were convinced that the sums were far too little if the nation was to keep pace with major foreign powers. During that period, Arnold worked closely with his Navy counterparts on aviation matters. He and Towers became lifelong friends.[55]

Continuing tense relations between the U.S. and Mexico boiled over in a major incident at the port city of Tampico on 9 April 1914 when Mexican police arrested a Navy shore party. Wilson was incensed. Determined to defend American honor and teach the Mexicans a lesson, he sent warships to blockade the ports of Tampico and Veracruz. Due to lobbying by naval aviators, several aircraft were embarked on the scout cruiser USS *Birmingham* and the battleship USS *Mississippi*. One contingent sailed for Tampico. The other aviation detachment made for Veracruz. The fleet assaulted and occupied Veracruz to prevent a German merchantman from landing arms there. Mexicans fired on two unarmed Navy floatplanes scouting inland, hitting one of them at least twice. Enraged, the pilot responded by hurling a large bar of laundry soap at his assailants.

U.S. naval aviation had made its combat debut. Although that episode had a decidedly comedic character, the deployment had demonstrated the potential operational value of aviation to the fleet.[56]

A far more serious clash occurred in March 1916. On that date, Pancho Villa, the Mexican bandit and erstwhile revolutionary, raided Columbus, New Mexico. His forces killed 17 American citizens. Wilson ordered Brig Gen John J. Pershing to chase Villa into Mexico and capture or kill him. The Army's First Aero Squadron was dispatched to New Mexico to support Pershing's expedition. Commanded by Foulois, the unit was equipped with 8 Curtiss JN-3 s powered by 90 hp. engines. Each of those fully equipped aircraft had cost the Army $5,000. The unit arrived at Columbus on 15 March 1916 and deployed to Casas Grandes, Mexico several days later. Although its mission was not clearly spelled out at the time, it soon became evident that the squadron would be used for reconnaissance and courier duties.[57]

The operational debut of the first formally constituted Army tactical air unit to be tested under realistic field conditions was a fiasco. Although not a shot was fired, six of the squadron's eight aircraft were rendered useless within two weeks. The Army reported that two of them collapsed in the relentlessly dry desert heat (that dried out their wooden frames and glued joints) and badly needed repairs. Their engines were too weak to fly over Mexico's Sierra Madres mountains. Moreover, the JN-3's lacked the range needed to support Pershing's cavalry columns as they searched for Villa in the vast wastelands south of the border. The squadron returned to Columbus on 22 April 1916 to await new aircraft. The *New York Times* referred to the surviving operational aircraft as "suicide machines" and the rest as "piles of junk." Pershing praised the aviators' courage and persistence but dismissed their aircraft as useless. He blamed the Signal Corps for failing to support his aviators and predicted that, in the future, an army without aircraft would lose to one that employed them for reconnaissance and courier duties.[58]

During March 1916, Congress, responding to the crisis with Mexico and the war in Europe, had passed additional military appropriations including $500,000 for Army aviation. A Technical Aero Advisory and Inspection Board, consisting of three officers and two civilian consulting engineers, was appointed to test aircraft being purchased for the Army and inspect the work underway at private aircraft plants. In April 1916, the board recommended to the CSO that the Army purchase 12 new aircraft and ship them immediately to the First Aero Squadron. They were the only planes immediately available and were never adequately tested before being sent to New Mexico. Ignoring formal bidding requirements, an Army Aviation Board let contracts to purchase the aircraft and shipped them to Foulois. Like their predecessors, the replacement aircraft proved to be totally unsuited to field conditions. Only 1 of the 12 ever got off the ground. They were then transferred to San Diego to train fledgling aviators. Secretary of War Newton D. Baker ordered eight more aircraft but manufacturers were slow in delivering them. Pershing had no operational aircraft at all for over two months.

To deal with the Mexican situation, an Army Technical Advisory and Inspection Board suggested purchasing 12 Curtiss R-2 aircraft at a cost of $12,000 each. By the time they reached the First Aero Squadron, a shaky truce existed between American and Mexican forces. The new planes proved to be only a marginal improvement over their predecessors. The First Aero Squadron stayed at Columbus, New Mexico until early 1917. Most of its energies were devoted to aircraft experimentation and testing.[59]

Military Preparedness

World War I stimulated a major reconsideration of American military policy. The resulting "Preparedness Movement" produced far-reaching defense reform legislation. Military historians Allan R. Millett and Peter Maslowski observe that "... it represented the first time that defense policy in peacetime influenced American politics and involved more people than a limited policy-making elite."[60] They added that "On the other hand, its legislative products came too late to have any substantial impact on American military readiness, either to fight World War I or to avoid intervention by imposing a peace before American entry into the war."[61]

To strengthen the nation's military posture in response to the stalemated European land war and German submarine operations, influential American citizens outside the government had formed pro-preparedness lobbies and launched propaganda campaigns to convince Wilson and Congress to significantly increase defense spending. An important result of that pressure was a new five-year naval building program designed to produce a fleet second only to the Royal Navy in size and superior to it in quality. Congress approved a General Board plan in August 1916 to make the U.S. Navy "second to none" in three years and authorized a $300 million shipbuilding program.[62]

The Fiscal Year 1917 Navy appropriation also called for the creation of a Naval Flying Corps of 150 officers and 350 enlisted ratings in addition to such personnel already provided for by law elsewhere in the sea service. Although the Navy had requested $13 million, the legislation provided only $3.5 million for aviation in the Naval Appropriations Act of 29 August 1916. Contracts were let for 47 planes, a number that was soon raised to 60. Negotiations had begun with Curtiss to supply the service with 30 "hydro airplanes" (i.e., airplanes equipped with floats). That process led to the delivery of 30 Curtiss N-9 tractor design training planes with enclosed cockpits to the Navy between November 1916 and February 1917. Since work on developing new aircraft engines at the Washington Navy Yard had been disappointing, American firms began working to purchase more powerful ones abroad.[63]

Responding to the Great War, Congress passed the National Defense Act of 1916. Millett and Maslowski, observe that the legislation "... represented the most comprehensive effort to organize a land force structure for future mobilization, but it made no special provisions for a crash preparedness program."[64] Nevertheless, its main provisions were significant. The regular Army was slated to grow to 175,000 personnel, expansible to 286,000 in wartime. The National Guard retained its prized role as the Army's first-line combat reserve and was projected to grow to 400,000. For the first time, the legislation granted the federal government emergency powers to compel private firms to support the needs of a massive wartime land force.[65]

The troubles with Mexico, strong American economic ties with the Allies, and increasingly effective British propaganda contributed to that change in public and elite political opinion. Moreover, private groups like the National Security League and the American Defense Society, prominent Republican public officials including Theodore Roosevelt and former Army Chief of Staff, Maj Gen Leonard Wood as well as active duty Army and Navy officers all lobbied for stronger defense measures. Wilson found it increasingly difficult for the U.S. to remain neutral after the sinking of the British ocean liner *Lusitania* by a German submarine in May 1915 killed 128 of the 139 U.S. citizens aboard the vessel.[66]

The 1916 legislation authorized eight Army aero squadrons in wartime. After years of ignoring warnings that American military aeronautics was falling dangerously behind other powers, the legislators appropriated $13,281,666—an enormous sum by the American standards of that day. In addition, the act authorized the training of reservists and the development of more powerful aircraft engines.[67]

Finally, the Signal Corps had the funds that it needed to develop a real air program, not just a modest experimental activity. The War Department authorized an air organization composed of four aero squadrons in the U.S. and three more in its overseas possessions. Each unit would be equipped with 12 aircraft. All of those planes would be superior in performance to those that had failed their operational tests over Mexico. New specifications for such aircraft had been issued by the CSO earlier in 1916.[68]

On 6 December 1916, the Signal Corps let preliminary orders for 96 twin-engine seaplanes at a cost of $2,007,000. They were intended to defend the coast of the continental U.S. as well as the Philippians, Hawaii and the Panama Canal. The orders were awarded to four firms: the Standard Aero Corporation (32); the Burgess Company (32); the Aeromarine Sales and Engineering Company (16); and the Curtiss Aeroplane and Motor Corporation (16). Their engines would be built by the Hall-Scott Motor Car Company, the Curtiss firm and the B.F. Sturtevant Company.[69]

Secretary Baker had also taken significant administrative actions to help prepare the Army's fledgling air arm for the possibility of American entry into the European conflict. Under his auspices, a technical board of officers and engineers conducted a thorough study of the American aircraft industry during Fiscal Year 1916 to assess its productive capacity and encourage improved U.S. military aircraft designs. In August 1916, the War Department announced that it had completed (unspecified) plans for the complete mobilization of the aircraft industry with the ambitious goal of making the Army's modest air arm the world's best.

In May 1916, as part of his effort to revamp the Army's aeronautical program, Baker placed Squier in charge of the Aviation Section of the Signal Corps. Although not a pilot himself, Squier, had just spent four years in Europe as the U.S. military attaché in London. Having secretly observed the air war over the western front firsthand, he knew that American military aeronautics was well behind that of the major European combatants.[70]

Squier understood that aeronautical progress required enormous amounts of money. Consequently, he carried his crusade for aeronautical research and development to Congress. Testifying before the House Military Affairs Committee in January 1917, Squier stressed the necessity for

Table 2.1 U.S. aircraft production 1909 to 1916

Year	*Total*	*Military*	*Civil*
1909	Not available	1	Not available
1910	Not available	–	Not available
1911	Not available	11	Not available
1912	45	16	29
1913	43	14	29
1914	49	15	34
1915	178	25	152
1916	411	142	269

Source Aerospace Industries Association of America (AIA), Inc., *Aerospace Facts and Figures, 1925* (Los Angeles, CA: 1964) 24, www.aia-aerospace.org

a high level of government support so that the U.S. could develop the same quality of aviation materiel as the warring European powers.[71] He indicated that the Aviation Section had "... distributed the orders [for aeronautical materiel] among a large number of manufacturers in order to encourage them, and thus build up the industry [in the United States]."[72] Squier suggested that, unless the government was able to provide stable funding over several years, manufacturers would be reluctant to become involved in developing and producing advanced military aircraft."[73]

Although Congress had appropriated the then unprecedented sum of nearly $19.5 million for Army aeronautics in 1916 and early 1917, the nation's aviation industry was able to deliver only 314 aircraft during that period. Manufacturers were continually pleading for more time to fulfill their contracts. The Army found it virtually impossible to penalize them because government policy was to withdraw penalties if the manufacturers claimed that extenuating circumstances had caused delays (Table 2.1).[74]

Charting Naval Aviation's Future

Despite its long-term implications, the immediate response to the Chambers board's report was extremely disappointing to naval aviation advocates. Historian Stein stresses that "... despite his public statements, Secretary Daniels was not willing to fight hard for aviation, especially since the Wilson administration was worried about a budget deficit caused by lower than expected tariff revenue.... When Daniels presented the Navy's proposed budget to Congress, he asked for only $350,000 for aviation and he did not fight hard to get it. Congress only appropriated the same low amount it had the previous two years—$65,000...."[75]

During December 1913, CAPT Mark L. Bristol replaced Chambers as the head of the Navy's aeronautics program. Devoted to aviation, Chambers had turned down opportunities for more sea duty and was slated for retirement by a "plucking board" instituted by Daniels for senior captains who hadn't had sufficiently punched that critical ticket for promotion to flag rank.

Although a non-flyer like his predecessor and not known as an aviation enthusiast, Bristol proved to be an effective advocate who was a good administrator with a strong personality. Bristol and Chambers worked well together fighting to advance the cause of naval aviation. During Bristol's tenure, naval aviation was officially recognized on 1 July 1914 with the creation of the Office of Naval Aeronautics in the Division of Operations under the Secretary of the Navy. He was officially designated as the "Director of Naval Aviation" in November 1914. Political scientist Vincent Davis concludes that those actions reflected the first official Navy commitment to aviation as a continuing activity and its initial step toward organizational autonomy within the service's bureau structure.

Aided by the war in Europe and the success of naval aviators at Veracruz and Tampico, Bristol played a key role in convincing Congress to appropriate $1 million for naval aviation during Fiscal Year 1915. He used the appropriation to purchase three new tractor planes from Burgess-Dunne ignoring the advice of naval aviators who actually had practical experience in aircraft design and flying. The aircraft failed to meet all of the Navy's specifications and didn't represent a significant increase in performance. That episode underscored the continuing frustration of naval aviators serving under the command of non-flyers. Nevertheless, Bristol made significant contributions directly or indirectly to naval aviation. He helped win congressional approval of a new organization that became the National Advisory Committee for Aeronautics (NACA). Congress authorized its creation on 3 March 1915 as a rider to the Naval Appropriations Act. A month later, Wilson appointed the committee's members, including senior military officers.[76]

Created in response to perceptions that the U.S. lagged far behind wartime technology, the NACA's explicit purpose was the "scientific study of problems of flight with a view to their practical solution."[77] It served a valuable advisory role during the early years of its existence but didn't actually make substantial contributions to the development of American aeronautics until it inaugurated its first full-scale program of aeronautical research in 1920 at its new Langley Memorial Aeronautical Laboratory in Hampton, Virginia.[78]

Meanwhile, to develop in-house expertise on aircraft technology and production, the BuC&R began designing a naval plane during Fiscal Year 1915. Work was begun on an airplane at the Washington Navy Yard based on the design and specifications drawn up by naval constructors. By 30 June 1916, the aircraft had been built and test flown at that installation. The experience gained by BuC&R during that process was seen as necessary for the Navy to deal effectively with the myriad of problems that would surface if the service needed to procure large numbers of aircraft from private firms during World War I.

Bristol gained a valuable ally in Rear Adm Taylor. While primarily concerned with ships, BuC&R's chief strongly believed in the future of naval aviation. He pushed for the study of aeronautics while building the first modern wind tunnel in the U.S. at the Washington Navy Yard and establishing an aerodynamics laboratory there. Taylor displayed a strong spirit of cooperation in dealing with the common problems he and Bristol faced. But, progress slowed after Daniels selected Adm William S. Benson to be the first Chief of Naval Operations (CNO) in 1915. After Bristol was placed under the CNO's office, Benson gradually stripped him of his authority and virtually ignored naval aviation.[79] Following his return to sea duty in March 1916, Bristol's old position as "Director of Naval Aeronautics," was eliminated. Aviation administrative duties were retained in the Office of the Chief of Naval Operations and placed in the hands of CAPT Josiah McKean who had been "long known as a frosty sailorman but not one with... [a] great interest in an air arm...."[80]

While most admirals remained focused on building more battleships, aviation's importance was gaining some traction in the Navy's secretariat. Secretary Daniels was a pacifist, tee-totaling former newspaper editor and publisher from North Carolina. He received his cabinet post as a political reward for advocating Wilson's presidential candidacy. While later known for several key initiatives including expanding the sea service for World War I and enlisting women in it, Daniels wrote (or at least signed) an article published in the *Washington Post* on 22 February 1914 which stressed that "With the Panama Canal, as well as our coast fortifications to safeguard, not only from land and water, but from the air, steps must and should be taken to muster an air fleet *absolutely second to none* [italics added] in either hemisphere."[81]

In July 1915, Daniels authorized the conversion of a building at the Washington Navy Yard to be used for testing aeronautical machinery. The facility evolved into the Aeronautical Engine Laboratory. During that same summer, he endorsed the General Board's recommendation of the urgent necessity of obtaining a fleet of 200 aircraft for the sea service. With war raging in Europe, the prospect that Congress would appropriate the necessary funds for those aircraft or additional ones appeared to be excellent. In September 1916, Daniels signed an order establishing formal flight testing and establishing procedures to determine whether operational aircraft were safe to fly. Those initiatives would be the basis for accepting new aircraft for operational duties in the sea service.[82]

Notes

1. "Cannon, Joseph Gurney, 1836–1926," Office of the Historian, Office of Art and Archives, Office of the Clerk, *History, Art and Archives, United States House of Representatives*, undated, history.house.gov; Benjamin D. Foulois, Maj Gen (Ret.), and Carroll V. Glines, Col (Ret.), *From the Wright Brothers to the Astronauts: The Memoirs of Major General Benjamin Foulois* (New York, NY: McGraw-Hill, 1968), 1.
2. Charles De Forest Chandler, Col (Ret.), and Frank Lahm, Brig Gen (Ret.), *How Our Army Grew Wings: Airmen and Aircraft Before 1914* (New York, NY: Ronald Press, 1943), 139; Charles J. Gross, *American Military Aviation: The Indispensable Arm* (College Station, TX: Texas A&M University Press, 2002), 16; Thomas W. McFadden, "Building Industries: Collective Action Problems and Institutional Solutions in the Development of the U.S. Aviation Industry, 1903–1938," Ph.D. dissertation, University of Arizona, 1999, 47, University of Arizona Libraries, lib.arizona.edu.
3. Henry H. Arnold, General of the Air Force (Ret.), *Global Mission* (New York, NY: Harper and Brothers, 1949), 8, 9; Dik Alan Daso, *Hap Arnold and the Evolution of American Airpower* (Washington, D.C.: Smithsonian Institution Press, 1 January 2001), 30–31, 37; Thomas M. Coffee, *HAP: The Story of the U.S. Air Force and the Man Who Built It, General Henry H. "Hap" Arnold* (New York, NY: The Viking Press, 1982), 19; Gross, *American Military Aviation: The Indispensable Arm*, 16.

4. Early aircraft fuselages were constructed using frames, typically of wood or steel tubing, which could then be covered with fabric such as cotton or Irish linen. Subsequently, three main types of fuselages evolved: Monocoque, Semi-Monocoque and Truss Frame. Monocoque fuselages featured an exterior skin that bears primary loads, minimizing the need for internal frameworks and benefiting aircraft with its lightweight and aerodynamically efficient design. Truss Frame fuselages used a lightweight steel alloy tube framework to ensure strength and rigidity while keeping weight low. Semi-Monocoque fuselages, merged Monocoque and Truss Frame features, providing a balanced approach that optimizes strength, weight and durability.
5. McFadden, "Building Industries: Collective Action Problems and Institutional Solutions in the Development of the U.S. Aviation Industry, 1903–1938," 68; T.H.G. Megson, *Aircraft Structures for Engineering Students* (London, UK: Edward Arnold Publishers LTD, 1972), 198; Bruce Robertson, *WWI British Aeroplane Colours and Markings* (Berkhampstead, UK: Albatros Publications Inc., 1996), 1–2; Sean Walsh, "Types of Aircraft Fuselages: Monocoque, Semi-Monocoque, and Truss Frame Explained," updated 23 February 2024, *PILOTPASSION*, pilotpassion.com.
6. Robert F. Futrell, *Ideas, Concepts, Doctrine: A History of Basic Thinking in the United States Air Force, 1907–1964* (Maxwell AFB, AL: The Air University, 1971, 1974), 8; Alfred Goldberg, Ed., *A History of the United States Air Force, 1907–1957* (New York, NY: Van Nostrand, 1957), 1–2; Laurence M. Burke II, *At the Dawn of Airpower: The U.S. Army, Navy, and Marine Corps' Approach to the Airplane, 1907–1917* (Annapolis, MD: Naval Institute Press, 2022), xvii.
7. Futrell, *Ideas, Concepts, Doctrine: A History of Basic Thinking in the United States Air Force, 1907–1964*, 8; Goldberg, *A History of the United States Air Force, 1907–1957*, 2; Chandler and Lahm, *How Our Army Grew Wings: Airmen and Aircraft Before 1914*, 144–145; *Juliette A. Hennessy, The United States Army Air Arm, April 1861–April 1917. Air Force Historical Study No. 98* (Maxwell AFB, Al.: USAF Historical Division, Research Studies Institute, Air University, May 1958, new imprint 1985), 20–21; Burke, *At the Dawn of Airpower: The U.S. Army, Navy, and Marine Corps' Approach to the Airplane, 1907–1917*, xviii, 11.

8. The General Board was an advisory body established on 13 March 1900 by Secretary Long because of inadequate staff work during the war with Spain. Composed of senior admirals, most nearing the ends of their careers, the board had no executive authority but its advice was highly influential because of its distinguished membership.
9. Arnold D. Turnbull, and Clifford L. Lord, *History of United States Naval Aviation* (New Haven, CT: Yale University Press, 1949), 2–3; A. Hunter Dupree, *Science in the Federal Government* (New York, NY: Arno Press, 1980, reprint of 1957 Harvard University Press publication), 304; Vincent Davis, *The Admirals Lobby* (Chapel Hill, NC: The University of North Carolina Press, 1967), 49–50; Matthew M. Oyos, "Theodore Roosevelt and the Instruments of War," *The Journal of Military History*, October 1996, 640; Clarke Van Fleet, Historian, Office of the Deputy Chief of Naval Operations (Air Warfare) and William J. Armstrong, Naval Air Systems Command, *United States Naval Aviation, 1910–1980.NAV AIR 00-80P-1. Supersedes NAVWEPS 00-80P-1 dated 1970* (Washington, D.C.: USGPO, 1981), 2.
10. Harry Combs with Martin Cardin, *Kill Devil Hill: Discovering the Secret of the Wright Brothers* (Boston, MA: Houghton Mifflin, 1979), passim; Richard Hallion, ed., *The Wright Brothers: Heirs Prometheus* (Washington, D.C.: Smithsonian Institution, 1978), passim; Gibbs-Smith, *A History of Flying*, 20–30, 34–36, 47, 77–82; C. Fayette Taylor, Professor of Automotive Engineering Emeritus, MIT, "Aircraft Propulsion: A Review of the Evolution of Aircraft Powerplants," *Annual Report of the Board of Regents of the Smithsonian Institution. Publication 4518* (Washington, D.C.: USGPO, 1963), 252; Gross, *American Military Aviation: The Indispensable Arm*, 16.
11. Chandler and Lahm, *How Our Army Grew Wings: Airmen and Aircraft Before 1914*, 141–142; Hennessy, *Army Air Arm*, 25; Marvin McFarland, Ed., *The Papers of Wilbur and Orville Wright, 2 Vols* (New York, NY: McGraw-Hill, 1953), Vol. I, 514–515, 518; Donald M. Pattillo, *Pushing the Envelope: The American Aircraft Industry* (Ann Arbor, MI: The University of Michigan Press, 1998), 6.
12. "War Department Annual Reports [Fiscal Year] 1909. Volume I" (Washington, D.C.: GPO, 1909), 50–51, Hathi Trust Digital

Library, "War Department Annual Reports [Fiscal Year] 1911. Volume I" (Washington, D.C.: GPO, 1912), 739, Hathi Trust Digital Library; McFarland, *Wright Papers*, Vol. II, 754–755, 757, 761–763; Chandler and Lahm, *How Our Army Grew Wings: Airmen and Aircraft Before 1914*, 142–143; Hennessy, *Army Air Arm*, 25–26; Herbert A, Johnson, "The Aero Club of America and Army Aviation, 1907–1916," *New York History*, October 1985, 375–377; Paul Wilson Clark, "Major General George Owen Squier: Military Scientist," PhD dissertation, Case-Western Reserve University, January 1974, 69, 136–138; Matthew Oyos, "Theodore Roosevelt and the Instruments of War," *The Journal of Military History*, October 1996, 640.

13. Clark, "Major General George Owen Squier: Military Scientist," 69, 110–120, 133, 136–138; Maj Gen Adolphus Washington Greely, *Reminiscences of Adventures and Service: A Record of Sixty-Five Years* (New York, NY: Charles Scribner's Sons, 1927), 162–163; M. Leroy Spearman, "Design Trends for Army/Air Force Planes in the United States," National Aeronautics and Space Administration (NASA) Technical Memorandum 4179, Langley Research Center, June 1990, 1, Hathi Trust Digital Library; Flint O. DuPre, *U.S. Air Force Biographical Dictionary* (New York, NY: Franklin Watts, 1965), 221.
14. Hennessey, *The United States Army Air Arm, April 1861–April 1917. Air Force Historical Study No. 98*, 26, 225; Chandler and Lahm, *How Our Army Grew Wings: Airmen and Aircraft Before 1914,* 146–150; "Airship to Carry Two: Gen. Allen Settles Point as to Test of Aeroplane," *Washington Post*, 27 June 1909, 5, ProQuest Historical Newspapers, Pentagon Library; Chandler and Lahm, "Signal Corps Specification, No. 486."
15. Clarence H. Danhoff, *Government Contracting and Technological Change* (Washington, D.C.: The Brookings Institution, 1968), 18.
16. Spearman, "Design Trends for Army/Air Force Planes in the United States," NASA Technical Memorandum 4179, 1; Foulois and Glines, *From the Wright Brothers to the Astronauts: The Memoirs of Major General Benjamin Foulois*, 53–54; Chandler and Lahm, *How Our Army Grew Wings*, 145, 150; Hennessy, *Army Air Arm*, 26–28; "Report of the Chief Signal Officer, War Department Annual Reports [Fiscal Year] 1908," Vol. II, 211, ProQuest Congressional Documents, Pentagon Library; C. de F. Chandler,

"The Birth of Military Aviation," *Current History*, December 1923, 378, 379, Pentagon Library; "Three Aeroplanes for Signal Corps," *New York Times*, 9 February 1908, 5, ProQuest Historical Newspapers, Anne Arundel County Public Library (AACPL) Maryland, www.aacpl.net.

17. Hennessy, *Army Air Arm*, 28, 33; Chandler and Lahm, *How Our Army Grew Wings: Airmen and Aircraft Before 1914*, 155; George Owen Squier, "Aeronautics in the United States at the Signing of the Armistice, November 11, 1918. Address Before the American Institute of Electrical Engineers," New York, NY, January 10, 1919, 3, George Owen Squire Papers, Box 6, Folder 2, United States Air Force Academy Library (USAFAL); Turnbull and Lord, *History of United States Naval Aviation*, 4.
18. Hennessy, *Army Air Arm*, 34; "U.S. War Department Annual Reports [Fiscal Year] 1911, Vol. I" (Washington, D.C.: GPO, 1912), 739, ProQuest Congressional Documents, Pentagon Library; "Wright Flies Among Cheers of Thousands: Three Successful Flights Made in Aeroplane at Fort Myer," *Atlanta Constitution*, 2 July 1909, 1, ProQuest Historical Newspapers, Pentagon Library; Robert Casari, "The First 59," *American Aviation Historical Society Journal*, Spring 1971, 4; Chandler, "The Birth of Military Aviation," *Current History*, December 1923, 381.
19. Leo Deluca, "How Ida Holdgreve's Stitches Helped the Wright Brothers Get Off the Ground," *Smithsonian Magazine*, 15 March 2021, www.smithsonianmag.com.
20. "U.S. War Department Annual Reports [Fiscal Year] 1913, Volume I" (Washington, D.C.: GPO, 1914), 786–787, Hathi Trust Digital Library; Richard P. Hallion, *Test Pilots: The Frontiersmen of Flight* (Garden City, NY: Doubleday and Company, 1981), 37–38; "U.S. War Department Annual Reports [Fiscal Year] 1912, Volume I" (Washington, D.C.: GPO, 1913), 966, Pentagon Library; "U.S. War Department Annual Reports [Fiscal Year] 1913, Vol. I" (Washington, D.C.: GPO, 1914), 782, Pentagon Library; J.W. Mitchell, "Army Tests Curtiss and Wright Planes," *Aero: America's Aviation Weekly*, 25 May 1912, 195, Hathi Trust Digital Library; Daniel L. Haulman, *One Hundred Years of Flight: USAF Chronology of Significant Air and Space Events 1903–2002* (Maxwell AFB, AL: Air Force History and Museums Program in association with the Air University Press, 2003), 7.

21. Roy A. Grossnick, *United States Naval Aviation, 1910–1995* (Washington, D.C.: Naval Historical Center, Department of the Navy, 1997), 2; Van Fleet and Armstrong, *United States Naval Aviation, 1910–1980,* 2; Burke, *At the Dawn of Airpower: The U.S. Army, Navy, and Marine Corps' Approach to the Airplane, 1907–1917*, 35, 36.
22. Burke, *At the Dawn of Airpower: The U.S. Army, Navy, and Marine Corps' Approach to the Airplane, 1907–1917*, 38–39; Turnbull and Lord, *History of United States Naval Aviation*, 7–8; Julius Augustus Furer, *Administration of the Navy Department in World War II* (Washington, D.C.: Naval History Division, USGPO, 1959), 353, ibiblio.org/hyperwar/USN/Admin-Hist/USN; Grossnick, *United States Naval Aviation, 1910–1995*, 3.
23. Stephen K. Stein, "Washington Irving Chambers: Innovation, Professionalization, and the New Navy, 1872–1919," PhD dissertation, The Ohio State University, 1999, 286, 287, 288, OhioLINK, Electronic Theses & Dissertations Center, www.ohiolink.edu; Grossnick, *United States Naval Aviation, 1910–1995*, 2; Burke, *At the Dawn of Airpower: The U.S. Army, Navy, and Marine Corps' Approach to the Airplane, 1907–1917*, 38–39; Stephen K. Stein, *From Torpedoes to Aviation: Washington Irving Chambers and Technological Innovation in the New Navy, 1876–1913* (Tuscaloosa, AL: The University of Alabama Press, 2007), 159–166.
24. U.S. Congress, House, "Hearings Before the Committee on Naval Affairs of the House of Representatives on Estimates Submitted by the Secretary of the Navy [Fiscal Year] 1914" (Washington, D.C.: GPO, 1914), 8 December 1913, 19, Hathi Trust Digital Library; Stein, *From Torpedoes to Aviation*, 159–166; Stein, "Washington Irving Chambers: Innovation, Professionalization, and the New Navy, 1872–1919" 288; Grossnick, *United States Naval Aviation, 1910–1995*, 2, 3; Burke, *At the Dawn of Airpower: The U.S. Army, Navy, and Marine Corps' Approach To The Airplane, 1907–1917*, 39.
25. Stein, "Washington Irving Chambers: Innovation, Professionalization, and the New Navy, 1872–1919", 295; Burke, *At the Dawn of Airpower: The U.S. Army, Navy, and Marine Corps' Approach to the Airplane, 1907–1917*, 41, 42; Lt Cdr William O. Shanahan, *Procurement of Naval Aircraft, 1907–1939. Volume XVII, Deputy*

Chief of Naval Operations (Air). Monographs in the History of Naval Aviation (Naval Aviation History Unit, Op-519B. Prepared under the direction of the Director of Naval History, 1946), 36, Navy Department Library, NHHC.
26. "Annual Reports of the Navy Department for the Fiscal Year 1910" (Washington D.C.: GPO, 1910), 23, Hathi Trust Digital Library; Shanahan, *Procurement of Naval Aircraft, 1907–1939*, 37; Van Fleet and Armstrong, *United States Naval Aviation, 1910–1980*, 3, 4; Turnbull and Lord, *History of United States Naval Aviation*, 17; Stein, "Washington Irving Chambers: Innovation, Professionalization, And the New Navy, 1872–1919," 290, 291, 292, 293, 296.
27. Ailerons are one of the primary flight control surfaces on aircraft. They are movable surfaces that control movement about the longitudinal axis—called roll. Ailerons are used along with the rudder to make turns.
28. McFadden, "Building Industries: Collective Action Problems and Institutional Solutions in the Development of the U.S. Aviation Industry, 1903–1938," 56–60; Richard F. Baker, "Glenn H. Curtiss," *U.S. Centennial of Flight Commission*, undated, centennia lofflight.net.
29. "Annual Report of the Navy Department for the Fiscal Year 1910" (Washington D.C.: GPO, 1910), 23, Hathi Trust Digital Library; Shanahan, *Procurement of Naval Aircraft, 1907–1939*, 37; Van Fleet and Armstrong, *United States Naval Aviation, 1910–1980*, 3, 4; Turnbull and Lord, *History of United States Naval Aviation*, 17; Stein, "Washington Irving Chambers: Innovation, Professionalization, and the New Navy, 1872–1919," 290, 291, 292, 293, 296; Furer, *Administration of the Navy Department in World War II*, 354; Turnbull and Lord, *History of United States Naval Aviation*, 11, 12; C.R. Roseberry, *Glenn Curtiss: Pioneer of Flight* (Garden City, NY: Doubleday & Company, Inc., 1972), 308–309; Alden Hatch, *Glenn Curtiss: Pioneer of Naval Aviation* (New York, NY: Julian Messner, Inc., 1942), 204–205, 206, 213–214; Capt. W. H. Sitz, USMC, *A History of U.S. Naval Aviation. Technical Note No. 18. Series of 1930* (Washington, D.C.: Bureau of Aeronautics, USGPO, 1930), 5, Internet Archive; Grossnick, *United States Naval Aviation, 1910–1995*, 3–6, "USS Birmingham (CL-2)," NHHC, undated, www.history.navy.mil.

30. Burke, *At the Dawn of Airpower: The U.S. Army, Navy, and Marine Corps' Approach to the Airplane, 1907–1917*, 40; Grossnick, *United States Naval Aviation, 1910–1995*, 3; Karl Jack Bauer and Stephen S. Roberts, *Register of Ships of the U.S. Navy, 1775–1990: Major Combatants* (Santa Barbara, CA: Greenwood Publishing Group, 1991), 135.
31. "Annual Reports of the Navy Department for the Fiscal Year 1910" (Washington D.C.: GPO, 1910), 23, Hathi Trust Digital Library; Shanahan, *Procurement of Naval Aircraft, 1907–1939*, 37; Van Fleet and Armstrong, *United States Naval Aviation, 1910–1980*, 3, 4; Turnbull and Lord, *History of United States Naval Aviation*, 17; Stein, "Washington Irving Chambers: Innovation, Professionalization, and the New Navy, 1872–1919," 290, 291, 292, 293, 296; Furer, *Administration of the Navy Department in World War II*, 354; Turnbull and Lord, *History of United States Naval Aviation*, 11, 12; Roseberry, *Glenn Curtiss: Pioneer of Flight*, 308–309; Hatch, *Glenn Curtiss: Pioneer of Naval Aviation*, 204–205, 206, 213–214; U.S. Congress, House of Representatives, Committee on Naval Affairs, 9 January 1913, 544, ProQuest Congressional Documents, Pentagon Library; Sitz, *A History of U.S. Naval Aviation. Technical Note No. 18. Series of 1930*, 5; Grossnick, *United States Naval Aviation, 1910–1995*, 3–6; "Wright Flying Machines," *Army and Navy Register*, 17 October 1910, 8, Nimitz Library, USNA; Bowers, *Curtiss Aircraft, 1907–1947*, 47; Burke, *At the Dawn of Airpower: The U.S. Army, Navy, and Marine Corps' Approach to the Airplane, 1907–1917*, 40, 43; Davis, *The Admirals Lobby*, 63–64.
32. Clark G. Reynolds, *Admiral John H. Towers: The Struggle for Naval Air Supremacy* (Annapolis, MD: Naval Institute Press, 1991), 4, 5, 6, 7, 24, 28–29, 32; Grossnick, *United States Naval Aviation, 1910–1995*, 10, 11; "John Henry Towers: 30 January 1885–30 April 1955," NHHC, history.navy.mil.
33. "Annual Reports of the Navy Department For Fiscal Year 1911" (Washington, D.C.: GPO, 1912), 58–59, Hathi Trust Digital Library; "Annual Reports of the Navy Department for Fiscal Year 1912" (Washington, D.C.: GPO, 1913), 156–157, Hathi Trust Digital Library; Burke, *At the Dawn of Airpower: The U.S.*

Army, Navy, and Marine Corps' Approach to the Airplane, 1907–1917, 47–49, 84, 88; Grossnick, *United States Naval Aviation, 1910–1995*, 6, 7, 9.

34. Allan R. Millett, *Semper Fidelis: The History of the United States Marine Corps. The Macmillan Wars of the United States*, Louis Morton, Gen. Ed. (New York, NY: Macmillan Publishing Co., 1980), 277; Edward C. Johnson and Graham A. Cosmas, Ed., *Marine Corps Aviation: The Early Years, 1912–1940* (Washington, D.C.: History and Museums Division, Headquarters, USMC, 1977, reprinted 1991), 2, 3, Internet Archive; "Annual Report of the Major General Commandant of the United States Marine Corps to the Secretary of the Navy for the Fiscal Year 1912" (Washington, D.C.: GPO, 1912), 12, Pentagon Library.
35. "Annual Reports of the Navy Department For Fiscal Year 1911" (Washington, D.C.: GPO, 1912), 58–59, Hathi Trust Digital Library; "Annual Reports of the Navy Department for Fiscal Year 1912" (Washington, D.C.: GPO, 1913), 156–157, Hathi Trust Digital Library; "Annual Reports of the Navy Department for Fiscal Year 1913" (Washington, D.C.: GPO, 1914), 142, Hathi Trust Digital Library; David W. Taylor, "Navy Airplane Development," *The Journal of the Society of Automotive Engineers*, December 1918, 373, Hathi Trust Digital Library; Sitz, *A History of U.S. Naval Aviation. United States Navy Department Bureau of Aeronautics Technical Note No. 18, Series of 1930*, 6, 7; Turnbull and Lord, *History of United States Naval Aviation*, 23; Stein, "Washington Irving Chambers: Innovation, Professionalization, and the New Navy, 1872–1919," 297–298, 300–301, 303; Van Fleet, and Armstrong, *United States Naval Aviation, 1910–1980*, 5, 6; Reynolds, *Admiral John H. Towers: The Struggle for Naval Air Supremacy*, 47; Thomas C. Hone and Curtis A. Utz, *History of the Office of the Chief of Naval Operations, 1915–2015* (Washington, D.C.: NHHC, Department of the Navy, 2020), 28, history.navy.mil; Stein, *Torpedoes to Aviation*, 156–173; Grossnick, *United States Naval Aviation, 1910–1995*, 7, 8, 9, 10, 18; "Committee After Military Plane," *San Francisco Chronicle*, 7 December 1910, 18, ProQuest Historical Newspapers, Pentagon Library; "Curtiss in Airship Rises From Water," *New York Times*, 27 January 1911, 1, ProQuest Historical Newspapers, Pentagon Library; Robert G. Skerrett, "Our Navy has the

Best Seaplane Catapult," *New York Times*, 2 April 1916, SM5, ProQuest Historical Newspapers, Pentagon Library; Wendell W. Hammer, "Shipboard Aviation," *Navy Air Pilot and Aeronautic Review*, April 1918, 12, Hathi Trust Digital Library; "Catapult for Aeroplanes," *New York Times*, 13 November 1912, 19, ProQuest Historical Newspapers, Pentagon Library; Burke, *At the Dawn of Airpower: The U.S. Army, Navy, and Marine Corps' Approach to the Airplane, 1907–1917*, 87–88; Cdr H. C. Richardson, BuAer, "Developments in Naval Aeronautics," *SAE [Society of Automotive Engineers] Transactions*, Vol. 19, PART II, 1924, 671, JSTOR, Pentagon Library; Van Fleet, and Armstrong, *United States Naval Aviation, 1910–1980,* 14; "Annual Reports of the Navy Department for the Fiscal Year 1917" (Washington, D.C.: GPO, 1917), 36, Hathi Trust Digital Library; Robert J. Cressman, "North Carolina II (Armored Cruiser No. 12)," published 8 May 2022, NHHC, history.navy.mil.

36. Russell F. Weigley, *History of the United States Army*: *The Macmillan Wars of the United States*, Louis Morton, Gen. Ed. (New York, NY: Macmillan, 1967), 289–290, 311, 319, 322, 558, 559; Goldberg, *A History of the United States Air Force, 1907–1957*, 4, 6, 8; "To Urge Aeros for Army: Secretary Dickinson Greatly Impressed by Flight in Machine," *Washington Post*, 8 November 1910, 2, ProQuest Historical Newspapers, Pentagon Library; Joseph J. Corn, *The Winged Gospel. America's Romance with Aviation, 1900–1950* (New York: Oxford University Press, 1983), passim.
37. Edgar F. Raines, Jr., *Army Historical Series. Eyes of the Artillery: The Origins of Modern U.S. Army Aviation in World War II. CMH Pub 70-31-1* (Washington, D.C.: Center of Military History (CMH), U.S. Army, 2000), 8, history.army.mil.
38. Hennessy, *Army Air Arm*, 79–85, 86, 122; Edward O. Purtee, *History of the Army Air Service, 1907–1926, Air Force Logistics Command (AFLC) Historical Study No. 279* (Wright-Patterson AFB, OH: Office of History, Air Materiel Command, May 1948), 15–18; Elretta Sudsbury, *Jackrabbits to Jets: The History of North Island, San Diego, California* (San Diego, CA: Neyenesch Printers, Inc., North Island Historical Committee, 2d Ed., October 1967), 27–29; Chandler and Lahm, *How Our Army Grew Wings: Airmen and Aircraft Before 1914*, 250, 258, 265, 266; Goldberg, *A*

History of the United States Air Force, 1907–1957, 7; "Naval Aviation Camp," *New York Times*, 30 December 1911, 8, ProQuest Historical Newspapers, Pentagon Library; "North Island Now Busiest Camp," *Aero: America's Aviation Weekly*, 17 February 1912, 394, Hathi Trust Digital Library; Grossnick, *United States Naval Aviation, 1910–1995*, 8.

39. See note above; H.H. Arnold to Grover C. Loening, Subj. Congratulations on New Job With Wrights, 29 July 1913, Grover C. Loening Papers, Box 1, 1913 Folder, Manuscript Division, Library of Congress (LOC); Grover C. Loening to Samuel Reber, Signal Corps, Subj.: Possible Job With Signal Corps, 15 June 1914, Grover C. Loening Papers, Box 1, 1914 Folder, Manuscript Division, LOC; Samuel Reber, Office of the Chief Signal officer, to Grover C. Loening, Subj.: Signal Corps Job, 13 July 1914, Grover C. Loening Papers, Box 1, 1914 Folder, Manuscript Division, LOC; War Department, Subj.: Order Appointing Grover C. Loening as a civil service "Aero Engineer" at $3600.00 Per Annum Salary in San Diego, Grover C. Loening Papers, Box 1, 1914 Folder, Manuscript Division, LOC; O. A. Brindley to Grover C. Loening, Subj.: Reber Eager to have Much Experimental Aviation Work Done At San Diego, 8 June 1914, Grover C. Loening Papers, Box 1, 1914 Folder, Manuscript Division, LOC; Grover C. Loening, Aeronautical Engineer, to the Chief Signal Officer of the Army, Subj.: Request for Salary Increase in Light of Growing Responsibilities, 21 January 1915, Grover C. Loening Papers, Box 2, January–May 1915 Folder, Manuscript Division, LOC; Grover C. Loening, *Our Wings Grow Faster* (Garden City, NY: Doubleday, Doran & Company, 1935), 48–50; Weigley, *History of the United States Army*, 559; Burke, *At the Dawn of Airpower: The U.S. Army, Navy, and Marine Corps' Approach to the Airplane, 1907–1917*, 136.
40. "Annual Reports of the Navy Department for the Fiscal Year 1915" (Washington D.C.: GPO, 1916), 42, 64th Cong., 1st sess., House of Representatives, Document No. 20, Hathi Trust Digital Library; Gordon Swanborough and Peter M. Bowers, *United States Navy Aircraft Since 1911* (London, UK: Putnam Aeronautical Books, 1990), 122–124; "The America Proves its Ability on Initial Trials," *Aero and Hydro*, 27 June 1914, 153, Hathi Trust Digital Library; "America Passes Tests – Prepared for Voyage," *Aero and Hydro*,

4 July 1914, 165, 166, Hathi Trust Digital Library; Grover C. Loening, *Takeoff Into Greatness: How American Aviation Grew So Big So Fast* (New York, NY: G.P. Putnam's Sons, 1968), 76–77.

41. Foulois and Glines, *From the Wright Brothers to the Astronauts: The Memoirs of Major General Benjamin Foulois*, 115; "Seek Safe Aeroplane," *Washington Post*, 23 March 1914, 3, ProQuest Historical Newspapers, AACPL, www.aacpl.net.
42. Foulois and Glines, *From the Wright Brothers to the Astronauts: The Memoirs of Major General Benjamin Foulois*, 115; Hennessy, *Army Air Arm*, 117; War Department, Office of the Chief Signal Officers, Order No. 16, 28 September 1914, Subj.: Appointing Grover C. Loening to Board to Conduct an Aeroplane Contest at the Signal Corps School, San Diego on or About 15 October 1914, Grover C. Loening Papers, Box 1, 1914 Folder, Manuscript Division, LOC; "Army Offers Prizes for Best Biplanes," *New York Times*, 29 June 1914, 4, ProQuest Historical Newspapers, Pentagon Library; Grover C. Loening, "Origin of Air Service Engineering," *The Airpower Historian*, October 1964, 97, 99, 100.
43. "Report of the Chief Signal Officer, United States Army, to the Secretary of War [Fiscal Year] 1914" (Washington, D.C.: GPO, 1914), 3, Hathi Trust Digital Library; "U.S. War Department Annual Reports [Fiscal Year] 1915. Volume I" (Washington, D.C.: GPO, 1916), 742, Congressional Documents, House of Representatives, 64th Cong., 1st sess., Document No. 92, Pentagon Library; "Appendix 16. Creation Of Aviation Section, Signal Corps, Sixty-Third Congress. Sess. II," in Chandler and Lahm, *How Our Army Grew Wings: Airmen and Aircraft Before 1914*, 316–319; Hennessy, *Army Air Arm*, 108–110; Mooney and Layman, *Military Aeronautics, 18*; Burke, *At the Dawn of Airpower: The U.S. Army, Navy, and Marine Corps' Approach to the Airplane, 1907–1917*, 132–133.
44. "Recent Strides in Aeroplane Branch of U.S. Army Must Be Kept Up Until Uncle Sam's Aircraft Hold Dominion of the Skies, Military Experts Insist," *Washington Post*, 4 October 1914, 6, ProQuest Historical Newspapers, Pentagon Library.
45. Lori A. Henning, *Harnessing the Airplane: American and British Cavalry Responses to a New Technology, 1903–1939* (Norman, OK: University of Oklahoma Press, 2019), 37.

46. Ibid.
47. Martin Claussen, *Materiel Research and Development in the Army Air Arm, 1914–1945. AAF (Army Air Forces) Historical Studies No. 50* (Washington, D.C.: Historical Office, Headquarters, U.S. Army Air Forces, November 1946), 12; Clark, "Major General George Owen Squier: Military Scientist," 155–158.
48. Herbert A. Johnson, *Wingless Eagle: U.S. Army Aviation Through World War I* (Chapel Hill and London: The University of North Carolina Press, 2001), 42.
49. Gross, *American Military Aviation: The Indispensable Arm*, 19–21, 53.
50. "Navy Aviators Busy at Guantanamo Bay," *Fly Magazine*, April 1913, 25, Hathi Trust Digital Library; "Aviation," *The Navy*, April 1913, 131, Hathi Trust Digital Library; Stein, *Torpedoes to Aviation*, 174–179; "Scouts Fly at Night," *New York Times*, 31 January 1913, 5, ProQuest Historical Newspapers, Pentagon Library; Stein, "Washington Irving Chambers: Innovation, Professionalization, and the New Navy, 1872–1919," 316; Reynolds, *Admiral John H. Towers: The Struggle for Naval Air Supremacy*, 56, 57, 58, 59, 61–62.
51. Grossnick, *United States Naval Aviation, 1910–1995*, 11.
52. Stein, "Washington Irving Chambers: Innovation, Professionalization, and the New Navy, 1872–1919," 329.
53. "Annual Reports of the Navy Department for the Fiscal Year 1915" (Washington D.C.: GPO, 1916), 40, 64th Cong., 1st sess., House of Representatives, Document No. 20, Hathi Trust Digital Library.
54. Grossnick, *United States Naval Aviation, 1910–1995*, 12, 13; Stein, "Washington Irving Chambers: Innovation, Professionalization, and the New Navy, 1872–1919", 306; Graham A. Cosmos, "The Formative Years of Marine Corps Aviation, 1912–1939," *Aerospace Historian*, Summer, June 1977, 83.
55. U.S. Congress, House Committee on Military Affairs, "Aeronautics in the Army," 12–16 August 1913, 74, 90–92, 63rd Cong., 1st sess., HR 5304, Reports Miscellaneous, 1913 Folder, Box 244, Henry H. Arnold Papers, Manuscript Division, LOC; Daso, *Hap Arnold and the Evolution of American Airpower*, 63–64; Goldberg, *A History of the United States Air Force, 1907–1957*, 7–9; Maurer Maurer, "The 1st Aero Squadron – 1913–1917," *Air Power Historian,* October 1957, 207, 208; Hennessy, *Army Air Arm*, 74;

Foulois and Glines, *From the Wright Brothers to the Astronauts: The Memoirs of Major General Benjamin Foulois*, 107–110; Arnold, *Global Mission*, 41–42; Weigley, *History of the United States Army*, 335; Gross, *American Military Aviation: The Indispensable Arm*, 25; "Aeros Off for Texas: First Full Army Squadron Begins Mobilization," *Washington Post*, 28 February 1913, 9, ProQuest Historical Newspapers, Pentagon Library; John T. Correll, "The First of the Force [Chandler, et. al.]," *Air & Space Forces Magazine*, 1 August 2007, airandspaceforces.com.

56. "Annual Reports of the Navy Department for the Fiscal Year 1914" (Washington, D.C.: GPO, 1915), 12, Hathi Trust Digital Library; "U.S. Navy Flying Boats to See Service [In Mexico]," *Aero and Hydro: America's Aviation Weekly*, 25 April 1914, 37, Hathi Trust Digital Library; "[Navy] Pilots Fly Over Vera Cruz." *Aero and Hydro: America's Aviation Weekly*, 2 May 1914, 58, Hathi Trust Digital Library; "[Mexican Troops] Fire At Navy Hydro," *Aero and Hydro: America's Aviation Weekly*, 16 May 1914, 81, Hathi Trust Digital Library; Reynolds, *Admiral John H. Towers: The Struggle for Naval Air Supremacy*, 77, 78, 80; Grossnick, *United States Naval Aviation, 1910–1995*, 14–15; Vice Adm Patrick N. L. Bellinger (Ret.), excerpt from unpublished memoir, "Naval Aviation Goes to War: Patrick Bellinger's Memoir of Veracruz," 40–47, *Journal of the Naval Aviation Foundation*, Spring 2015, 45; "He Did Not Curry Favor [Henry C. Mustin]," *Aerospace Historian*, September 1983, 193; Van Fleet, and Armstrong, *United States Naval Aviation, 1910–1980*, 10.
57. Maurer, "The 1st Aero Squadron – 1913–1917," *Air Power Historian,* October 1957, 209; Hennessy, *Army Air Arm*, 145–167; Goldberg, *A History of the United States Air Force, 1907–1957*, 10; James L. Crouch, "Wings South: The First Foreign Employment of Air Power by the United States," *Aerospace Historian*, Spring, March 1972, 28.
58. Maurer, "The 1st Aero Squadron – 1913–1917," *Air Power Historian*, October 1957, 209; Hennessy, *Army Air Arm*, 145–167; Foulois and Glines, *From the Wright Brothers to the Astronauts: The Memoirs of Major General Benjamin Foulois*, 122–137; Goldberg, *A History of the United States Air Force, 1907–1957*, 10; Elsbeth E. Freudenthal, *The Aviation Business: From Kitty Hawk to Wall*

Street (New York, NY: Vanguard Press, 1940), 22–23; "6 Aeroplanes Lost in Mexico Campaign," *New York Times*, 23 April 1916, 6, ProQuest Historical Newspapers, Pentagon Library; Crouch, "Wings South: The First Foreign Employment of Air Power by the United States," *Aerospace Historian*, Spring, March 1972, 51–55, Pentagon Library.

59. Goldberg, *A History of the United States Air Force, 1907–1957*, 10; Freudenthal, *The Aviation Business: From Kitty Hawk to Wall Street*, 22–23; Hennessy, *Army Air Arm*, 172–173; Maurer, "The 1st Aero Squadron - 1913–1917," *Air Power Historian*, October 1957, 209; Daso, *Hap Arnold and the Evolution of American Airpower*, 81–82; "Complains of Aeroplanes," *New York Times*, 28 April 1916, 5, ProQuest Historical Newspapers, Pentagon Library; "Army Men Aiding Aviation Industry," *New York Times*, 3 September 1916, 8, ProQuest Historical Newspapers, AACPL, www.aacpl.net.
60. Allan R. Millett and Peter Maslowski, *For the Common Defense: A Military History of the United States of America* (New York, NY: The Free Press, A Division of Macmillan, Inc., 1984), 321.
61. Ibid.
62. Ibid., 323; Turnbull and Lord, *History of United States Naval Aviation*, 81.
63. Grossnick, *United States Naval Aviation, 1910–1995*, 19, 21; Turnbull and Lord, *History of United States Naval Aviation*, 80–85; Adrian O. Van Wyen, Historian, Deputy Chief of Naval Operations (Air) and the Editors of Naval Aviation News, *Naval Aviation in World War I* (Washington, D.C.: Office of the Chief of Naval Operations, USGPO, 1969), 6; Van Fleet, and Armstrong, *United States Naval Aviation, 1910–1980*, 15. "The Hydro-aeroplane," *Scientific American*, 2 December 1911, www.scientificamerican.com; Burke, *At the Dawn of Airpower: The U.S. Army, Navy, and Marine Corps' Approach to the Airplane, 1907–1917*, 173, 185.
64. Millett and Maslowski, *For the Common Defense: A Military History of the United States of America*, 322–324.
65. Ibid., 324–325.
66. Weigley, *History of the United States Army*, 342–354; Walter Millis, *Arms and Men: A Study in American Military History* (New York, NY: G.P. Putnam's Sons, 1956), 214–226.

67. Weigley, *History of the United States Army*, 348–349; Goldberg, *A History of the United States Air Force, 1907–1957*, 10; Hennessy, *Army Air Arm*, 154; Edwin L. Williams, Jr., "Legislative History of the Air Arm," *Military Affairs*, Summer 1956, 82.
68. Hennessy, *Army Air Arm*, 157.
69. "Contracts Let By Army for $2,007,000 Worth of Seaplanes," *Aviation and Aeronautical Engineering*, 15 December 1916, 321, Hathi Trust Digital Library.
70. Telegram, Subj.: Squier Sent Secretly to Western Front, 6 January 1915, "The George Owen Squier Papers, War Diary," November 1914–January 1915, U.S. Army Military History Institute (USAMHI); "Annual Report of the Secretary of War [Fiscal Year] 1916" (Washington, D.C.: GPO, 1916), 8, 39, Pentagon Library; Foulois and Glines, *From the Wright Brothers to the Astronauts: The Memoirs of Major General Benjamin Foulois*, 138–141; Hennessy, *Army Air Arm*, 157; "Squier Commands Aviation Section," *San Francisco Chronicle*, 13 April 1916, 10, ProQuest Historical Newspapers, Pentagon Library.
71. U.S. Congress, House, Committee on Military Affairs [Fiscal Year] 1918 Army Appropriations Bill, Hearings Before the Committee, 64th Cong., 2d sess., 1917, 1009, Pentagon Library.
72. Ibid.
73. Ibid., 1009–1011.
74. Ibid., 1010; Goldberg, *A History of the United States Air Force, 1907–1957*, 10–11; Hennessy, *Army Air Arm*, 197, 252; "U.S. War Department Annual Reports [Fiscal Year] 1915 (In Three Volumes), Volume I" (Washington, D.C.: GPO, 1916), 76, Congressional Documents, House of Representatives, 64th Cong., 1st sess., Document No. 92, Pentagon Library; Glenn L. Martin, "The First Half Century of Flight in America: The Seventeenth Wright Brothers Lecture," *Journal of the Aeronautical Sciences*, February 1954, 75; "Army Men Aiding Aviation Industry," *New York Times*, 3 September 1916, 8, ProQuest Historical Newspapers, Pentagon Library.
75. Stein, "Washington Irving Chambers: Innovation, Professionalization, and the New Navy, 1872–1919," 331.
76. "First Annual Report of the National Advisory Committee for Aeronautics, 1915, Third Edition" (Washington, D.C.: GPO, 1916), 7, Hathi Trust Digital Library; Jerome C. Hunsaker,

"Forty Years of Aeronautical Research," *Annual Report of the Smithsonian Institution, 1955* (Washington, D.C.: USGPO, 1956), 243–247; "President Names Aviation Board," *New York Times,* 3 April 1915, 4, ProQuest Historical Newspapers, Pentagon Library; Grossnick, *United States Naval Aviation, 1910–1995,* 16; U.S. Congress, House of Representatives, Committee On Naval Affairs, 21 February 1916, 1815, ProQuest Congressional Documents, Pentagon Library; Van Fleet, and Armstrong, *United States Naval Aviation, 1910–1980,* 11.

77. George W. Gray, *Frontiers of Flight: The Story of NACA Research* (New York, NY: Alfred A. Knopf, 1945, 1948), 10–13.
78. Hunsaker, "Forty Years of Aeronautical Research," *Annual Report of the Smithsonian Institution, 1955,* 247–251; Goldberg, *A History of the United States Air Force, 1907–1957,* 11; Martin Claussen, *Comparative History of Research and Development Policies Affecting Air Materiel, 1915–1944, AAF Historical Study No. 20* (Washington, D.C.: Army Air Forces Historical Division, June 1945), 24–25.
79. "Annual Reports of the Department of the Navy for the Fiscal Year 1916" (Washington, D.C.: GPO, 1917), 309, 311, U.S. Congress, House, House Documents, VOL. 30, 64th Cong., 2d sess., Hathi Trust Digital Library; Stein, "Washington Irving Chambers: Innovation, Professionalization, and the New Navy, 1872–1919," 332, 333, 335–336; Furer, *Administration of the Navy Department in World War II,* 353; Turnbull and Lord, *History of United States Naval Aviation,* 36, 37, 47, 51, 55; Tom D. Crouch, *Wings: A History of Aviation from Kites to the Space Age* (Washington, D.C.: Smithsonian National Air and Space Museum, in association with W. W. Norton & Company, New York, NY, 2003), 185; Grossnick, *United States Naval Aviation, 1910–1995,* 12, 15;. George F. Pearce, "He Did Not Curry Favor [Henry C. Mustin]," *Aerospace Historian,* September 1983, 193–194; "Chiefs of Naval Operations," NHHC undated, history.navy.mil; Van Fleet, and Armstrong, *United States Naval Aviation, 1910–1980,* 11; Sitz, *A History of U.S. Naval Aviation. United States Navy Department Bureau of Aeronautics Technical Note No. 18, Series of 1930,* 6; U.S. Congress, House, *Estimates Submitted by the Secretary of the Navy [For Fiscal Year] 1915, Hearings Before the Committee on Naval Affairs of the House of Representatives* (Washington, D.C.: GPO,

1915), 278, 287–290, 300, Hathi Trust Digital Library; "Annual Report of the Chief of the Bureau of Construction and Repair to the Secretary of the Navy for the Fiscal Year 1915" (Washington, D.C.: GPO, 1915), 7, 319, Hathi Trust Digital Library; "Annual Report of the Chief of the Bureau of Construction and Repair to the Secretary of the Navy for the Fiscal Year 1916" (Washington, D.C.: GPO, 1916), 11, Hathi Trust Digital Library; "Annual Reports of the Navy Department for the Fiscal Year 1915" (Washington, D.C.: GPO, 1916), 40, Hathi Trust Digital Library; "Annual Reports of the Navy Department for the Fiscal Year 1917" (Washington, D.C.: GPO, 1917), 35, Hathi Trust Digital Library; Burke, *At the Dawn of Airpower: The U.S. Army, Navy, and Marine Corps' Approach to the Airplane, 1907–1917*, 163–164, 230; Davis, *The Admirals Lobby*, 65–66.

80. Grossnick, *United States Naval Aviation, 1910–1995*, 17; "Progress of American Aviation," *Aeronautics*, 7 February 1917, 110, Hathi Trust Digital Library; "Notes On American Aviation," *Aeronautics*, 9 May 1917, 341, Hathi Trust Digital Library; Pearce, "He Did Not Curry Favor [Henry C. Mustin]," *Aerospace Historian*, September 1983, 195.
81. Regina T. Akers, *The Navy's First Enlisted Women: Patriotic Pioneers. World War I Commemorative Series* (Washington, D.C.: NHHC, Department of the Navy, 2019), 17–18, history.navy.mil; James Tertius de Kay, *Roosevelt's Navy: The Education of a Warrior President, 1882–1920* (New York, NY: Pegasus Books, 2012), 95; Josephus Daniels, Secretary of the Navy, "High Time for an Air Navy," *Washington Post*, 22 February 1914, 51, ProQuest Historical Newspapers, Newspapers: Pentagon Library; Stein, "Washington Irving Chambers: Innovation, Professionalization, and the New Navy, 1872–1919," 318–319, 322–326; Turnbull and Lord, *History of United States Naval Aviation*, 34–35.
82. Grossnick, *United States Naval Aviation, 1910–1995*, 17; "Seek 200 Navy Aeros," *Washington Post*, 12 July 1915, 5, ProQuest Historical Newspapers, Pentagon Library; U.S. Congress, House of Representatives, Committee on Naval Affairs, 21 February 1916, 1815–1816, 1818, 1824–1825, 1826, 1837, 1857, 1858, ProQuest Congressional Documents, Pentagon Library; Van Fleet, and Armstrong, *United States Naval Aviation, 1910–1980,* 15.

CHAPTER 3

The Great War: Enlisting Science and Engineering in American Aviation, 1917–1918

Introduction

During World War I, the airplane rapidly evolved from a fragile technological curiosity into a significant military weapon. Driven by the hope that control of the air might end the mindless stalemated slaughter in the trenches below, major European powers raced to expand the role of aviation from observation to air superiority over the battlefield and close support of their own ground troops. Airplanes also went to sea. There were rudimentary attempts at strategic bombing. Those innovative developments also stimulated a frantic race to achieve technological superiority in the air which forced the major belligerents to expand their substantial aeronautical research and development programs.[1]

After it declared war in April 1917, the U.S. was quickly swept up by this enthusiasm for air power. Although woefully unprepared for a modern war against a major industrialized nation, especially its aeronautical dimension, America quickly committed itself to the bizarre idea that building a huge air force virtually from scratch might tip the scales of battle against the Central Powers. But, when that aerial force failed to materialize quickly and play the decisive role predicted for it, there was an enormous political backlash. Many observers judged that America's first wartime air program had been a dismal failure.

C. J. Gross, *Technological Innovation and the Rise of Aviation, 1903–1941*, Palgrave Studies in the History of Science and Technology, https://doi.org/10.1007/978-3-031-71108-4_3

Traditionalists on Capitol Hill concluded that they must never again tolerate the graft, corruption and what they saw as abject failure of the wartime aircraft program. They were determined to reassert conventional competitive contracting practices and stricter congressional oversight. Senior civilian officials and military officers who had managed the program rejected the critics' conclusions. They were convinced that the program had been successful in the final analysis. They anticipated a strong cooperative postwar relationship between the military and the aviation industry. Army and Navy officers who had participated in the air war in Europe were convinced that it was the coming next big thing. Some like Army flyer Brig Gen William "Billy" Mitchell were convinced that air power might render conventional ground and naval forces nearly obsolete.[2]

World War I radically altered the American aviation industry and the technological foundations it rested upon. The war transformed it from a small, struggling cottage industry dominated by a handful of creative designers into a huge enterprise based upon professional science and engineering. While its mass production capacity temporarily disappeared after 1918, the importance of science and engineering did not. Some of the young men like Donald W. Douglas who had been drawn into aviation by the conflict, later built the aircraft manufacturing firms that made possible the hugely successful American aircraft production program during World War II.

The war demonstrated that superior technology and mass production were key elements of aerial victory. The conflict led to the creation of several strong research organizations in the federal government designed specifically to foster aeronautical progress. As noted previously, the NACA had already been established. McCook Field in Dayton, Ohio, home of the Army's Engineering Division established in August 1917, played a key role in that service's technological development during the war. Opened that same year, the Naval Aircraft Factory (NAF) in Philadelphia, Pennsylvania played a similar role for the sea service.[3]

Improvising Intervention

When President Wilson read his war message to Congress in April 1917, the U.S. was woefully unprepared for war against a major industrial power. The nation had continued to rely on her geographic isolation and a formidable Navy for security while Europe consumed itself in total war.

Like most of his fellow citizens, Wilson had believed that America had no business intervening militarily in Europe's bloody quarrel. However, a combination of circumstances—growing economic ties with the Allies, skillful British propaganda, the anglophile attitudes of key American elites and heavy-handed diplomacy by the German government—gradually drew the U.S. toward the fray. The immediate cause of American belligerency was the German decision to resume unrestricted submarine warfare against merchant shipping, including that of neutral nations including the U.S.[4]

Military historians Allan R. Millett and Peter Maslowski conclude that "The American role in World War I derived its character less from strategic thinking in the United States than from the geopolitical notion that the future well-being of the United States depended upon the balance of power in Europe and the outcome of the war."[5] They added that "Discarding the hallowed assumption that Europe's affairs did not involve the United States and the security of the western hemisphere, the Wilson administration decided that the nation had a critical stake in Allied victory. American involvement stemmed from economic self-interest as well as an emotional commitment to support 'democracy' (France and Great Britain) against 'autocracy' (Germany)."[6]

America's government was unready for war on a grand scale. It had no system in place to mobilize the manpower needed for a massive land force. Not until after the U.S. declaration of war did the Wilson administration submit proposal to Congress for a draft. The result was the "Selective Service Act" of 18 May 1917.[7]

Secretary of War Newton D. Baker—a lawyer and former progressive mayor of Cleveland, Ohio—was an avowed pacifist. He ran his department independently of the Navy and other federal agencies. Although the National Defense Act of 1916 had made it possible to begin to address the economic issues posed by modern total war, little had actually been accomplished when America entered the fight. The legislation had given the President the authority to place mandatory orders for military and naval supplies with private firms. If businesses failed to comply, they could be seized by the federal government. The act also required the War Department to survey the American arms and munitions industries. A rider to the Army appropriations act in August 1916 established a Council of National Defense consisting of the Secretaries of War, Navy, Interior, Agriculture, Commerce and Labor. The Council, assisted by an advisory body of businessmen and other public figures, was responsible for dealing with the problems of procurement and economic mobilization in an emergency.[8]

While the War Department was ignoring the problems of economic mobilization, the Council had established a Munitions Standards Board in March 1917. Soon after the U.S. declared war the following month, the Board was replaced by a General Munitions Board that included military officers. The new organization exercised responsibility for coordinating military procurement. But, it soon became apparent that, like its predecessor, it lacked adequate authority. The Council of National Defense replaced it with the War Industries Board in July 1917. To provide centralized direction and planning for aircraft production, the Council of National Defense had established the Aircraft Production Board on 16 May 1917 following a recommendation by a NACA advisory committee. But, the War Department was unwilling and unable, because of fierce resistance by the Army's various bureaus, to cooperate with the War Industries Board. Consequently, none of this institutional machinery proved to be effective and had to be energized by new leadership and organizational changes after the near collapse of America's economic mobilization during the winter of 1917–1918.[9]

America's military was no more prepared than its civilian government to meet the challenges posed by modern war. The Army had approximately 200,000 men under arms when the U.S. went to war.[10] Although the nation possessed rudimentary strategic planning capabilities—the General Board of the Navy, established in 1900 as well as the General Staff of the Army and the Joint Army and Navy Board, both created in 1903—its military contingency plans proved to be irrelevant to the situation that the nation faced in 1917.[11]

In early 1917, the American armed forces lacked practical experience in organizing, equipping, training and leading massive wartime forces. The Civil War had provided them with the last real test of those capabilities. The Army's tiny General Staff consisted of only 19 officers during April 1917. It was quickly overwhelmed by the challenges of preparing a massive land force for war.

The Army itself was not ready to face the battle-hardened German force. Most of its strength was scattered along the nation's southwestern border in small units as a result of the 1916 crisis with Mexico. It possessed no unit larger than a regiment and was poorly equipped with machine guns, artillery and mortars. John J. Pershing, then a major general, created a headquarters for American forces in France after his arrival there on 13 June 1917. His army became known as the American Expeditionary Forces (AEF).[12]

Pershing's airmen had to build an air force from scratch while securing aircraft and other equipment from the Allies. Brig. Gen. Mitchell was assigned to command of the AEF's front line units after a short stint as the Air Service's overall commander. Col Raynal C. Bolling was given responsibility for the rear areas. The absence of an overall air commander generated an enormous amount of tension and conflict. That was intensified by the fact that most of the AEF's aviators were young, inexperienced and ill-prepared to assume senior military posts. They were in frequent conflict with ground officers. Neither group had a firm grasp of the needs and capabilities of their respective military arms.

To help rectify that situation and revive his floundering aviation program, Pershing appointed Brig. Gen. William L. Kenly, an experienced artillery officer, as the AEF's chief air commander in September 1917. Bolling was given control of supplies but clashed with the volatile Mitchell. Pershing reorganized his Air Service again later that year and brought in now Brig. Gen. Benjamin D. Foulois and his staff from Washington, D.C. Mitchell clashed with the newcomers, especially Foulois. He blasted them as "carpetbaggers."[13] Bickering, infighting, personality conflicts and slow progress continued to plague the Air Service. Pershing, disgusted with the situation, assigned his old West Point classmate and friend Brig. Gen. Mason M. Patrick during May 1918 to rectify that situation. Although a non-flyer at that stage of his career, Patrick was a seasoned officer known for his impressive administrative and organizational skills. Graduating second in his 1886 class at the U.S. Military Academy, Patrick had been commissioned in the Army Corps of Engineers, an assignment usually reserved for the cream of the crop at the Military Academy. Appointed to the temporary rank of major general in June 1918, Patrick wasted little time in sorting things out in the AEF's Air Service. After several organizational reshuffles, Mitchell was assigned as the air leader of the First Army where he earned a reputation the outstanding U.S. tactical air commander of the war.[14]

The Navy was in better shape. Drawing upon the late nineteenth-century technological innovations that produced steam-powered steel warships with fast-firing long-range cannons, the sea power ideas of Navy CAPT Alfred Thayer Mahan and the enormous prestige that it had achieved during the Spanish-American War, the sea service had solidified its claim as the nation's first line of defense. Its battle fleet was being modernized by a massive shipbuilding program launched in 1917.[15] But, that proved to be the wrong program for the war that the U.S. actually found itself engaged in. With the German High Seas Fleet bottled up by Britain's Grand Fleet after the battle of Jutland in 1916, U-boats were the real menace the Navy faced. The sea service was compelled to launch a crash program to build vessels that could safely escort the expected flood of American troops and materiel to Europe against German submarines. Rear Adm William S. Sims was sent to London in April 1917 to serve as the senior American naval representative to the British Admiralty. Sims was subsequently given command of the nation's naval forces operating in the European area and promoted to the temporary rank of vice admiral. Early on he called for construction of aircraft to fight U-boats but not at the expense of building more small American warships.[16]

Economic Mobilization

The improvised U.S. military intervention was accompanied by a general mishandling of America's economic mobilization. Prior to 1917, the U.S. government had only taken halting steps to prepare for the economic challenges of modern war against other powers. Much of that situation was simply due to naivety. But, some of it was rooted in the Wilson's administration's desire to maintain a neutral and non-provocative stance toward the belligerents. When the U.S. officially entered the fray, planners and political decision-makers confronted four basic economic problems. First, the Allies were already making heavy purchases of military equipment and supplies in the U.S. The modern American munitions industry had largely sprung up in response to those needs. Second, the American economy was already heavily supporting the civilian needs of France and Great Britain. Third, the American people had usually been unwilling (and not called upon) to make serious personal sacrifices for past wartime efforts. With the exception of the Confederacy during the American Civil War, "guns and butter" had been the nation's dominant wartime economic theme. Public officials were reluctant to break that historic pattern and ask the nation's consumers to severely tighten their collective belts. No plan had been prepared or even conceived to meet the overall materiel requirements of a massively increased armed force intended for a huge overseas military intervention. Finally, the Army and Navy were engaged in a competitive procurement free-for-all to secure the materiel they needed for a massive wartime expansion. Within the War Department, there were over 150 separate and uncoordinated purchasing committees vying against each other to obtain scarce goods from America's private economic sector by July 1917.[17]

By the winter of 1917–1918, the Wilson administration's ad hoc approach had produced a massive crisis. The American economy was near collapse. Military programs were plagued by serious shortages of supplies and equipment. Congress and the War Industries Board pressured Baker to consolidate the Army's chaotic procurement and supply operations.[18]

The economic mess was not straightened out until the War Industries Board was reorganized and placed under the vigorous leadership of Wall Street investor Bernard Baruch. It was given unprecedented authority by Wilson in March 1918. After taking the board's helm, Baruch established a system of mandatory priorities that determined the order in which contracts could be filled by manufacturers. The board's independent Price Fixing Committee also set the prices of various industrial products including steel, coke, iron and rubber. The War Industries Board became the federal government's chief economic mobilization agency, coordinating the work of the military services and civilian wartime agencies.[19]

Innovating Wartime Aviation

The American aviation program had stalled along with the rest of the nation's wartime economic mobilization. Patent infringement lawsuits had slowed technical progress in the aviation industry even before the U.S. declared war. The Wright-Martin Company and the Curtiss Airplane and Motor Company controlled most of the patents needed to build aircraft in the U.S. Those two firms were involved in patent infringement lawsuits against each other. Neither could build an airplane without violating the patent rights of the other. No other American manufacturer could produce airplanes without obtaining licenses from both Wright-Martin and Curtiss. That messy legal situation threatened to cripple American aviation production and technical innovation.

Franklin D. Roosevelt, the politically astute and ambitious acting Secretary of the Navy, had called upon the NACA to help break that legal logjam. After several months of study and quiet political negotiations, that body finally convinced aircraft producers to form a manufacturers association and establish a cross-licensing agreement. It required all aircraft producers to join the newly formed Aircraft Manufacturers Association (originally the Manufacturers Aircraft Association) and pay it a set fee for each plane they produced. The proceeds would then be divided between, Wright-Martin, Curtiss and the association. Members of the association agreed to reimburse Wright-Martin and Curtiss a royalty of $200.00 for each aircraft they sold. Of that sum, a ceiling of $2 million was placed upon payments to Wright-Martin and Curtiss as a means of leveling the airframe manufacturing situation. A three-man board of directors of the association, headed by Professor Joseph Ames of Johns Hopkins University, a physicist and highly respected NACA charter member, administered the pact.[20]

Professor Ames' involvement with NACA was multifaceted and long-lasting. Mary Catherine Bitter of the National Aeronautics and Space Administration (NASA)[21] writes that Ames' scientific training enabled "... him to contribute significantly to the organization's research initiatives and that his diligence made him a natural choice for an administrative leader as well. Ames became chairman of the executive committee in 1919, a role that handled concerns essential for the NACA's internal operations, such as managing the various aspects of the budget, overseeing the construction of facilities, and deciding upon specific research initiatives."[22] After Ames became chairman of the organization's main committee in 1927, that "... gave him great influence over research priorities, and also made him the key player in the NACA's relationship with other agencies of the federal government. Ames' influence long outlived him. By handpicking and mentoring young, promising engineers and scientists, like Hugh Dryden (who would later head the NACA) Ames defined the future of the NACA. Under his guidance in these roles, the NACA became a world-renowned, pioneering institution in aeronautics research."[23]

The cross-licensing agreement and a ruling by Attorney General Thomas Wyatt Gregory that it was not in restraint of trade ended the patent troubles that had plagued the American aviation industry.[24] Historian Alex Roland concludes that "In effect, the cross-licensing agreement of 1917 established that the American aviation industry would operate without major patents. Small royalties would be paid for certain contributions within the Manufacturers Associations but in general the ideas and techniques of aircraft manufacturers were to be shared openly among the members."[25]

The Army and Navy aviation programs were woefully short of trained personnel, modern aircraft and airships as well as money. Army aviation was manned by only 131 men and 1,087 enlisted men. In April 1917, only 26 of those officers were considered qualified military aviators. None of them had ever trained for combat. Furthermore, most of its approximately 250 aircraft were obsolete by European standards. When the U.S. declared war, naval aviation—Navy and Marine Corps combined—was far smaller than its Army counterpart. It consisted of 48 officers and 239 enlisted men. They were equipped with 54 airplanes, 1 airship and 3 balloons. Every one of its aircraft was obsolete by the standards of the warring European powers. Like Army fliers, naval aviators had not been trained for combat. With a few exceptions the senior uniformed leadership of the American armed forces didn't think that aviation would play important role in the war.[26]

During the early stages of America's' direct military participation in the war, Squier, now a major general, was arguably the most important military officer involved in the fledgling American aviation program. A contemporary observed that he had a "... great gift for brilliant analysis. Short, dapper, keen-eyed, he talked and walked decisively. He was what people called a 'born leader.' Men followed him implicitly because of his passionate sincerity. He gave his subordinates unlimited authority which inspired them to act with vigor, promptness, and initiative."[27]

Notwithstanding its tiny military aviation programs and small aircraft industry, the American public embraced the fantastic notion that one of its key contributions to victory could be a gigantic air force. Overwhelmed by the problems of raising a massive land force, the Army's General Staff largely ignored aviation during the spring of 1917. Moreover, it was unprepared to deal effectively with the multitude of issues, both large and small, raised by the development of military aviation.[28] Aviation enthusiasts inside and outside the Army had moved to fill the vacuum created by the General Staff's inattention. In March 1917, the NACA proposed that the U.S. spend $54 million on wartime aviation and produce some 20,000 military aircraft. However, Squier was initially unable to convince the secretaries of War and Navy to support that. Nevertheless, his tiny office continued to promote the concept of a large wartime air force. The Aviation Section's proposal was quite modest. It called for $54 million to equip a force of 16 reconnaissance and the same number of balloon squadrons. The plan was submitted to Secretary Stimson just before President Wilson asked Congress for a declaration of War. During June 1917, Congress appropriated $43.45 million for that aviation program.[29]

While the Aviation Section was struggling to win approval for a relatively modest aerial program, French Premier Alexander Ribot dropped a bombshell on the American government. Ribot wired Wilson in May 1917 asking the U.S. to create a gigantic air force.[30] He requested an American flying program of 4500 airplanes with their required personnel and support equipment that would be sent to the French front in time for the 1918 campaign (Illustration 3.1). The monthly production goal for American factories would be 2000 airplanes and 4000 aircraft engines.[31]

General Staff officers were appalled by Ribot's stunning and enormously costly proposal. His cable was passed from the President to the secretaries of War and Navy and subsequently to the recently created Army-Navy Technical Board. That body, established to deal with the standardization of aircraft and associated material, sent the message to Squier. He directed Foulois and his staff "... to develop a program that would truly correspond to the needs of the Allies, a much different program than their previous experience had led them to pursue."[32] Several weeks later, Squier and Foulois presented their draft program to Baker. They bypassed the General Staff, a bold move for career military officers. Baker

Illustration 3.1 Maj Gen George Owen Squier (*Photo* Maj Gen George Owen Squier, Chief Signal Officer, U.S. Army. United States Army images: www.arlingtoncemetery.net/gosquier.htm)

approved it and obtained President Wilson's support before sending it to Congress. In response, the legislators appropriated $10,800,000 for aviation on 12 May 1917. During the following month, Congress allocated $43,450,000 for the Signal Corps' Aviation Section—the largest appropriation for aviation to date. But that amount was a relative trifle compared to the $707,541,000 request Foulois was preparing to meet the estimated cost of Ribot's request. On 24 July 1917, President Wilson signed the slightly pared down $640,000,000 appropriation Congress had approved in response to the Aviation Section's proposal.[33]

Official Navy Department thinking on aviation's importance changed significantly during the first half of 1917. As of February, it had asked that less than 1%, or $5.3 million of its budget request to Congress, be allocated to aviation. On 30 June 1917, Daniels submitted an estimate of $45,000,000 for the immediate needs of naval aviation to the Treasury Department which transmitted it to the Speaker of the House of Representatives on 12 July 1917.[34]

Meanwhile, Howard E. Coffin, an automotive engineer and industrialist, had been recruited to chair the Aircraft Production Board in May 1917. The board's key mission was to bring the aircraft and the federal government into close cooperation to stimulate the production of greater quantities of more advanced military flying machines for the war. To help expedite the process of obtaining wartime aviation contracts, American airplane makers formed the Aircraft Manufacturers' Association that month. Coffin and his staff had begun to put together an extensive public relations campaign to build support for the massive American aviation program. The industrialist courted some of the most influential newspaper editors in the country. Basically, he convinced them that the program could end the war quickly and relatively little loss of life. High-profile industrialists and officers from Allied military missions in the U.S. echoed that same narrative. Coffin arranged to have senior government officials issue press releases touting the program. Across the nation, newspaper editors soon published pieces urging Congress to support a huge Yankee aviation program. It was very effective propaganda that played to Americans' growing, if unrealistic, faith in advanced technology and the nation's huge industrial capacity.[35]

To speed production, the federal government relied upon negotiated cost-plus-percentage-of-cost agreements as recommended by the Aircraft Production Board. Before that production contracts were fixed price. On

4 March 1917, Congress changed the law to allow fixed-price incentive fee contracts. Those new arrangements covered all of a contractor's labor, materials, plant depreciation and overhead expenses. In addition, they were guaranteed a fixed profit—12.5% for aircraft and 15% for engines—of the estimated cost to build that equipment. Bonuses would be paid if expenses were held below projected costs. A business would get 25% of the savings from building an aircraft or engine below the estimated while the rest went to the government. Beyond accelerating the procurement process, those contracting practices were justified on the grounds that most manufacturers could not accurately predict the cost of building aircraft, engines and other aviation equipment. Critics charged that they contributed to personal favoritism, incompetence, corruption and widespread waste in America's wartime aviation program.

To obtain more detailed information about the specific equipment that the U.S. would need for its aviation program, a special aeronautical commission was sent to Europe in June 1917. The commission was headed by Maj Raynal C. Bolling, an Army Reserve officer and former general counsel of the United States Steel Corporation who had played a prominent role in aviation circles for several years. His commission, which sailed for Europe on 17 June 1917, had five major objectives. First, it wanted to establish agreements permitting the U.S. to manufacture aviation equipment of foreign design. Second, the commission was determined to secure sample aircraft, engines and accessories from the Allies and send them to the U.S. Third, it would conclude preliminary arrangements to purchase airplanes in Europe for U.S. flyers to use in combat. Fourth, the commission would arrange to have American aviation personnel trained by the Europeans. Finally, it would help the air programs of the allies by arranging for the allocation of American raw materials for European use.[36]

The commission concluded that the U.S. would have to depend heavily on the Allies at least initially to fulfill its wartime requirements. Despite stateside enthusiasm for building a huge air force, Bolling's group recognized that America had a tiny aircraft industry compared to the major European combatants as well as little organized capability then for aeronautical research and development. Under the immense operational pressures generated by air combat, pursuit aircraft—the key to air supremacy in the war—were changing rapidly. The group concluded that it would be very difficult for the U.S. to design, produce and ship new versions of such aircraft across the Atlantic before they became obsolete. Consequently, it recommended that Americans should concentrate on producing training aircraft and engines as well as certain European-designed service planes and engines whose designs were not changing nearly so rapidly. The latter category included observation aircraft and bombers. It urged that single-seat pursuit (i.e., fighter) aircraft should be purchased directly from the Europeans. Although it emphasized that the U.S. must first obtain aircraft that would support the ground forces, the commission also called for development of a strategic air force that could conduct independent bombing operations.[37]

Bolling's group identified specific European designs that it believed would be best suited to American needs and mass production methods. It negotiated preliminary agreements to produce those aircraft in the U.S. and ensured that prototypes as well as related technical information would be sent west across the Atlantic. The group established draft arrangements to purchase European aircraft directly and exchange critical raw materials with the Allies. It recommended that certain European engines should be produced in America. The commission also suggested that the Liberty engine, then under development in the U.S. might be used in some foreign aircraft.[38] Naval historian William F. Trimble observes that its naval representatives "... saw a substantial role for large, long-range flying boats in suppressing the German U-boat offensive and called an expansion of navy patrols in the eastern Atlantic."[39]

The situation facing U.S. military planners after they received the Bolling commission's report was fluid and confusing. Messages had continued to arrive almost daily in Washington, D.C. from the AEF concerning its aircraft and engine needs. Cable communication links across the Atlantic were poor resulting in slow and sometimes garbled transmission of vital technical, programmatic and operational data. The technical staff of the AEF's Air Service was woefully undermanned. It suffered from frequent leadership changes and a shortage of technically qualified personnel. There were no clear-cut organizational channels and procedures to transmit vital data from France to the U.S. Stateside, air planners had to rely on a variety of sources including allied military missions in Washington, D.C., impressions of individual American officers plus additional American missions that were dispatched to Europe after Bolling's.[40]

By early 1918, it had become apparent that America's huge vaunted air armada was not going to materialize any time soon in Europe's skies.

Army pilots, equipped with second rate Allied land-based aircraft, didn't begin flying combat missions until April 1918. The first American-made plane, a British-designed DH-4, didn't reach France until May 1918.

Primarily due to production problems stateside, plans for a large American Air Service in Europe were dramatically reduced. The initial U.S. aviation plan approved by Pershing authorized 263 U.S. flying squadrons for service in the war zone. At the time of the Armistice on 11 November 1918, only 45 American squadrons were assigned to combat duties. Meanwhile, stateside aircraft production had fallen well behind schedule. The American public, Congress and the press sensed an enormous scandal. Acrimonious charges of waste, mismanagement, corruption and sabotage by German spies and sympathizers were rampant. Wilson's congressional opponents, sensing political vulnerability, tried unsuccessfully to strip the chief executive of his powers to manage the entire war effort.[41]

Institutional confusion and mismanagement within the War Department hobbled the aviation production program. Baker, convinced that only Detroit possessed the management expertise and the manufacturing capacity to mass produce aircraft, had turned to auto magnates including Howard E. Coffin and Edward A. Deeds to run the aviation production program. However, they had no practical experience in aviation. The result was even more confusion. Baker compounded his error by not exercising strong supervision over the inexperienced men he had picked (Illustration 3.2).[42]

America's ambitious aviation production program was also undermined by the fragile nature of the nation's technical-industrial base in aeronautics. There was no functioning national program of aeronautical research and development when the U.S. went to war. NACA's laboratories at Langley Field, Virginia were not completed until after the Armistice. Neither the military services nor universities had established significant aeronautical research and development programs compared to those of the major warring European powers. During 1915, only 25 universities reported that they had aviation courses.

MIT had offered the nation' first aeronautical engineering course. That initiative was launched by the school's president, Richard C. Maclaurin. He enlisted the help of Lt Jerome C. Hunsaker—a young naval officer, aviation enthusiast and student in naval construction at that school—to help turn that idea into reality. Maclaurin convinced the Navy to send him to survey European aviation laboratories in 1913. Hunsaker, who

Illustration 3.2 DH-4 aircraft (*Photo* A DH-4 aircraft inside a Dayton-Wright Airplane Company plant in 1918. Photo courtesy of Wright State University Libraries' Special Collections & Archives)

had graduated at the top of his Naval Academy Class in 1908, was accompanied by Dr. Alfred F. Zahm, who taught physics at Catholic University in Washington, D.C. Zahm had built the first significant wind tunnel in the U.S. in 1901. Upon their return from overseas, Hunsaker developed an aerodynamics curriculum which was approved for the 1914–1915 MIT school year. He also built a wind tunnel at that institution and was assisted by Donald Douglas, a young aviation enthusiast who enrolled there in 1912 after dropping out of the Naval Academy.[43]

Historian Trimble emphasizes that "It is impossible to overstate the significance of Hunsaker and Zahm's European trip. For the first time, Americans had access to the latest aerodynamic research, firmly grounded in theory and mathematical principles.... Before 1913 there was little attention paid to the subject of theoretical aerodynamics in the United

States, the work of Hunsaker and Zahm being the sole exception."[44] It represented a basic shift in emphasis from practitioners like the Wright brothers to theoreticians including Zahm.[45]

There were very few trained aeronautical engineers and designers in the U.S. in 1917. To help alleviate that critical problem, the War Department appealed to the Allied powers to send experienced flyers, aeronautical engineers and designers to the U.S. to help build a strong aviation program. Shortly after that, a number of foreign military officers were dispatched to America for that purpose and assigned to the CSO's staff. Foreign officers with the latest operational experience were continually rotated to Washington, D.C. during the war. Aeronautical engineers from Great Britain, France and Italy were brought to America to help design U.S. flying machines.[46]

Initial War Department plans had identified no more than 10 or 11 American plants capable of wartime production of military aircraft and engines. Their annual productive capacity was set at an exaggerated 7,200 planes. The aviation industry's capitalization was $15 to $20 million. It only employed some 8,000 workers. With the exception of seaplanes, none of those firms had ever designed or built combat aircraft. To help alleviate that situation, early on the government turned to automobile firms that, although they had no previous experience in building aircraft, possessed the skilled workers and machinery needed for mass production.

The nation's aircraft manufacturing industry was still largely in the handicraft stage. The standard practice was to concentrate all of a firm's resources on manufacturing a single aircraft until it was completed. Then resources were shifted to the next machine on order. After those firms began receiving large wartime military contracts, aircraft components were built by subcontractors, and many of these subcomponents—including engines, propellers, radios, weapons, wheels, landing gear assemblies—were put together on moving assembly lines. But final aircraft assembly still largely resembled the pre-Fordism practice of bringing the various parts to a stationary work station and using teams of skilled workers to complete the final assembly process.[47]

Curtiss was the only significant exception to the grim portrait of industrial inadequacy. His firm had a maximum rated production capacity of about 1,500 aircraft per year. By April 1917, it was already building trainers for Canada and seaplanes for Great Britain. Early in the war, Curtiss had received $364,000 in orders for its aircraft including large flying boats and trainers from the British government. Funded by an

August 1916 congressional appropriation, Curtiss had delivered 12 JN-4 and 72 JN-4B trainers to the U.S. military services by early 1917. The JN-4, or *Jenny* as it came to be known, was designed by British-born engineer B. Douglas Thomas who Curtiss recruited from the United Kingdom to work for him. Most American Army aviators were trained in it during the war. Many naval aviators were trained on a seaplane variant designated the N-9.[48]

The ambitious wartime aviation production program propelled a huge expansion of America's aircraft manufacturing capabilities. Lured by one of the largest pork barrels in the nation's history, the government was inundated by unsolicited aircraft proposals from charlatans and dreamers as well as legitimate businessmen. Official policy, however, focused on awarding contracts to the limited number of companies that either had aircraft production experience or the general capacity for mass production. Curtiss and the newly established Dayton-Wright firm got the lion's share of aircraft orders. Several automakers, including Fisher Body and Packard, also produced substantial numbers of aircraft during the war.[49]

Motivated by several factors including national pride as well as the desire to field a new and more powerful engine superior to the many existing European models, the Aircraft Production Board and the Joint Army and Navy Production Board approved a project to develop a standard American power plant that could be mass produced. Experienced industry engine developers Elbert J. Hall and Jesse G. Vincent shut themselves up for five days in a room in Washington's Willard Hotel to design it. The former was from the Hall-Scott Motor Car Company which had been producing aircraft engines since the 1913. The latter had been the vice president of engineering for the Packard Motor Car Company. Both men were serving as wartime reserve Army officers when they tackled that task. On 4 June 1917, they published plans for a new engine with 8-cylinder and 12-cylinder versions. Most of the engine's features had already been demonstrated by experiments and manufacturing experience at their firms. Based on design features pioneered in the 1915 German Mercedes engine, the new American power plant did not incorporate any radical new technologies. Rather, it represented a superb synthesis of the existing state of the art. The government allocated $250,000 to develop the engine.

The first standard eight-cylinder engine was tested at that plant on 23 July 1917 and flown the following August 29th. Contracts to produce the engine at scale were let beginning in August. Driven by growing demands for increased horsepower, the problem-plagued 8-cylinder model was deemphasized in favor of a 12-cylinder model. The first contracts to build 22,500 12-cylinder power plants were signed during August 1917. Originally named the "United States Standard 12-cylinder Aviation Engine," Rear Adm Taylor suggested early in its large-scale production phase that it be rechristened the "Liberty Engine." The name stuck.[50]

By 29 November 1918, 15,572 Liberty engines had been produced in the U.S. The Navy received 3,472 for its seaplanes while 5,323 went to American plants manufacturing airplanes. The remainder was sent to several different entities including: training fields in the U.S., the AEF in France and the Allied powers. The War Department reported that over 64,000 Liberties would have been delivered under contract if production had not been cut short by the Armistice on 11 November 1918 (Illustration 3.3). That output was considered an enormous achievement especially in the face of production delays caused by a variety of problems including: the need to construct factories, coal shortages, railroad tie-ups and the scarcity of special alloy steels.[51]

By the Armistice, over 300 plants were involved in American aircraft and engine production. It had taken an enormous amount of resources to achieve that immense expansion of manufacturing capacity. Many of the firms that became involved in the wartime aviation were not used to the fine tolerances, specific manufacturing processes and demanding inspection methods required to produce military aircraft and power plants. Skilled, experienced labor was in short supply. The available stock of machine tools had been totally inadequate to deal with surging demand.[52]

Fundamental differences between European and American mass production methods hindered aircraft production in the U.S. Pioneered by the automobile industry, American large-scale manufactures relied on standardized (i.e., interchangeable) components of finished products. That approach required tedious preparation of detailed drawings based on highly accurate measurements to produce parts, dies, tools, jigs and fixtures. The European practice was quite different, stressing the talents of highly skilled mechanics who virtually handcrafted airframes. European parts were measured in the metric system which was largely unused in the U.S. As a result, American firms found it extremely difficult to mass produce European designs in a timely manner. The latter's blueprints and technical specifications were usually not detailed enough to be readily accommodated to Yankee mass production methods. That mismatch required American firms to undertake the time-consuming process of

Illustration 3.3 Liberty engine (*Photo* Army Maj Henry H. Arnold with the first Liberty engine produced for wartime use courtesy of the National Museum of the United States Air Force (NMUSAF), undated, nationalmuseum.af.mil)

disassembling European aircraft, making meticulous measurements of them, completing detailed drawings based on those measurements and then fabricating new tools to build them.[53]

Turbulence

The general failure of American economic mobilization as well as the inexperience and a skimpy aeronautical industrial base contributed to the near breakdown of aviation production by December 1917. The resulting public controversy and partisan political attacks led to several investigations of the flagging aircraft program.[54] Spurred by growing public disclosures of the aircraft program's shortcomings, the Senate Military Affairs Committee appointed a subcommittee to investigate it. During the summer of 1918 hundreds of witnesses including Secretary Baker testified before the panel. However, its members could not agree on their findings. The majority concluded that combat aircraft production had been a failure. They attributed the program's troubles to ignorance and mismanagement. The majority chided the government for misrepresenting the program's progress. To straighten out the mess, they urged the President to centralize control of the program in the hands of one executive who would report to him. The Senators also demanded that the aviation program be removed from the Signal Corps. The committee's minority differed sharply. They concluded that the program was doing well and that it had accomplished much given the difficult circumstances it had confronted.[55]

Fueled by conflicting reports from the Senate and other investigations, public controversy continued to swirl around the aviation program, President Wilson turned to his last Republican rival for the presidency, former Supreme Court Associate Justice Charles Evans Hughes, to help the Department of Justice investigate the situation. The appointment of the highly regarded jurist helped calm the political turbulence surrounding the situation. Hughes' report, issued on the eve of the Armistice, concluded that aviation program's "... delays and waste were chiefly... matters for administrative correction through unification of effort under competent control."[56]

Hughes dismissed as unfounded charges of widespread corruption and outright theft. He criticized the confusion and poor management he found in the program. He reported that Deeds, now an Army colonel assigned to the aircraft program, had provided confidential advice on the aircraft program to a former business associate at the Dayton-Wright firm and had provided "false and misleading information" to the public about the program's progress.[57] Hughes recommended that "The evidence with respect to Colonel Edward A. Deeds should be presented to the Secretary of War to the end the Colonel Deeds may be tried by court-martial..."[58] However, no action was taken against Deeds. Instead, Baker exonerated him based upon the recommendation of a Special War Department Board of Inquiry.[59]

A committee headed by Howard Coffin had also been formed to investigate the aircraft program. On 12 April 1918, it submitted a preliminary report urging the Army to strip the Signal Corps of its aviation responsibilities. Over Squier's strenuous objections, Baker and Wilson agreed. The President had already begun moving haltingly to straighten out the aviation mess. Later that month he appointed John D. Ryan, railroad man, financier and head of the Anaconda Copper Company, as Director of Aircraft Production. The appointment was reportedly based on the advice of the committee headed by Hudson Snowden Marshall, the U.S. Attorney for the Southern District of New York. It had been established earlier in the year to investigate problems in the aircraft program. Although lacking any experience in aircraft production, Ryan proved to be an excellent choice. He had a solid business management background and was untainted by any hint of corruption. Ryan promptly called upon pre-war aviation experts for advice.[60]

President Wilson issued an executive order removing the Aviation Section from the Signal Corps on 20 May 1918. His order created two new organizations in the War Department: the Division of Military Aeronautics (DMA) and the Bureau of Aircraft Production (BAP). The former, headed by Army flyer Brig Gen (later Maj Gen) William L. Kenly, was responsible for aviation organization and training. It also developed requirements for aircraft and associated equipment. The latter, under Mr. Ryan, managed the aircraft production program. Squier, in his capacity as the CSO, would continue to focus on the administration of Army signals programs including the design and manufacture of aircraft radios.[61]

Despite the turbulence at the top, the aircraft production gained momentum. The War Department's Fiscal Year 1918 report noted that "In reporting upon the accomplishments of the year it must not be overlooked that the greatest thing done, in a fundamental way, was the creation of an industry capable of supplying the needs of the future for aircraft and aircraft accessories, there being now engaged in this work about 350 concerns, employing an industrial army of 200,000 men and women."[62]

Some firms considered women more effective than men in making wings and wing surfaces for airplanes. Harry Bowers, president of the Standard Aero Corporation, stressed that "Preparing the wing surfaces and making the wings... is light work requiring deft fingers and close application. It is ideal work for women—the very characteristics of their sex make them superior to men.... [Before employing them] We found that women were extensively used in Britain and France in making wings and that they had proved to be better at this work than men."[63] After employing them in his New Jersey plant, Bowers discovered that "In in a very short time we discovered that the women were better workmen than the men.... We found the women to be painstaking and fast in their work, and, in addition, they are precise, and careful manipulation of sensitive machines, women are admirably adapted to much of the light machine work done on airplanes...."[64] The U.S. Women's Bureau in the Department of Labor reported that 186 of each 1,000 employees working in the 34 airplane plants surveyed in October 1918 were women.[65]

While American women helped fulfill the surging wartime manpower requirements in industry and the armed forces, the DMA and the BAP initially acted independently of each other. Both reported directly to the Secretary of War. There was no overall head of the Army's air organization. That arrangement worked poorly. Finally, during August 1918, the President combined them under Ryan whom he appointed Second Assistant Secretary of War and Director of the Air Service. However, that move came too late to have any significant impact on the war effort.

It had taken until the end of July 1918 before the DMA and the BAP worked out an effective arrangement for deciding which new aircraft designs would be produced. Essentially the DMA's Technical Section would write the performance specifications which the BAP's engineers would use. The Technical Section would also set priorities for developing new aircraft and modifying existing designs. DMA and BAP officials would share responsibility for production decisions. The Technical Section would also maintain the power to veto production changes. However, that new planning and decision-making structure was weakened by the failure of Pershing's Air Service to develop an effective process to collect, assess and transmit timely information on the AEF's rapidly changing aviation requirements. The Armistice arrived on 11 November 1918 before the Air Service's revamped stateside headquarters organization could be fully tested.[66]

Despite mismanagement, delay organizational turbulence and the public controversy associated with it, aircraft production in the U.S. finally began to gather steam during 1918. The trickle of military aircraft delivered to the armed forces became a flood. Only 750 planes were completed during January of that year. By the following October, that figure reached 1,651 aircraft. During the 12 months before April 1917, American industry had produced fewer than 800 aircraft, most of them trainers for export. The Allies provided 5,198 aircraft to the U.S. during the war. From April 1917 through December 1918, the U.S. produced 12,571 aircraft. Over 8,500 of them were trainers, a prudent first step in building a large wartime air force. Altogether, the U.S. spent over $848 million on military aeronautics during Fiscal Year 1917 through Fiscal Year 1919.[67]

In terms of its exaggerated original unrealistic claims, the wartime U.S. aircraft program was a clear failure. However, it transformed American aircraft production from a handicraft operation into a large-scale modern manufacturing enterprise that depended upon professional science and engineering as well as the practical skills of pioneer designers. There were 24 aircraft and engine companies—large, small and experimental—in America by the Armistice. They employed some 175,000 men and women and represented a total capitalization of $23 million (Table 3.1). Some boasted excellent design staffs. The war captured the imagination of men like Douglas who were destined to play crucial roles in American aviation during the interwar period and its vast military aviation production program in World War II.[68]

The career of James H. "Dutch" Kindelberger illustrated the importance of the talented and ambitious men World War I attracted to aviation. The son of a steelworker, he had studied engineering at the Carnegie Institute of Technology and worked as a draftsman before the U.S. declared war. During the conflict, he enlisted in the Army and served as a pilot instructor. Following the Armistice, Kindelberger became the chief draftsman and assistant chief engineer for the Glenn L. Martin Company. Later, while employed as chief engineer for Douglas Aircraft, Kindleberger engineered the design of the famed DC-3 airliner. In 1934, he became president of the General Aviation Manufacturing which was subsequently renamed North American Aviation. Under his leadership, the latter was responsible for developing a long line of famous aircraft including the P-51 *Mustang* fighter, F-86 *Sabre Jet* fighter, F-100 *Super Sabre* fighter and the X-15 research plane.[70]

Table 3.1 U.S. Aircraft production during World War I[69] (number of aircraft)

Year	*Total*	*Military*	*Civil*
1914	49	15	34
1915	178	26	152
1916	411	142	269
1917	2148	2013	135
1918	14,020	13,991	29

Centralizing Army Aeronautical Research and Development

While the Aircraft Production Board had fumbled to establish a coherent and practical national wartime effort, Maj Gen Squier was initially responsible for actually implementing the Army's program including organizing and training air units. Assigned as the CSO in February 1917, he was also responsible for planning, organizing and directing the wartime buildup of the rest of the Signal Corps.[71] On 22 October 1917, Squier established a Science and Research Division in the Signal Corps under the leadership Lt Col Robert. A. Millikan, an experimental physicist and holder of a doctorate from Columbia University. He would be awarded the Nobel Prize in physics in 1923. Much of Millikan's work during the war was devoted to solving practical problems in America aviation. Working primarily with industry, his organization made some significant contributions to the war effort. Among those were: establishing all specifications for aircraft instruments; developing a new aerial compass; creating a new aerial camera and film; and developing a self-sealing gas tank for combat aircraft. Millikan's wartime experience was part of a larger pattern employed by the armed forces to apply scientific expertise to some of their problems by offering scientists commissions and assignments in military laboratories.[72]

But the Science and Research Division offered only a partial solution to the Signal Corps' aviation needs. Additional engineering work was being conducted at several locations across the nation. Squier and Deeds concluded that the slow progress of the aviation program was being worsened by the absence of a strong centralized aeronautical engineering organization in the Signal Corps. They were determined to establish a temporary organization where the Army could conduct such activities for the remainder of the war. Deeds suggested that such a facility be established in Dayton, Ohio.[73]

The Army chose a Dayton site known as South Field as a temporary location to consolidate its wartime aeronautical research and development activities until NACA's experimental station at Langley Field, Virginia became operational. Named McCook Field, it honored a local Dayton family whose members had distinguished themselves during the Civil War. After a crash construction program, McCook officially opened on 4 December 1917 as the Airplane Experimental Station of the Signal Corps. During the war, most of the Army's aeronautical engineering and test responsibilities were gradually concentrated there. An aeronautical laboratory was established there to conduct testing and research activities. Squier assigned it responsibility for all technical and experimental work on aircraft and engine designs that had earlier been conducted at Langley Field, Virginia. Later, McCook was directed to conduct aircraft and engine development, install cameras on experimental aircraft and work to synchronize machine gun fire through rotating propeller blades. During January 1918, McCook was made responsible for the design of all Army aircraft and accessories. By executive order, all civilian employees at McCook Field were exempted from competitive civil service exams. From the outset, it was operated as more of a business than a military installation. At the war's end, McCook was manned by 58 officers, 267 enlisted men and 1915 civilians.[74]

Frequent leadership and organizational changes hampered the work at McCook Field during 1918. Between October 1917 and November 1918, five different officers commanded it. The chief issue facing McCook was its relationship with the Production Engineering Division, a rival Signal Corps organization formed in Washington, D.C. that January. The Production Engineering Division was responsible for preparing established European equipment and experimental American-designed aircraft for production in the U.S. Inevitably, the Production Engineering Division found itself involved in design and experimental activities supposedly the province of McCook Field. Competition, overlap and poor communications between the two organizations were not overcome until the two were merged in August 1918 at McCook as the Airplane Engineering Division. During January 1919, the Airplane Engineering Division was renamed the Technical Division. The following March, it became the Engineering Division. The latter title was maintained until its functions were merged with the supply, maintenance and procurement functions of the Army's air arm in 1926.[75]

McCook Field's primary contribution during World War I was to adapt European aircraft designs to U.S. requirements and production methods. Its staff redesigned the British DH-9 bomber and Bristol fighter. It also contracted with private firms for engineering on numerous aircraft including a LePere two-seat fighter, Loening monoplane and Martin bomber. Those contractors also worked on engines, carburetors, radiators, propellers and other equipment. In addition, McCook's staff conducted extensive flight tests and developed ancillary equipment including propellers, self-sealing fuel tanks, visual signaling devices and a system to control synchronized machine gun fire through rotating propeller blades.

For basic and applied research, McCook turned to universities, the Federal Bureau of Standards, private individuals and industry. Its engineers designed and built a wind tunnel in 1918. The wind tunnel was used to calibrate airspeed instruments and test airfoils. McCook's staff also widely disseminated technical information on aeronautics to industry, universities and other interested parties. Working from district offices, its engineers monitored the production of aircraft, engines and accessories by various manufacturers.[76]

McCook Field also served as a training ground for some of the men who later rose to leading technical and managerial position in American aviation. Because there had been very few aeronautical engineers and other professionally trained personnel in the Army in April 1917, the Signal Corps had been forced to recruit such talent from the private sector. For instance, Alfred V. Verville, who came to Dayton after working for Glenn Curtiss, became widely respected during the interwar period as one of the world's leading aircraft designers. In the early 1920s, he designed the first American internally braced, low-wing monoplane with retractable landing gear, a design that set the direction for the future development of pursuit aircraft.

Don R. Berlin, another McCook alumnus, held a bachelor's degree in mechanical engineering from Purdue University. After conducting wind tunnel tests at McCook during World War I, he worked for the Douglas, Northrop and Curtiss-Wright firms. At Curtiss-Wright, Berlin designed the Curtiss Hawk 75 series of Army aircraft that entered the Army's pursuit aircraft inventory in the late 1920s. They evolved into the Curtiss P-40 series which was the mainstay of the Army's pursuit strength in early World War II. After the war, he held executive positions with several aircraft firms including Boeing and McDonnell before retiring.[77]

Army aviation grew to approximately 183,000 personnel during the war, including some 12,000 were pilots. Approximately 58,000 Air Service members served in Europe. Some 1500 AEF aviators in 45 flying squadrons carried the deadly burdens of aerial combat starting during the spring of 1918. They focused on observation and air superiority missions shooting down 781 German aircraft and 73 enemy balloons. In the process, the Air Service suffered 583 casualties and lost 289 aircraft.[78]

Despite its promising accomplishments, historian Henning concludes that aviation hadn't fulfilled the exaggerated predictions made for it during the Great War nor had the cavalry and, by extension other combat arms, become obsolete. She stresses that "The horse cavalry continued to play a vital role... well after the invention of motorized vehicles. The caution shown by cavalrymen and government officials... about aviation prior to the Great War was a skeptical but rational response to an unproven innovation. It would take time for aviation to become the revolutionary technology many claimed it would be.[79]

Wings for the Navy

The Navy's wartime aviation program was far smaller, more limited in scope, cheaper and relatively uncontroversial compared to the Army's. Because the surface fleets of the Central Powers had been reduced to minor raids by the time America entered the war, German submarines were the main threat the Navy faced. That fact, plus the established expertise of Curtiss, helped shape the decision to concentrate on seaplanes and lighter-than-air craft instead of planes that could go to sea with the fleet. The Navy didn't attempt to develop land planes. Instead it used those provided for the Army.[80]

Historian Trimble stresses that the enormous size of the total U.S. aviation program "... coupled with confusion over the types of airplanes to be manufactured and high-level decisions regarding civilian control of the production effort, led to concern that the needs of the navy would be subordinated to huge army orders at private plants."[81] In June 1917, Rear Adm Taylor ordered his staff to examine the existing American aircraft industry and formulate recommendations for a naval aircraft manufacturing facility. They concluded that a government factory would help assure the numbers of aircraft needed by the Navy, aid its development programs and check "excessive costs."[82] They stressed that approximately $1 million would need to be appropriated "... for the prompt construction of a plant having the capacity of 1,000 school machines per year, or an equivalent of other types: this plant to include... sufficient equipment for experimental work to permit the Department to push the development of its designs with the greatest rapidity."[83]

Taylor approved the report and sent it to Secretary Daniels. The latter greatly admired Taylor and saw such a plant as a way to hold down costs and promote efficiency. Furthermore, he was determined to restrict wartime profiteering by businessmen. On Taylor's recommendation, Daniels allocated $1 million from the department's aviation appropriation on 27 July 1917 to build the NAF in the Philadelphia Navy Yard.[84] Contract award orders were sent out on August 6th. Actual construction was begun that month and was completed that November.

Rear Adm Taylor selected LT Cdr Frederick G. Coburn to be the NAF's first manager. A 1904 Naval Academy graduate with a master's degree in engineering from MIT, Coburn was a naval constructor and an outstanding manager.[85] Under his guidance, the factory was completed in five months. The first aircraft produced there, an H-16 flying boat, made its maiden flight on 27 March 1918. It was equipped with two, 230-pound bombs and five Lewis machine guns.[86]

Coburn found that his most difficult initial task was to recruit, train and direct a productive civilian workforce. With the aircraft industry still in its infancy, skilled aircraft workers were in extremely short supply. After the factory opened, no more than 25 individuals who had worked in the aircraft industry were employed in there in a workforce that grew to 3,700. Instead, individuals were recruited from auto factories, machine shops and other factories doing similar work. In a significant departure from established practice, the Navy stressed that "The employment of women on this work was begun at an early date, the maximum number reached being about 900. Women are admirably adapted to this class of work, and those in the factory did their part most effectively."[87]

Near the end of 1918, when the NAF reached its highest employment of 3,640 workers, 24.5% were women. Historian Trimble finds that their primary tasks involved "... cutting and sewing fabrics, covering wings and other aerodynamic surfaces, and painting; a few women's jobs demanded considerable skill, such as drill press operation and other machinist positions. Only three women achieved supervisory status."[88] Women were paid less than their male counterparts (Illustration 3.4). Trimble observes that "Although women put in the same forty-nine hour workweek as did men, the assumption was that 'their output had not equaled' that of their male counterparts and consequently, they received only helpers' wages."[89]

The NAF was originally designed to produce about 1,000 small aircraft annually or their equivalent in larger ones; however, from the outset, it focused on producing large flying boats for overseas use. Driven by an enormous expansion of naval aircraft requirements, Daniels in January 1918 approved "... an expansion in the nature of an assembly plant where hulls, wings, and other parts constructed under contract by private concerns could be assembled.... early in June, 1918, the full [productive] capacity of the original factory was reached."[90]

Illustration 3.4 Woman worker at the NAF (*Photo* Unidentified woman worker at the NAF files a metal part on 7 November 1918, Catalog #: NH 2659, NHHC, history.navy.mil)

At a cost of approximately $3.5 million, additional facilities were constructed at the NAF to accommodate the expanded workload. The enlarged plant began production on July 6th of that year. To build seaplane hulls, the NAF obtained the services of seven small yacht building shipyards. To fabricate aircraft wings, metal parts, tanks, engine foundation assemblies, tail sections and other items, the NAF contracted with approximately a dozen firms. Those firms were placed under the control of the NAF's manager who maintained a branch office in each of them. He directed them as if they were part of his own factory.[91]

Like the Army, naval aviation's wartime experience demonstrated the importance of centralized control of a strong aeronautical research and development program to combat aviation. It firmly established the organizational foundation of those functions in the sea service.

The Joint Technical Board on Aircraft had recommended the initial wartime Navy aircraft production program during May 1917. It consisted of 300 trainers, 200 service seaplanes, 100 speedy scouting aircraft and 100 large seaplanes. The board listed the Curtiss N-9 and R-6 as the best models for training and operational service seaplanes. The Curtiss N-9 served as the standard Navy primary and advanced seaplane trainer during World War I. Originally developed by Curtiss as a private venture, 560 were produced for the Navy during the conflict. The Curtiss "R" series seaplanes were widely used by the U.S. Army and Navy as well as the Royal Navy for scouting, observation and training. Although only two R-6s were built for the U.S. Navy, an improved version was the first American built aircraft to serve overseas with U.S. forces during the war. A squadron equipped with this model began anti-submarine patrols from the Azores on 17 January 1918.[92]

On 25 August 1917, Rear Adm Taylor initiated the development of a long-range flying boat. He wanted an aircraft that could deal with German submarines and be able to fly across the Atlantic easing ship delivery problems to the war zone. During the project's early stages, his staff collaborated with Curtiss engineers to design the flying boat. While not operational during the war, the jointly designed Navy/Curtiss (NC) aircraft made history in 1919 when they became the first planes to fly across the Atlantic, albeit in stages.

Daniels signed a contract with Curtiss in December 1917 for the construction of four NC-type planes allowing a 10% profit margin. Following many changes as the work progressed, NC-1's costs reached $150,000. To speed work on the highly secretive project, subcontracts were let to various other firms including three of the four hulls (Illustration 3.5). Meanwhile, Baker had taken a significant step to assure the success of the naval aircraft production program when he approved a recommendation in November 1917 "... that priority be given by the War Department to naval needs for aviation material necessary to equip and arm seaplane bases."[93]

Demands for naval aircraft production grew steadily in early Fiscal Year 1918. During the summer of 1917, a large naval aircraft program had been launched, primarily for coastal patrol and hunting enemy submarines. In September 1917, a board of officers returned to the U.S. after investigating the naval aircraft situation abroad. They concluded that there were no fully satisfactory types of foreign seaplanes suitable for coastal patrol. The board reported that an American-designed flying boat, the Curtiss H-16, modified and improved by the British, was operating successfully with two English engines. It recommended that it be

Illustration 3.5 Seaplane Hull under construction (*Photo* Seaplane Hull under construction at the NAF, 3 June 1918. Catalog #: NH 2649. Copyright Owner: NHHC, history.navy.mil)

modified to suit American manufacturing conditions and be equipped with U.S. engines. During that period, a single-engine, flying boat built by Curtiss was tested with the Liberty engine that greatly improved the aircraft's performance. Consequently, a production program was adopted in October 1917 that featured both essentially American seaplanes, one equipped with the single Liberty engine and the other equipped with twin Liberty engines.

As previously noted, Curtiss had pioneered the construction of large flying boats starting in 1914. In 1916, the Navy ordered a variant of his first large flying boat designated the H-12. During October 1917, the sea service let contracts for the H-16, a modified version of the flying boat which Curtiss had sold to the Royal Navy. Although their production lagged initially, it soon exceeded established requirements. Based on operational experience, those machines were modified in ways which didn't slow production. The NAF built 150 H-16s for the U.S. Navy while Curtiss constructed an additional 124.[94]

In response to a meeting with French naval officers during the spring of 1917, Towers cobbled together a token aviation unit of 7 officers and 122 mechanics and dispatched them to the continent. In accordance with agreements made earlier with the French mission in Washington, D.C., they used French airplanes. The initial increment of the Navy's First Aeronautic Detachment arrived in Pauillac, France on June 5th. Operating a French Le Tellier flying boat from Le Croisic Naval Air Station, they flew their first coastal patrol on November 18th.[95]

The first shipment of those American-built naval aircraft arrived in France on 24 May 1918. After a modification that increased the aircraft's wingspan from 62 to 74 feet, the HS-1 was redesignated the HS-2L. That change was made so that the aircraft could carry more powerful bombs. On 13 June 1918, the first HS-1 assembled in France made its initial flight. Approximately 1,178 of the aircraft were produced. Since the numbers ordered were too large for Curtiss to produce alone, many were constructed under license by other American firms.[96]

Hunsaker, returned to duty in the Navy Department, had proposed a large aircraft procurement program in September 1917. It included 100 Curtiss H-1 twin Liberty engine flying boats, 825 Curtiss HS-1 single Liberty engine flying boats and 640 Curtiss R-6 trainers. Soon afterward, Vice Adm Sims requested an additional 630 twin-engine flying boats. That enormous increase threw Hunsaker's production plans into disarray. Neither the NAF nor any of the existing civilian plants could handle the expanded workload. Historian Trimble writes that the solution "... was to assign 480 of the airplanes to the Naval Aircraft Factory on the assumption that the factory would rely on outside contractors for major components and carry out final assembly only."[97]

During the summer of 1918, Daniels requested an increase of 1,300 flying boats. He wanted them delivered by 1 July 1919. That enormous increase had to be divided up among several different firms by Hunsaker's staff. It intensified the problems of coordinating the delivery of raw materials and spare parts to the various factories involved in the program. Finding the right contract arrangements for the situation was another challenge. Because some private firms were concerned that the Navy might not reimburse them for the needed expansion of their facilities, the Navy junked its established flat-price contract. Trimble writes that instead the sea service adopted cost-plus-a-fixed-fee contract which guaranteed them a 15% profit "... based on a 'bogey' [i.e., benchmark] price agreed upon by both parties. In addition, the navy contributed a quarter of any saving from the 'bogey' price [to the firm], thus encouraging the manufacturer to come in under that price."[98]

Naval aviation had grown enormously by the Armistice. Daniels reported that "On, November 11, 1918, there were 282 officers and 2180 men in the Marine section; 2835 officers and 132,873 sailors in the Naval Flying Corps, with 10,579 men of other naval ratings assigned to aviation duty. The Navy's aviation equipment included 1,865 flying boats and seaplanes and 242 land planes. A total of 570 aircraft had been sent abroad before the end of hostilities."[99] Daniels observed that the war had emphasized "... the necessity of developing aviation as an integral part of our military and naval forces..... Our aviation force must be made a part of the fleet as well as an effective coast patrol-capable not only of operating from shore, but also at sea, accompanying our ships wherever they may go."[100]

Notes

1. Richard P. Hallion, *The Rise of Fighter Aircraft, 1914–1918* (Annapolis, MD: Nautical & Aviation Publishing Company of America, 1 January 1988), passim; Maurer Maurer, Ed., "Final Report of the Chief of Air Service, AEF," *The U.S. Air Service in World War I. Volume I* (Maxwell AFB, AL: The Albert F. Simpson Historical Research Center and Washington, D.C.:

The Office of Air Force History, 1978), 82; R. D. Layman, *Naval Aviation in the First World War: Its Impact and Influence* (Annapolis, MD: Naval Institute Press, 1996), 10–12.

2. Irving Brinton Holley, Jr., *Ideas and Weapons: Exploitation of the Aerial Weapon by the United States in World War I: A Study in the Relationship of Technological Advance, Military Doctrine, and the Development of Weapons* (Hamden, CT: Archon, 1971), passim; Charles F. O'Connell, Jr., "The Failure of the American Aeronautical Production and Procurement Effort During the First World War I," master's thesis, The Ohio State University, 1978, passim; Goldberg, *A History of the United States Air Force, 1907–1957*, 15; "America's Munitions, 1917–1918: Report of Benedict Crowell, The Assistant Secretary of War, Director of Munitions" (Washington, D.C.: GPO, 1919), 235–267; Senate Report No. 380, 65th Cong., 2nd sess., 1918, "Investigation of the War Department – Aircraft Production," passim; John K. Barnes, "The Vindication of Squier and Deeds, What Really Happened to the Billion Dollar Aircraft Appropriation," [article originally appeared in the July 1921 issue of the *World's Work* magazine and was privately reprinted by Edward Deeds in booklet form, 1936], 1–25; Maj Gen George Owen Squier, "Aeronautics in the United States at the Signing of the Armistice, November 11, 1918," presented at the 354th meeting of the American Institute of Electrical Engineers, New York, City, New York, 10 January 1919, 3, 62–63, George Owen Squier Papers, Series Six, Box 6, Folders, USAFAL; R.M. Jones, Air Service, for the Office of the Assistant Secretary of War, Planning Branch, Procurement Division, Rooms 25 41-2551 Munitions Building, Washington, D.C., Subj.: Wartime Aircraft Production, 10 February 1923, 1, National Archives, Record Group (RG) 107, Records of the Office of the Secretary of War, Entry 144, Box 1; Benedict Crowell and Robert Forrest Wilson, *How America Went to War. Demobilization. Our Industrial and Military Demobilization After the Armistice* (New Haven, CT: Yale Press, 1921), 200.

3. Maurer Maurer, "McCook Field, 1917–1927," *The Ohio Historical Quarterly*, January 1958, 21–22; Grossnick, *United States Naval Aviation, 1910–1995*, 28; William F. Trimble, *Wings for the Navy: A History of the Naval Aircraft Factory, 1917–1956* (Annapolis, MD: Naval Institute Press, 1990), 13; Hallion, *Test*

Pilots: The Frontiersmen of Flight, 57; Gross, *American Military Aviation: The Indispensable Arm*, 28, 35, 47; Wayne Biddle, *Barons of the Sky: From Early Flight to Strategic Warfare, the Story of the American Aerospace Industry* (New York, NY: Simon & Schuster, 1991), 87–88.

4. Millis, *Arms and Men: A Study in American Military History*, 213–238; Weigley, *History of the United States Army*, 342; Warren Cohn, Ed., *Intervention 1917: Why America Fought* (Boston, MA: Heath Co., 1966), passim; Millett and Maslowski, *For the Common Defense: A Military History of the United States of America*, 321, 337.
5. Millett and Maslowski, *For the Common Defense: A Military History of the United States of America*, 328.
6. Ibid.
7. Millis, *Arms and Men: A Study in American Military History*, 225; Weigley, *History of the United States Army*, 342, 348–350, 354, 360–361.
8. Millis, *Arms and Men: A Study in American Military History*, 225; Weigley, *History of the United States Army*, 342, 348–350, 354, 360–361; Daniel R. Beaver, *Newton Baker and the American War Effort, 1917–1919* (Lincoln, NE: University of Nebraska Press, 1966), 53–54; Weigley, *History of the United States Army*, 558.
9. "Resolution of the Council of National Defense Creating Aircraft Production Board," 16 May 1917, AAF Records, RG 18, Entry # 84–85, Box 2, National Archives, Minutes of the Aircraft Production Board, May–October 1917; "Aeronautics: Third Annual Report of the National Advisory Committee for Aeronautics [Fiscal Year] 1917" (Washington, D.C.: GPO, 1918), 17, Hathi Trust Digital Library; Weigley, *History of the United States Army*, 348–350, 360–361; Paul C. Koistinen, "The Industrial Military Complex in Historical Perspective: The Interwar Years," *Journal of American History*, March 1970, 821, 822; Goldberg, *A History of the United States Air Force, 1907–1957*, 15; Phillip S. Dickey, III, *The Liberty Engine, 1918–1942, Smithsonian Annals of Flight*, Vol. I, No. 3 (Washington, D.C.: Smithsonian Institution Press, 1968), 234–335.
10. Millis, *Arms and Men: A Study in American Military History*, 234–235.

11. Weigley, *History of the United States Army*, 200–201; James E. Hewes, Jr., *From Root to McNamara: Army Organization and Administration, 1900–1963* (Washington, D.C.: Center of Military History, U.S. Army, 1975), 19.
12. "War Department Annual Reports, 1918 (In Three Volumes), Volume I" (Washington, D.C.: GPO, 1919), 8, U.S. Congress, House of Representatives, 65th Cong., 3rd sess., Document No. 1432, ProQuest Congressional Documents, Pentagon Library; Secretary of War to Maj Gen John J. Pershing, Subj." "Command Authority and Duties in Europe," 26 May 1917, "War Department Annual Reports, 1919 (In Three Volumes), Volume I" (Washington, D.C.: GPO, 1919), 7, U.S. Congress, House of Representatives, 66th Cong., 2nd sess., Document No. 426, ProQuest Congressional Documents, Pentagon Library; "Final Report of Gen. John J. Pershing," 1 September 1919, "War Department Annual Reports [Fiscal Year] 1919, Volume I" (Washington, D.C.: GPO, 1919), 551, U.S. Congress, House of Representatives, 66th Cong., 2nd sess., Document No. 426, ProQuest Congressional Documents, Pentagon Library; Millis, *Arms and Men: A Study in American Military History*, 237; Weigley, *History of the United States Army*, 335–356, 358–359.
13. "War Department Annual Reports [Fiscal Year] 1917. Volume I" (Washington, D.C.: GPO, 1918), 51, U.S. Congress, House of Representatives, 65th Cong., 2nd sess., Document No. 856, ProQuest Congressional Documents, Pentagon Library; Robert P. White, *Mason Patrick and the Fight for Air Service Independence* (Washington, D.C.: Smithsonian Institution Press, 2001), 18–19; Gross, *American Military Aviation: The Indispensable Arm*, 37–38.
14. White, *Mason Patrick and the Fight for Air Service Independence*, 20–21; "New Chief of Overseas Air Service [Brig Gen Mason M. Patrick]," *Air Service Journal*, 11 July 1918, 56, Hathi Digital Trust Library; James P. Tate, *The Army and Its Air Corps: Army Policy Toward Aviation, 1919–1941* (Maxwell AFB, AL: Air University Press, July 1998), 19; "Major General Mason M. Patrick," prepared by the USAF, undated, www.af.mil; Roger J. Spiller, Ed., "PATRICK, Mason Mathews," *Dictionary of American Military Biography Volume II* (Westport, CT: Greenwood Press, 1984), 826–827; Gross, *American Military*

Aviation: The Indispensable Arm, 38; Louis A. Peake, "Major General Mason M. Patrick, Military Aviation Pioneer," *American Aviation Historical Society (AAHS) Journal*, Fall 1981, 200, 202; Pershing, *My Experiences in the World War*, Vol. I, 333. Marcosson, *Colonel Deeds: Industrial Builder*, 220.

15. Millis, *Arms and Men: A Study in American Military History*, 226; Millett and Maslowski, *For the Common Defense: A Military History of the United States of America*, 321–322; Kenneth J. Hagan, *This People's Navy: The Making of American Sea Power* (New York, NY: The Free Press, A Division of Macmillan, Inc., 1991), 252.
16. Millis, *Arms and Men: A Study in American Military History*, 235–236; Millett and Maslowski, *For the Common Defense: A Military History of the United States of America*, 339; Hagan, *This People's Navy: The Making of American Sea Power*, 255; "What Was the Battle of Jutland?," British Imperial War Museum, undated, www.iwm.org.uk; Turnbull and Lord, *History of United States Naval Aviation*, 97.
17. Hewes, *From Root to McNamara: Army Organization and Administration, 1900–1963*, 29; Weigley, *History of the United States Army*, 358–362; Beaver, *Newton Baker and the American War Effort, 1917–1919*, 50–78; "Curtiss Has Nearly $7,000,000 of Foreign Orders," *Aviation and Aeronautical Engineering*, 1 December 1916, 204, Hathi Trust Digital Library; H. D. Norris, "The Industrial War Load and Its Distribution," May 1941, 1,4, Historical Office, Air Materiel Command (AMC); Crouch, *Wings: A History of Aviation from Kites to the Space Age*, 187.
18. Beaver, *Newton Baker and the American War Effort, 1917–1919*, 80–81.
19. Bernard M. Baruch, Chairman, War Industries Board, "American Industry in the War: A Report of the War Industries Board" (Washington, D.C.: GPO, 1921), 22–29, Internet Archive; Hewes, *From Root to McNamara: Army Organization and Administration, 1900–1963*, 23, 44, 48; Beaver, *Newton Baker and the American War Effort, 1917–1919*, 80–81; Weigley, *History of the United States Army*, 361.
20. Planning Branch, Office of the Assistant Secretary of War, "A Survey of the American Aircraft Industry," 1935, 3, Franklin D. Roosevelt Presidential Library and Museum (FDRPL&M),

PSF, Box 14, War Department 1933–1941 Folder; Alex Roland, *Model Research: The National Advisory Committee for Aeronautics, 1915–1918. Volume I* (Washington, D.C.: NASA, 1975), 39–41; *The Aircraft Yearbook For 1919* (New York City, NY: Manufacturers Aircraft Association Inc., 1919), 35, 37, AIA, aia-aerospace.org; 1st Lt Clayton Bissell, "Brief History of the Air Corps and Its Late Developments," A.S. Mimeo U 746, 1 January 1927, 27, Office of History, Headquarters Air Force Systems Command (AFSC/HO) archives; Clark, "Major General George Owen Squier: Military Scientist," 273–280; "Joseph Sweetman Ames," NASA, 10 December 2013, nasa.gov; "Nation not in Patent Deal: Cross-Licensing in Airplane Industry Between Companies Only," ***New York Times***, 27 July 1917, 16, ProQuest Historical Newspapers, AACPL, www.aacpl.net.

21. NACA's successor organization founded in 1958.
22. Mary Catherine Bitter, "Joseph Ames," NASA History Program Office, 20 March 2013, last updated 7 August 2017, nasa.gov.
23. Ibid.
24. Planning Branch, Office of the Assistant Secretary of War, "A Survey of the American Aircraft Industry," 1935, 3, FDRPL&M, PSF, Box 14, War Department 1933–1941 Folder; Roland, *Model Research: The National Advisory Committee for Aeronautics, 1915–1918, Volume I*, 39–41; Clark, "Major General George Owen Squier: Military Scientist," 273–280; "Aeroplane Agreement Legal," *Wall Street Journal,* 11 October 1917, 2, ProQuest Historical Newspapers, Pentagon Library.
25. Roland, *Model Research: The National Advisory Committee for Aeronautics, 1915–1918. Volume I*, 41.
26. Goldberg, *A History of the United States Air Force, 1907–1957*, 13; Holley, *Ideas and Weapons: Exploitation of the Aerial Weapon by the United States in World War I: A Study in the Relationship of Technological Advance, Military Doctrine, and the Development of Weapons*, 37; Grossnick, *United States Naval Aviation, 1910–1995*, 24; Layman, *Naval Aviation in the First World War: Its Impact and Influence*, 49; "Army Lacking in Air Equipment Says Officer: Not One Machine Suitable for Field Service in Present War," *Chicago Tribune*, 4 May 1917, 7, ProQuest Historical Newspapers, Pentagon Library.

27. "Brig. Gen. George O. Squier is Promoted," *Air Service Journal*, 11 October 1917, 438, Hathi Trust Digital Library; Isaac F. Marcosson, *Colonel Deeds: Industrial* Builder (New York, NY: Dodd, Mead and Company, 1 January 1947), 220.
28. Gross, *American Military Aviation: The Indispensable Arm*, 29; Hiram Bingham, *An Explorer in the Air Service* (New Haven, CT: Yale University Press, 1920), 23, Hathi Trust Digital Library.
29. "History of Aviation Section Signal Corps and Division of Military Aeronautics From April 1917, to October 1, 1918," ca 1918, 9, H.H. Arnold Papers, Box 222, File "Air Service 1918", Manuscript Division, LOC; Foulois and Glines, *From The Wright Brothers to the Astronauts: The Memoirs of Major General Benjamin Foulois*, 141; Arnold, *Global Mission*, 53; Harry Howe Ransom, "The Air Corps Act of 1926: A Study of the Legislative Process," Ph.D. dissertation, Princeton University, August 1953, 27–29; Jerold E. Brown, "From the Ground Up: Air Planning in the Office of the Chief Signal Officer 1917–1918," *Aerospace Historian*, Fall, September 1986, 177.
30. Brig Gen Benjamin D. Foulois, "Air Service Lessons Learned During the Great War," 29 January 1919, Benjamin D. Foulois Papers, Folder 1, USAFAL; "History of the Air Service: Lecture Delivered By Major F. P. Lahm, United States Army, at the United States Military Academy, West Point, New York," 13 February 1920, 9, U-370, A.S. RG 18, Records of the Army Air Forces (AAF), Entry 211 Air Corps U Stencils, Box 4, Vol. VIII, National Archives; Arnold, *Global Mission*, 50; Goldberg, *A History of the United States Air Force, 1907–1957*, 14; Foulois and Glines, *From the Wright Brothers to the Astronauts: The Memoirs of Major General Benjamin Foulois*, 143.
31. Maj Gen W.L. Kenly, Director of Military Aeronautics, "History of Aviation Section Signal Corps and Division of Military Aeronautics From April 1917, to October 1, 1918," ca 1918, 8–9, H.H. Arnold Papers, Box 222, File "Air Service 1918", Manuscript Division, LOC; Lahm, "History of the Air Service: Lecture Delivered By Major F. P. Lahm, United States Army, at the United States Military Academy, West Point, New York," 13 February 1920, 9; Arnold, *Global Mission*, 50; Foulois and Glines, *From the Wright Brothers to the Astronauts: The Memoirs of Major General Benjamin Foulois*, 143.

32. Brown, "From the Ground up: Air Planning in the Office of the Chief Signal Officer 1917–1918," *Aerospace Historian*, Fall, September 1986, 179.
33. Ibid., 179–180; "History of Aviation Section Signal Corps and Division of Military Aeronautics From April 1917, to October 1, 1918," ca 1918, 14–15, H.H. Arnold Papers, Box 222, File "Air Service 1918", Manuscript Division, LOC; John F. Victory, "Early Aeronautical Research," address at the Air Research and Development Command, USAF officers' open mess at Baltimore, Maryland, 24 April 1953, John F. Victory Papers, Series 2, Box 19, Folder 2, USAFAL; Ransom, "Air Corps Act of 1926: A Study of the Legislative Process,"36–37 "Huge American Air Fleet, Costing $600,000,000 Assured," *New York Times*, 29 July 1917, 69, ProQuest Historical Newspapers, Pentagon Library; Hallion, *The Rise of Fighter Aircraft*, iii-iv; "$640,000,000 Appropriated for Aviation: Senate Passes and President Signs Bill Creating Army Air Service," *Air Service Journal*, 26 July 1917, 82–83, Hathi Trust Digital Library.
34. "Asks Bigger Air Fund," *New York Times*, 19 February 1917, 2, ProQuest Historical Newspapers, Pentagon Library; "The Navy Appropriations," *Aviation and Aeronautical Engineering*, 15 July 1917, 555, Hathi Trust Digital Library; "Navy's Need for $45,000,000 for Aviation Explained to Speaker Clark by Secretary Daniels," *Air Service Journal*, 26 July 1917, 83, Hathi Trust Digital Library.
35. Ransom, "Air Corps Act of 1926: A Study of the Legislative Process," 39, 40, 41; Gross, *American Military Aviation: The Indispensable Arm*, 31–32; "Americans Plan to Fight to Finish in Air Battles: Special Board Under Howard Coffin Announces Extensive Program," *Chicago Daily Tribune*, 21 May 1917, 5, ProQuest Historical Newspapers, Pentagon Library; "Aircraft Manufacturers Approve New Board," *New York Times,* 22 May 1917, 24, ProQuest Historical Newspapers, Pentagon Library; *Aircraft Yearbook for 1934* (Clinton, MA: Clinton Press, Inc., 1934), 110, AIA, aia-aerospace.org.
36. U.S. Congress, Senate, *Aircraft Production. Hearing Before the Subcommittee of the Committee on Military Affairs*, 65th Cong., 2nd sess., Vol. I, 3 June 1918, 82; 7 June 1918, 294–295; 19 June 1918, 316; 21 June 1918, 439, ProQuest Congressional

Documents, Pentagon Library; "History of Aviation Section Signal Corps and Division of Military Aeronautics From April 1917, to October 1, 1918," ca 1918, 24, H.H. Arnold Papers, Box 222, File "Air Service 1918", Manuscript Division, LOC; "Aircraft Production Board Recommended Cost Plus Fixed-Profit Contracts for Aircraft and Engines," 29 May 1917, AAF Records, RG 18, Entry # 84–85, Box 2, Minutes of the Aircraft Production Board, May–October 1917, National Archives; Hone and Utz, *History of the Office of the Chief of Naval Operations, 1915–2015*, 38; Lahm, "History of the Air Service: Lecture Delivered By Major F. P. Lahm, United States Army, at the United States Military Academy, West Point, New York," 13 February 1920, 10; J.H. Towers, "Operations of Naval Aircraft [During World War I]," *SAE [Society of Automotive Engineers] Transactions*, Vol. 14, Part 1, 1919, 373–374, JSTOR, Pentagon Library; O'Connell, "The Failure of the American Aeronautical Production and Procurement Effort During the First World War I," 69–70; Irving Brinton Holley, Jr., *Buying Aircraft: Materiel Procurement for the Army Air Forces. Special Studies. U.S. Army in World War II* (Washington, D.C.: Center of Military History, USA, 1964, reprinted 1975), 82; William F. Trimble, *Jerome C. Hunsaker and the Rise of American Aeronautics* (Washington, D.C.: Smithsonian Institution Press, 2002), 36; Bissell, "Brief History of the Air Corps and Its Late Developments," A.S. Mimeo U 746, 1 January 1927, 21.

37. Maurer Maurer, Comp. and Ed., *The U.S. Air Service in World War I. Volume II* (Maxwell AFB, AL: The Albert F. Simpson Historical Research Center and Washington, D.C.: The Office of Air Force History, 1978), 131–133; Lahm, "History of the Air Service: Lecture Delivered BY Major F. P. Lahm, United States Army, at the United States Military Academy, West Point, New York," 13 February 1920, 12; Historical Section, Army War College, "The Signal Corps and Air Service: A Study of Their Expansion in the United States 1917–1918," Monograph No. 16, July 1922 (Washington, D.C.: GPO, 1922), 74, Henry H. Arnold Papers, Box 244, Miscellaneous Reports File, 1922, Manuscript Division, LOC; O'Connell, "American Aeronautical Failure," 51–52.

38. Mauer, *The U.S. Air Service in World War I. Vol I, 59–61*; Claussen, *Materiel Research and Development in the Army Air Arm, 1914–1945. AAF Historical Studies No. 50*, 27–28; Historical Section, Army War College, "The Signal Corps and Air Service: A Study of Their Expansion in the United States 1917–1918," Monograph No. 16, July 1922, 74; Memorandum, R.M. Jones, Air Service, for the Office of the Assistant Secretary of War, Planning Branch, Procurement Division, Subj.: Wartime Aircraft Production, 10 February 1923, 6, RG 107, Records of the Office of the Secretary of War, Entry 144, Box 1, National Archives; "Order Airplanes Abroad: Our Aviation Service Places Large Contracts In Allied Countries," *New York Times*, 16 August 1917, 13, ProQuest Historical Newspapers, AACPL, www.aacpl.net.
39. Trimble, *Wings for the Navy: A History of the Naval Aircraft Factory, 1917–1956*, 16–17.
40. Ibid., 5–6; Foulois, "Lessons Learned in the Great War," 22; General John J. Pershing to the Adjutant General, U.S. Army, Subj.: Permission to Purchase French Aircraft, 16 July 1917, Papers of Benjamin D. Foulois, Box 10, "Aircraft Production (General)" file, Manuscript Division, LOC; General John J. Pershing, Commander-in-Chief, American Expeditionary Forces, *My Experiences In World War I* (New York, NY: Frederick A. Stokes Company, 1931), Vol. I, 286.
41. "History of Aviation Section Signal Corps and Division Of Military Aeronautics From April 1917, to October 1, 1918," 100–101, H.H. Arnold Papers, Box 222, File: "Air Service 1918," LOC; Foulois, "Air Service Lessons Learned During the Great War,"22; "De Havilland DH-4," NMUSAF, published 7 April 2015, nationalmuseum.af.mil; Historical Section, Army War College, "The Signal Corps and Air Service: A Study of Their Expansion in the United States 1917–1918," Monograph No. 16, 77; "Plane Inquiry Secret," *Washington Post*, 29 Mar 1918, 9, ProQuest Historical Newspapers, Pentagon Library.
42. O'Connell, "The Failure of the American Aeronautical Production and Procurement Effort During the First World War I," 17; "America Groping Says Wadsworth: War Machinery Has Broken Down Under Inadequate System, Senator Asserts," *New York*

Times, 6 February 1918, 5, ProQuest Historical Newspapers, Pentagon Library.

43. "Aeronautics: Third Annual Report of the National Advisory Committee for Aeronautics [Fiscal Year] 1917," 20; "Aeronautics: Fourth Annual Report of the National Committee for Aeronautics [Fiscal Year] 1918" (Washington, D.C.: GPO, 1920), 24, NACA Collection [University of North Texas] UNT Libraries, Government Documents Department, UNT Digital Library; Trimble, *Jerome C. Hunsaker and the Rise of American Aeronautics*, 13, 22–34, 97–98; Paul A. Hanle, *Bringing Aerodynamics to America* (Cambridge, MA: The MIT Press, 1982), 82–83; Howard S. Wolko, *In the Cause of Flight: Technologists of Aeronautics and Astronautics. Smithsonian Studies in Air and Space, No. 4* (Washington, D.C.: The Smithsonian Institution, 1981), 20–21.
44. Trimble, *Jerome C. Hunsaker and the Rise of American Aeronautics*, 30–31; Van Fleet, and Armstrong, *United States Naval Aviation, 1910–1980*, 9.
45. Trimble, *Jerome C. Hunsaker and the Rise of American Aeronautics*, 31.
46. "War Department Annual Reports [Fiscal Year] 1918. Volume I" (Washington, D.C.: GPO, 1919), 1074, U.S. Congress, House of Representatives, 65th Cong., 3rd sess., Document No. 1432, ProQuest Congressional Documents, Pentagon Library.
47. "War Department Annual Reports, 1918 (In Three Volumes), Volume I," 1073–1074, U.S. Congress, House of Representatives, 65th Cong., 3rd sess., Document No. 1432; Freudenthal, *The Aviation Business: From Kitty Hawk to Wall Street*, 37; Glenn L. Martin, "The First Half-Century of Flight in America," *Journal of Aeronautical Sciences*, February 1954, 75; Memorandum, R.M. Jones, Air Service, for the Office of the Assistant Secretary of War, Planning Branch, Procurement Division, Subj.: Wartime Aircraft Production, 10 February 1923, 5, National Archives, RG 107, Records of the Office of the Secretary of War, Entry 144, Box 1; Claussen, *Comparative History of Research and Development Policies Affecting Air Materiel, 1915–1944, AAF Historical Study No. 20*, 27.
48. Freudenthal, *The Aviation Business: From Kitty Hawk to Wall Street*, 37; Theodore M. Knappen, *Wings of War: An Account*

of the Important Contributions of the United States to Aircraft Invention, Engineering, Development and Production During the World War (New York City, NY: Putnam, 1920), 39; Hatch, *Glenn Curtiss: Pioneer of Naval Aviation*, 247–249; Roseberry, *Glenn Curtiss: Pioneer of Flight*, 397; Henry Woodhouse, *Textbook of Naval Aeronautics* (New York, NY: The Century Co., 1917), 144; "JN-4 'Jenny'," NHHC, undated, history.navy.mil.

49. "America's Munitions, 1917–1918: Report of Benedict Crowell, the Assistant Secretary of War, Director of Munitions," 243, 251–252; 33, 39; O'Connell, "The Failure of the American Aeronautical Production and Procurement Effort During the First World War I," 32–35.
50. Jesse G. Vincent, "Reasons Behind the Liberty Aviation Engine," *The Journal of the Society of Automotive Engineers*, January 1918, 26, Hathi Trust Digital Library; C. Fayette Taylor, *Aircraft Propulsion: A Review of the Evolution of Aircraft Piston Engines. Smithsonian Annals of Flight, Volume 1, Number 4* (Washington, D.C.: Smithsonian Press, 1971), 30, 33, repository.si.edu; "America's Munitions, 1917–1918: Report of Benedict Crowell, The Assistant Secretary of War, Director of Munitions," 53–54, 265–280; "War Department Annual Reports [Fiscal Year] 1918. Volume I," 1076–1077.
51. "Completion of Liberty Engine," *Air Service Journal*, 20 September 1917, 338, Hathi Trust Digital Library; "Progress in Production of Planes and Engines," *Air Service Journal*, 21 February 1918, 252, Hathi Trust Digital Library; Crowell, "America's Munitions, 1917–1918: Report of Benedict Crowell, the Assistant Secretary of War, Director of Munitions," 278, 280; Bissell, "Brief History of the Air Corps and Its Late Developments," A.S. Mimeo U 746, 1 January 1927, 33.
52. "War Department Annual Reports [Fiscal Year] 1918. Volume I" (Washington, D.C.: GPO, 1919), 51–52, U.S. Congress, House of Representatives, 65th Cong., 3rd sess., Document No. 1432, ProQuest Congressional Documents, Pentagon Library; "Lahm, "History of the Air Service: Lecture Delivered By Major F. P. Lahm, United States Army, at the United States Military Academy, West Point, New York," 13 February 1920, 12, 13; John B. Rae, *Climb to Greatness. The American Aircraft Industry, 1920–1960* (Cambridge, MA: MIT Press, 1968), 2; Knappen,

Wings of War: An Account of the Important Contributions of the United States to Aircraft Invention, Engineering, Development and Production During the World War, 139, 176–177; William Glenn Cunningham, *The Aircraft Industry: A Study in Industrial Location* (Los Angeles, CA: Lorrin L. Morrison, 1961), 37; Crowell, "America's Munitions, 1917–1918: Report of Benedict Crowell, the Assistant Secretary of War, Director of Munitions," 243, 245–249; Memorandum, R.M. Jones, Air Service, for the Office of the Assistant Secretary of War, Planning Branch, Procurement Division, Subj.: Wartime Aircraft Production, 10 February 1923, 5, 8, National Archives, RG 107, Records of the Office of the Secretary of War, Entry 144, Box 1; "Annual Report of the Bureau of Construction and Repair to the Secretary of the Navy for the Fiscal Year 1918" (Washington, D.C.: GPO, 1918), 15, Hathi Trust Digital Library.

53. "America's Munitions, 1917–1918: Report of Benedict Crowell, the Assistant Secretary of War, Director of Munitions," 266–267; Knappen, *Wings of War: An Account of the Important Contributions of the United States to Aircraft Invention, Engineering, Development and Production During the World War*, 69–70; O'Connell, "The Failure of the American Aeronautical Production and Procurement Effort During the First World War I," 77; Memorandum, R.M. Jones, Air Service, for the Office of the Assistant Secretary of War, Planning Branch, Procurement Division, Subj.: Wartime Aircraft Production, 10 February 1923, 6, National Archives, RG 107, Records of the Office of the Secretary of War, Entry 144, Box 1; Earle Remington, "The Future of American Aeronautics" *The North American Review*, Vol. 209, No. 758 (Jan., 1919), 53, JSTOR, Pentagon Library.
54. "History of Aviation Section Signal Corps and Division of Military Aeronautics From April 1917, to October 1, 1918," ca 1918, 90, H.H. Arnold Papers, Box 222, File "Air Service 1918", Manuscript Division, LOC; "Board Will Sift Aircraft Charges: Report Made to President by Borglum to Come in for Close Scrutiny," *New York Times*, 17 March 1918, 5, ProQuest Historical Newspapers, Pentagon Library; "Our War with Germany," XIII (March 5–March 31) *The North American Review*, Vol. 207, No. 750 (May, 1918), 783, JSTOR, Pentagon Library.

55. "History of Aviation Section Signal Corps and Division of Military Aeronautics From April 1917, to October 1, 1918," ca 1918, 90–91, H.H. Arnold Papers, Box 222, File "Air Service 1918", Manuscript Division, LOC; Ransom, "The Air Corps Act of 1926: A Study of the Legislative Process," 55; Historical Section, Army War College, "The Signal Corps and Air Service: A Study of Their Expansion in the United States 1917–1918," Monograph No. 16, 113–114; U.S. Congress, Subcommittee of the Senate Committee on Military Affairs, *Investigation of the War Department—Aircraft Production*, pt. 1, S. Rept. 380, 65th Cong. 2nd, 29 May 1918–12 July 1918, passim, ProQuest Congressional Documents, Pentagon Library; U.S. Congress, Subcommittee of the Senate Committee on Military Affairs, *Investigation of the War Department—Aircraft Production*, pt. 1, S. Rept. 380, pt. 2, 1, 5, 65th Cong. 2nd sess., 12 July 1918–16 August 1918, passim, ProQuest Congressional Documents, Pentagon Library.
56. President Woodrow Wilson to Charles Evans Hughes, Subj.: Request Hughes Participate in Investigation of U.S. Aircraft Production Program, 13 May 1918, Presidential Materials, ProQuest Congressional Documents, Pentagon Library; "History of Aviation Section Signal Corps and Division of Military Aeronautics From April 1917, to October 1, 1918," ca 1918, 91, H.H. Arnold Papers, Box 222, File "Air Service 1918," Manuscript Division, LOC; "Hughes Appointed as Investigator: Will Help Department of Justice on Borglum's Charges, at President's Request," *New York Times*, 16 May 1918, 3, ProQuest Historical Newspapers, Pentagon Library.
57. Ransom, "The Air Corps Act of 1926; A Study of the Legislative Process," 56–57; "Hughes Airplane Report Accuses Deeds and Others," *New York Times*, 1 November 1918, 1, 10, ProQuest Historical Newspapers, Pentagon Library.
58. "Hughes Airplane Report Accuses Deeds and Others," *New York Times*, 1 November 1918, 10, ProQuest Historical Newspapers, Pentagon Library.
59. Ransom, "The Air Corps Act of 1926: A Study of the Legislative Process," 57; Secretary of War Newton D. Baker to Chairman, Committee on Military Affairs, House of Representatives, Subj.: Col. Deeds, 16 January 1919, reprinted in *Air Service Newsletter*, War Department, Washington, D.C., 18 January 1919, 7, Air

Force Historical Support Division (AFHSD), www.afhistory.af.mil; "Col. E.A. Deeds Is Exonerated By Gen. S.T. Ansell," *Air Service Journal*, 25 January 1919, 2, Hathi Trust Digital Library.

60. "Report of the Director of Military Aeronautics, U.S. Army, to the Secretary of War, 1918" (Washington, D.C.: GPO, 1918), 4, Hathi Trust Digital Library; "War Department Annual Reports [Fiscal Year] 1918. Volume I" (Washington, D.C.: GPO, 1919), 51, U.S. Congress, House of Representatives, 65th Cong., 3rd sess., Document No. 1432, ProQuest Congressional Documents, Pentagon Library; "History of Aviation Section Signal Corps and Division of Military Aeronautics From April 1917, to October 1, 1918," ca 1918, 92, H.H. Arnold Papers, Box 222, File "Air Service 1918," Manuscript Division, LOC; "J.D. Ryan Chosen to Direct Making of Army Aircraft," *New York Times*, 25 April 1918, 1, 3, ProQuest Historical Newspapers, Pentagon Library; Clark, "Major General George Owen Squier: Military Scientist," 278–279; Freudenthal, *The Aviation Business*: *From Kitty Hawk to Wall Street*, 55; Hewes, *From Root to McNamara: Army Organization and Administration, 1900–1963*, 36.
61. "Report of the Director of Military Aeronautics, U.S. Army, to the Secretary of War [Fiscal Year] 1918," 4; "Report of the Bureau of Aircraft Production" (Washington, D.C.: GPO, 1918), 1, Hathi Trust Digital Library; "War Department Annual Reports [Fiscal Year] 1918. Volume I" (Washington, D.C.: GPO, 1919), 51, 1073, 1390, 1407, U.S. Congress, House of Representatives, 65th Cong., 3rd sess., Document No. 1432, ProQuest Congressional Documents, Pentagon Library; Bissell, "Brief History of the Air Corps and Its Late Developments," A.S. Mimeo U 746, 1 January 1927, 31–32; Holley, *Ideas and Weapons: Exploitation of the Aerial Weapon by the United States in World War I: A Study in the Relationship of Technological Advance, Military Doctrine, and the Development of Weapons*, 68–69; Ransom, "The Air Corps Act of 1926: A Study of the Legislative Process," 57; "Wilson Recasts Aviation Service: Takes All Control of Operation and Production Away From the Signal Corps," *New York Times,* 21 May 1918, 10, ProQuest Historical Newspapers, AACPL, www.aacpl.net; "Major General William Lacy Kenly," USAF, undated, www.af.mil.

62. "War Department Annual Reports [Fiscal Year] 1918. Volume I" (Washington, D.C.: GPO, 1919), 1075, U.S. Congress, House of Representatives, 65th Cong., 3rd sess., Document No. 1432, ProQuest Congressional Documents, Pentagon Library; Squier, "Aeronautics in the United States at the Signing of the Armistice, November 11, 1918," 20.
63. "Women Making Wings for Airplanes," *Air Services Journal*, 9 August 1917, 146, Hathi Trust Digital Library.
64. Ibid.
65. Bulletin of the Women's Bureau, No. 12, "The New Position of Women in American Industry," U.S. Women's Bureau, Department of Labor (Washington, D.C.: GPO, 1920), 74, Hathi Trust Digital Library.
66. "Report of the Director of Military Aeronautics, U.S. Army, to the Secretary of War [Fiscal Year] 1918," 7; "War Department Annual Reports [Fiscal Year] 1918. Volume I" (Washington, D.C.: GPO, 1919), 51, U.S. Congress, House of Representatives, 65th Cong., 3rd sess., Document No. 1432, ProQuest Congressional Documents, Pentagon Library; "Mr. John D. Ryan Appointed Second Assistant Secretary of War, and Given Full Charge of All Aircraft Work," *Air Power*, September 1918, 209, Hathi Trust Digital Library; Bissell, "Brief History of the Air Corps and Its Late Developments," A.S. Mimeo U 746, 1 January 1927, 32; O'Connell, "The Failure of the American Aeronautical Production and Procurement Effort During the First World War I," 25–26; Holley, *Ideas and Weapons: Exploitation of the Aerial Weapon By the United States in World War I: A Study in the Relationship of Technological Advance, Military Doctrine, and the Development of Weapons*, 69, 78–79; Maurer, *The U.S. Air Service in World War I. Volume I*, 131; Mooney and Layman, *Organization of Military Aeronautics, 1907–1935 (Congressional and War Department Action). U.S. Air Force Historical Study No. 25*, 35; "History of Aviation Signal Corps and Division of Military Aeronautics From April 1917, to October 1, 1918," undated, 112–113, H. H. Arnold Papers, Manuscript Division, LOC.
67. Crowell, "America's Munitions, 1917–1918: Report of Benedict Crowell, The Assistant Secretary of War, Director of Munitions," 235–237, 243, 250, 254, 291; O'Connell, "The Failure

of the American Aeronautical Production and Procurement Effort During the First World War I," 90–92.

68. Wolko, *In the Cause of Flight: Technologists of Aeronautics and Astronautics. Smithsonian Studies in Air and Space, No. 4*, 91–92; *The Aircraft Yearbook for 1934*, 112; Trimble, *Jerome C. Hunsaker and the Rise of American Aeronautics*, 34;" McDonnell Douglas Corporation: American Company," *Britannica*, undated, www.britannica.com; "Donald Douglas: American Aircraft Designer," *Britannica*, undated, www.britannica.com; Frank Cunningham, *Sky Master: The Story of Donald Douglas* (Philadelphia, PA: Dorrance and Company Publishers, 1943), 74–75, 88; Crouch, *Wings: A History of Aviation from Kites to the Space Age*, 183, 184.
69. *Aerospace Facts and Figures, 1964* (Los Angeles, CA.: Aerospace Publishers, Inc., 1964), 24, AIA, aia-aerospace.org.
70. Wolko, *In the Cause of Flight: Technologists of Aeronautics and Astronautics. Smithsonian Studies in Air and Space, No. 4*, 91–92; "James H. Kindelberger," San Diego Air & Space Museum, undated, sandiegoairandspace.org.
71. H.H. Blee, Army Air Service, "History of Organization and Activities of the Airplane Engineering, Bureau of Aircraft Production," ca 15 August 1919, 2–3, unpublished history in archives of the Office of History, Aeronautical Systems Division (ASD/HO), AFSC; Memorandum, R.M. Jones, Air Service, for the Office of the Assistant Secretary of War, Planning Branch, Procurement Division, Subj.: Wartime Aircraft Production, 10 February 1923, 1, 2, RG 107, Records of the Office of the Secretary of War, Entry 144, Box 1, National Archives.
72. Purtee, *History of the Army Air Service, 1907–1926, Air Force Logistics Command (AFLC) Historical Study No. 279*, 63; Squier, "Aeronautics in the United States at the Signing of the Armistice, November 11, 1918," 43–46; "Robert A. Millikan, Biographical," *The Nobel Prize, in Physics 1923*, 3 June 2024, www.nobelprize.org.
73. Paper, Subj.: Col Deeds Recommended Centralized Signal Corps Aviation Research Facility Be Established in Dayton, Ohio, 25 September 1917, AAF Records, RG 18, Entry # 84–85, Box 2, Minutes of the Aircraft Production Board, May–October 1917, National Archives; Holley, *Ideas and Weapons: Exploitation of the*

Aerial Weapon by the United States in World War I: A Study in the Relationship of Technological Advance, Military Doctrine, and the Development of Weapons, 104; Purtee, *History of the Army Air Service, 1907–1926, Air Force Logistics Command (AFLC) Historical Study No. 279*, 63; Claussen, *Materiel Research and Development in the Army Air Arm, 1914–1945, Army Air Forces Historical Studies No. 50*, 26; Maurer, "McCook Field, 1917–1927," *The Ohio Historical Quarterly*, January 1958, 22–23.

74. Historical Section, Army War College, "The Signal Corps and Air Service: A Study of Their Expansion in the United States 1917–1918," Monograph No. 16, July 1922 (Washington, D.C.: GPO, 1922), 64, Henry H. Arnold Papers, Box 244, Miscellaneous Reports File, 1922, Manuscript Division, LOC; Paper, Subj.: "Aircraft Production Board Recommended That the New Aviation Research Facility in Dayton, Ohio Be Named McCook Field, 1 October 1917," AAF Records, RG 18, Entry # 84–85, Box 2, National Archives, Minutes of the Aircraft Production Board, May–October 1917; Maurer, "McCook Field, 1917–1927," *The Ohio Historical Quarterly*, January 1958, 22–24; "History of the Air Corps Materiel Division [1917–1927]." Ca 1927, Office of Air Force History Micro # 200–1, v-8110, A.C, 4929–79 (page numbers blacked out); Albert E. Misenko and Phillip H. Pollock, *Engineering History, 1917–1918: McCook Field to the Aeronautical Systems Division, 4th ed.* (Wright-Patterson AFB, OH: ASD/HO, February 1979), iii, 3–4; Blee, "History of Organization and Activities of the Airplane Engineering, Bureau of Aircraft Production," 2–3; Albert Misenko, "Aero Propulsion and Power Directorate: The McCook Field Years (1917–1927)," AFRL-PR-WP-TR-1998–2020, USES, Inc., Propulsion Directorate, Air Force Research Laboratory, Air Force Materiel Command, Wright-Patterson, AFB, OH, February 1995, 34–35, Internet Archive; "Wright Field to Be Used for Aircraft Engineering and Experimenting," *Air Service Journal*, 11 October 1917, 443, Hathi Trust Digital Library.

75. Terrance M. Dean, 'The History of McCook Field; Dayton, Ohio, 1917–1927," unpublished Masters Thesis, University of Dayton, 1969, 9; "History of the Air Corps Materiel Division [1917–1927]." ca 1927, Office of Air Force History, Micro # 200-1, v-8110, A.C, 4929-79; Holley, *Ideas and Weapons:*

Exploitation of the Aerial Weapon by the United States in World War I: A Study in the Relationship of Technological Advance, Military Doctrine, and the Development of Weapons, 105; Hallion, *Test Pilots: The Frontiersmen of Flight*, 57; Misenko and Pollock, *Engineering History, 1917–1918: McCook Field to the Aeronautical Systems Division*, xiii–xiv; Memo, Al Misenko, Aeronautical Systems Division History Office, for Charles J. Gross, HQ AFSC/HO, Subj.: McCook Field Commanders, 4 October 1985, w/atch: "Notes on McCook Field, prepared for Dayton Chamber of Commerce," by Historical Division, HQ AMC [Headquarters Air Materiel Command], Wright- Patterson AFB, 23 September 1960.

76. Claussen, *Materiel Research and Development in the Army Air Arm, 1914–1945, Army Air Forces Historical Studies No. 50*, 28; Claussen, *Comparative History of Research and Development Policies Affecting Air Materiel, 1915–1944, AAF Historical Study No. 20*, 28; Blee, "History of Organization and Activities of the Airplane Engineering, Bureau of Aircraft Production," 86–91; Dean, "The History of McCook Field; Dayton, Ohio, 1917–1927," 46–55; "Bureau of Standards," *Annual Report of the Secretary of Commerce [Fiscal Year] 1918* (Washington, D.C.: GPO, 1918), 66–68, Hathi Trust Digital Library; Walt Boyne, "The Treasure Trove of McCook Field," pt. 1, *Airpower*, July 1975, 13; Maurer, "McCook Field, 1917–1927," 25.
77. Boyne, "The Treasure Trove of McCook Field," pt. 1, 15; Wolko, *In the Cause of Flight: Technologists of Aeronautics and Astronautics. Smithsonian Studies in Air and Space, No. 4*, 84; Swanborough and Bowers, *United States Military Aircraft Since 1911*, 201–203; History, "Against the Wind: 90 Years of Flight Test in the Miami [Ohio] Valley," History Office, Aeronautical Systems Center, Air Force Material Command, Wright-Patterson AFB, Ohio, February 1994, 6; Associated Press (AP), "Alfred Verville, Airplane Designer," *New York Times*, 14 March 1970, 31, ProQuest Historical Newspapers, Pentagon Library; Walter H. Waggoner, "Don R. Berlin, 83, a Designer of Aircraft for World War II," *New York Times*, 8 June 1982, D26, ProQuest Historical Newspapers, Pentagon Library.
78. Gross, *American Military Aviation: The Indispensable Arm*, 43.

79. Henning, *Harnessing the Airplane: American and British Cavalry Responses to a New Technology, 1903–1939*, 55–56.
80. "Annual Reports of the Navy Department for Fiscal Year 1917" (Washington, D.C.: GPO, 1918), undated, 35–37, U.S. Congress, House of Representatives, 65th Cong., 2nd sess., Document No. 618, ProQuest Congressional Documents, Pentagon Library.
81. Trimble, *Wings for the Navy: A History of the Naval Aircraft Factory, 1917–1956*, 8–9.
82. Ibid., 9; F. G. Coburn, NAF Manager, "Problems of the Naval Aircraft Factory During the War," *Transactions of the Society of Automotive Engineers, Inc., Part I, Volume XIV, 1919* (New York, NY: Offices of The Society, SAE, 1920), 304, Hathi Trust Digital Library.
83. Trimble, *Wings for the Navy: A History of the Naval Aircraft Factory, 1917–1956*, 10.
84. "Annual Reports of the Navy Department for Fiscal Year (Including Operations and Recommendations to December 1, 1918) 1918" (Washington, D.C.: GPO, 1918), 64, U.S. Congress, House of Representatives, 65th Cong., 3rd sess., Document No. 1450, ProQuest Congressional Documents, Pentagon Library; Trimble, *Wings for the Navy: A History of the Naval Aircraft Factory, 1917–1956*, 13; Grossnick, *United States Naval Aviation, 1910–1995*, 28; Trimble, *Jerome C. Hunsaker and the Rise of American Aeronautics*, 41–42.
85. Trimble, *Wings for the Navy: A History of the Naval Aircraft Factory, 1917–1956*, 14; Sitz, *A History of U.S. Naval Aviation. Technical Note No. 18. Series of 1930*, 15, Internet Archive; "Naval Aircraft Factory Completed," *Wall Street Journal*, 22 December 1917, 8, ProQuest Historical Newspapers, Pentagon Library; Frank A. Blazich, Jr., "United States Navy and World War I: 1914–1922," 108, NHHC, history.navy.mil; Coburn, "Problems of the Naval Aircraft Factory During the War," *Transactions of the Society of Automotive Engineers, Inc., Part I, Volume XIV, 1919*, 305; "The Naval Aircraft Factory," *Mechanical Engineering: The Journal of the American Society of Mechanical Engineers*, January 1919, 112, Hathi Trust Digital Library.

86. Trimble, *Wings for the Navy: A History of the Naval Aircraft Factory, 1917–1956,* 18; Grossnick, *United States Naval Aviation, 1910–1995*, 33; Van Fleet, and Armstrong, *United States Naval Aviation, 1910–1980,* 32; "Annual Report of the Chief of the Bureau of Construction and Repair to the Secretary of the Navy for the Fiscal Year 1919" (Washington, D.C.: GPO, 1919), 12, Hathi Digital Trust; Coburn, "Problems of the Naval Aircraft Factory During the War," *Transactions of the Society of Automotive Engineers, Inc., Part I, Volume XIV, 1919,* 305–306.
87. "Annual Reports of the Navy Department for Fiscal Year (Including Operations and Recommendations to December 1, 1918) 1918," 65–66; Sitz, *A History of U.S. Naval Aviation. Technical Note No. 18. Series of 1930,* 16, Internet Archive.
88. Trimble, *Wings for the Navy: A History of the Naval Aircraft Factory,* 31.
89. Ibid.
90. "Annual Reports of the Navy Department for Fiscal Year (Including Operations and Recommendations to December 1, 1918) 1918," 514–515; Trimble, *Wings for the Navy: A History of the Naval Aircraft Factory, 1917–1956*, 19, 20.
91. Sitz, *A History of U.S. Naval Aviation. Technical Note No. 18. Series of 1930*, 16.
92. Reynolds, *Admiral John H. Towers: The Struggle for Naval Air Supremacy*, 112, 122; Turnbull and Lord, *History of United States Naval Aviation,* 98; Grossnick, *United States Naval Aviation, 1910–1995*, 24, 25, 26; Swanborough and Bowers, *United States Military Aircraft Since 1911*, 114, 119–120; Van Fleet, and Armstrong, *United States Naval Aviation, 1910–1980,* 25, 26; Blazich, "United States Navy and World War I: 1914–1922," 34; Davis, *The Admirals Lobby*, 68.
93. Grossnick, *United States Naval Aviation, 1910–1995*, 29, 30; Roseberry, *Glenn Curtiss: Pioneer of Flight*, 410–415; "Annual Reports of the Navy Department for the Fiscal Year 1919" (Washington, D.C.:GPO, 1920), 28, U.S. Congress, House of Representatives, 66th Cong., 2nd sess., Document No. 729, ProQuest Congressional Documents, Pentagon Library; Trimble, *Jerome C. Hunsaker and the Rise of American Aeronautics,* 53–55.
94. "Annual Report of the Bureau of Construction and Repair to the Secretary of the Navy for the Fiscal Year 1918," 12; "Annual

Reports of the Navy Department for the Fiscal Year (Including Operations and Recommendations to December 1, 1918) 1918," 1, 511–512; "Annual Reports of the Navy Department for Fiscal Year 1917," 36; Trimble, *Wings for the Navy: A History of the Naval Aircraft Factory, 1917–1956*, 17; Swanborough and Bowers, *United States Military Aircraft Since 1911*, 122–123; Reynolds, *Admiral John H. Towers: The Struggle for Naval Air Supremacy*, 111; Roger E. Bilstein, *The American Aerospace Industry: From Workshop to Global Enterprise* (New York, NY: Twayne Publishers, 1996), 10–11.

95. Josephus Daniels, Secretary of the Navy, *Our Navy at War*, (Washington, DC: Pictorial Bureau, 1922), 228–229, Hathi Trust Digital Library; Roberts, "Origins of Navy Patrol Aviation, 1911 to 1920s," *Dictionary of American Naval Aviation Squadrons—Volume 2: The History of VP, VPB, VP(H) and VP(AM) Squadrons*, 3, NHHC, history.navy.mil; LT Kenneth Whiting, USN, First Aeronautic Detachment, to Secretary of the Navy Josephus Daniels, Subj.: French Bases, 7 July 1917, 1, "Documentary Histories WWI," NHHC, history.navy.mil; Reynolds, *Admiral John H. Towers: The Struggle for Naval Air Supremacy*, 111, 112, 115; Turnbull and Lord, *History of United States Naval Aviation*, 119–120; Grossnick, *United States Naval Aviation, 1910–1995*, 26; Van Fleet, and Armstrong, *United States Naval Aviation, 1910–1980*, 26, 30; Blazich, "United States Navy and World War I: 1914–1922," 37, NHHC, history.navy.mil; "Annual Reports of the Navy Department for the Fiscal Year 1917" (Washington, D.C.: GPO, 1918), 39, Hathi Trust Digital Library.
96. "New Curtiss Flying Boat in Trial Flight," *Air Service Journal*, 2 August 1917, 119, Hathi Trust Digital Library; "HS-1 and HS-2 Seaplanes," National Museum of the U.S. Navy, undated, NHHC, history.navy.mil; Grossnick, *United States Naval Aviation, 1910–1995*, 35; Swanborough and Bowers, *United States Navy Aircraft Since 1911*, 125–127; Van Fleet, and Armstrong, *United States Naval Aviation, 1910–1980*, 33; Bowers, *Curtiss Aircraft: 1907–1947*, 103.
97. Trimble, *Jerome C. Hunsaker and the Rise of American Aeronautics*, 49.
98. Ibid., 49–50.

99. "Annual Reports of the Navy Department for the Fiscal Year 1919" (Washington, D.C.: GPO, 1920), 41, U.S. Congress, House of Representatives, 66th Cong., 2nd sess., Document No. 729, ProQuest Congressional Documents, Pentagon Library.
100. Ibid., 40.

CHAPTER 4

Organizing Postwar Aviation, 1918–1926

Plagued by growing public disenchantment with the results of World War I, a return to relatively frugal congressional appropriations, the absence of any credible threat to the nation's security plus the skepticism of some political leaders and tradition-bound senior military officers, American military aviation struggled to find its place after the Armistice.[1] But despite aviation's promising record during the Great War and the enthusiasm of air power visionaries like Army Brig Gen William "Billy" Mitchell, the U.S. was unwilling to entrust its security to the bold new aeronautical technology and unproven theories about air power's war-winning potential.[2]

Aeronautical technology had made enormous strides during the war. While most of that had come from the European great powers, Glenn Curtiss and the U.S. Navy had accomplished much during the conflict with the former's seaplanes. Their NC (Navy, Curtiss) flying boats demonstrated their technological prowess in 1919 when they became the first aircraft to fly across the Atlantic.[3]

The American aircraft industry had to deal with outdated and inappropriate ideas about what kind of government procurement policies could best promote the development of aviation. Historian Jacob A. Vander Meulen concludes that "Until World War II, the congressional overseers

C. J. Gross, *Technological Innovation and the Rise of Aviation, 1903–1941*, Palgrave Studies in the History of Science and Technology, https://doi.org/10.1007/978-3-031-71108-4_4

tried to make the industry a model of traditional, laissez-faire capitalism, an ideology for which it was equally unsuited. Most congressmen concluded that the wartime deficiency resulted from a failure of free enterprise to offer a sufficient reign.... congressmen used their powers over military contracting to preserve a destructive, competitive regulatory environment for aircraft manufacture."[4] Nevertheless, the first postwar president, Warren G. Harding, recognized the necessity for close cooperation between military and commercial aviation in developing the infant science of aeronautics (Illustration 4.1). In his initial message to Congress, he stressed that "Aviation is inseparable from either the Army or the Navy, and the Government must, in the interest of national defense, encourage its development for military and civil purposes. The encouragement of the civil development of aeronautics is especially desirable as relieving the Government largely of the expense of development and maintenance of an industry almost entirely borne by the government through appropriations for the military, naval and postal air services."[5]

with 1919 transatlantic naval aviators (*Photo* Navy secretary Josephus Daniels (2nd from left front row) and assistant secretary Franklin D. Roosevelt (4th from left front row) with transatlantic naval aviators. H-030-2, NHHC, posted 6 July 2015, history.navy.mil)

After the Armistice, America rapidly demobilized most of its massive wartime military force and dismantled much of the formidable armaments industry which had sustained it. President Woodrow Wilson negotiated postwar settlements at the Versailles peace conference. Disillusioned with the results of the peace treaty, Congress rejected his League of Nations and eventually sought to return to "normalcy"—a new term for old, familiar ways of thinking popularized by Harding during his 1920 presidential campaign. Public and congressional opinion turned inward favoring a unilateralist (not isolationist) foreign policy. Strict economy became the watchword in government, especially on Capitol Hill. Americans were willing to rely on the nation's geographic isolation and a powerful Navy rather than a large standing Army or an independent air force to protect them from foreign foes.

Nothing complicated the development of military policy during the interwar period more than the emergence of aircraft as potentially formidable weapons. They blurred the established boundaries between the armed forces while precipitating enormous fights within them because of aviation's voracious appetite for resources. Those circumstances shaped the postwar debate over U.S. national defense policy including the role of

Illustration 4.1 Secretary Daniels and assistant secretary Roosevelt

military aviation. American political and military elites spent much of the 1918–1926 period arguing about how to organize the nation's military aviation programs as well as how (or if) to support the nation's struggling aircraft industry and promote the development of commercial aviation.

A small group of military aviators, plus their allies in the armed forces, aviation industry and political supporters, made some technical progress during the 1920s. Those gains were rationalized by its advocates as a response to the challenge posed by better funded European aviation programs rather than specific threats to the nation. The experimental work of the U.S. military, closely tied with industry and the National Advisory Committee for Aeronautics (NACA), made notable contributions to American aeronautics.

American military aviation procurement practices were tailored to keep a small but technically progressive segment of the nation's struggling aircraft industry alive to foster aeronautical progress and serve as a mobilization base for wartime expansion. Without adequate military orders and funding, aviation advocates feared that the best aircraft firms would go out of business. They realized that maintaining technological superiority and fostering the development of commercial aviation for wartime expansion had become key national defense requirements. Shaped by low budgets, a conservative approach to aeronautical innovation, squabbles about the uses and control of air power, as well as the lack of credible threats to American security, the pace of American technological innovation was modest during the early interwar period.

To a limited extent, the armed forces worked together to promote aeronautical progress and hold down costs. They established a Joint Army and Navy Board on Aeronautics (later, the Aeronautical Board) during 1919 to make recommendations to coordinate their aviation activities with other federal government agencies. The military air arms cooperated continually on matters of material supply, experimental work and specialized training.[6]

Austerity, Policy and Politics

President Wilson, engrossed in the peace process and incapacitated by serious illness, gave no real guidance to postwar defense planners. The armed forces moved to fill the resulting policy vacuum. Almost as if the Great War had never happened, the Navy sought to revive its ambitious 1916 warship construction program and reassert its claims as the nation's first line of defense. In essence, the sea service wanted to build itself into global naval supremacy based on battleships aided by a significant aviation component.

The Navy's only significant foreign rivals were the British and the Japanese. No responsible public official actually believed that the U.S. would ever fight Great Britain again. Japan was another matter. In 1919, the Navy moved its strongest units to the Pacific assuming that the western Pacific was the main region of potential conflict with Japan. Since the Harding administration failed to provide any guidance regarding that assumption, the Navy prepared and planned for its own Pacific war.

To save money and forestall a naval race with either of those powers, Harding's administration convened an arms limitation conference in 1921. The Washington Naval Conference of 1921–1922 produced several landmark treaties. The Four-Power Treaty abrogated the Anglo-Japanese alliance and, in effect, established zones of naval influence for major powers in the Pacific. The toothless Nine-Power Treaty pledged its signatories to respect China's independence and territorial integrity.

The most important outcome of the conference, the Five-Power Treaty, partially limited naval armaments. Under its key provision, the world's major naval powers agreed to a "naval holiday."[7] Capital ship and aircraft carrier displacement tonnages were established for each of the contracting parties, the British Empire, U.S., France, Italy and Japan.[8] The Five-Power Treaty scuttled the Navy's plans to revive its 1916 battleship construction program. Since there were few treaty limitations on other types of naval vessels, the sea service pivoted to concentrate on developing cruisers, submarines, anti-submarine warfare vessels and other ships. The agreement placed no restraints on naval aviation. A principal long-term result of the Five-Power Treaty was to force the Navy to concentrate significant portions of its resources on emerging technologies including aircraft and submarines at the expense of traditional ones.[9] Historians Allan R. Millett and Peter Maslowski observe that "... the treaty allowed two battle-cruiser hulls to be converted to aircraft carriers, USS *Lexington* and USS *Saratoga*. When the two mammoth 33,000-ton carriers joined the fleet, they were the best of their type in the world."[10]

Within the Navy, there was confusion and ardent controversy over which course aviation should take. Initially, based on the service's wartime experience, the dominant faction championed shored-based patrol work with seaplanes. Another argued that the fastest way to have aviation move to sea with the fleet was to put float planes on battleships and cruisers.

Those aircraft would scout for enemy fleets, spot the fall of long-range gunfire and defend the skies over friendly warships. Some officers advocated aircraft carriers that had been pioneered by the Royal Navy during World War I.

Naval aviation benefitted from the creation by Congress on 12 July 1921 of the Bureau of Aeronautics (BuAer). It was inspired unwittingly by Mitchell's campaign for a separate air force which motivated the sea service's senior leadership to resolve their differences of opinion in an effort to save naval aviation. That situation produced a concerted effort to place naval aviation's administration on a sounder basis. During February 1920, Rear Adm David W. Taylor had tasked his BuC&R staff to draft a bill that would create a new naval aviation organization. Wilson's Navy Secretary Josephus Daniels had strongly endorsed the need for that legislation. President Harding told a special session of Congress on 12 April 1921 that aviation was integral to the Army and Navy while advocating creation of BuAer. The organization was officially established by the Navy on 10 August 1921 and began operations the following September 1st, with CAPT William A. Moffett, Sr. as chief in the rank of rear admiral.[11]

Political scientist Vincent Davis stresses that BuAer "... tied the naval fliers more firmly and loyally to the Navy, and at the same time it provided them with greater freedom."[12] He added that "... this new degree of organizational autonomy was much greater than that enjoyed by any other body of specialists within the Navy, in that BuAer was endowed with responsibilities and authority enjoyed by none of the other bureaus."[13] In addition to obtaining equipment like the other bureaus, BuAer secured the critical personnel administration functions that LT Cdr John Towers had provisionally acquired during World War I. Davis concludes that "... the creation of BuAer was clearly the most important intra-Navy political development arising from the aviation issue in the decades between the two world wars."[14]

Gen Payton C. March, the Army Chief of Staff, sponsored an ambitious plan for a 500,000 man active-duty postwar establishment. It would be organized as an expansible force held at half strength in peacetime. The Air Service's portion of the proposed force would operate 87 flying squadrons and 42 balloon companies. In effect, the plan would require the purchase of several thousand new aircraft of types the Air Service did not possess including bombers, single-seat fighters and dual-seat fighters.[15]

Aviation enthusiasts including Brig Gen William "Billy" Mitchell championed a different postwar American defense policy. They were convinced that the airplane had revolutionized warfare necessitating a radical overhaul of the nation's defense policy and institutions. Mitchell was clearly the most outspoken air power radical. He was convinced that wartime breakthroughs in aeronautical technology had destroyed the assumption that the U.S. could rely on geography for its defense. To protect the nation against aerial attack, an independent air force was required. On an equal footing with the Army and Navy, the new military service would be free to develop the doctrines, tactics and technologies needed for successful aerial warfare. Although he recognized that aeronautical technology was not ready in 1919 for strategic bombing, Mitchell believed that an independent air force must possess its own offensive capabilities. He also understood that air power needed a strong industrial and personnel base that only commercial aviation could provide. Mitchell was convinced that the federal government must heavily subsidize that industry before it could evolve into a viable segment of the nation's economy.[16]

There was enormous official attention in the U.S. focused on aviation immediately after World War I. Numerous bills addressing the issue were introduced on Capitol Hill during the 1919–1926 period. Several official Army bodies were formed to grapple with the difficult issues raised by military aviation. For example, the "Tactics and Organization Board," led by Maj Gen Joseph T. Dickman, was appointed in 1919 by General Pershing to study World War I's lessons as seen by the AEF. The Dickman Board, composed entirely of ground officers, rejected the idea of a separate air force but acknowledged the important auxiliary role of aviation during the war.

In response to congressional pressure, the War Department convened a board of officers that year headed by Maj Gen Charles T. Menoher, a non-flyer who headed the Air Service, to examine the aviation issue. That body, composed of four major generals all artillerymen, opposed a separate air arm while endorsing the idea that aviation was growing in value as an auxiliary to conventional ground combat arms including the infantry, artillery and cavalry. Although it rejected the notion that the air arm could independently influence the outcome of a war, the board recommended that the Air Service be placed on a par with the Army's combat arms. Aviation's growing value as an auxiliary after World War I was demonstrated in repeated maneuvers supporting other combat arms as well as during patrols with cavalry units along the southwest border beginning in 1919 to protect the U.S. against hostile Mexican bands.

Meanwhile, Secretary of War Newton Baker had appointed a commission in 1919 to study European aviation developments. Headed by Assistant Secretary of War Benedict Crowell, the group reached conclusions that were fundamentally at odds with those of the Dickman and Menoher boards. Its members recommended creation of a separate air force as well as a new federal agency responsible for all civil and military aviation activities in the U.S. They urged the federal government to subsidize enough commercial production to serve as the nucleus of a wartime aircraft production industry.[17]

The National Defense Act of 1920 rejected the expansible army of Peyton March and the visionary air force of Billy Mitchell. Instead, it created a smaller active-duty land force supported by the National Guard and organized federal reserves.[18] The postwar sea service's future was based on the Naval Act of 1916 which had called for "a Navy second to none." Although Congress repeated that goal in the Naval Act of 1919, it failed to appropriate enough money to fully fund the ambitious 1916 shipbuilding program. Historians Millett and Maslowski conclude that "In sum, the concept of sea power remained largely intact. The instrument of that concept—the battleline of the U.S. Navy—might, however, be altered by a combination of international and organizational-technological change."[19]

In the event of war, the Navy, held in a high state of readiness for combat, would guard the nation's ocean approaches while the U.S. raised a massive land force of citizen soldiers including reservists, National Guardsmen and volunteers. The War Department's plan for a postwar standing Army of 500,000 was cut to 280,000 by Congress. Even that reduced manning level was never reached during the interwar years because of the nation's revulsion against war and a desire to return to the normal peacetime practice of very limited military expenditures.[20]

The National Defense Act of 1920 recognized the growing importance of military aviation. It made the Air Service a permanent combat arm of the Army for the first time. The legislation authorized 1,516 officers and 16,000 enlisted men (including 2,500 aviation cadets) for the air arm within the Army's overall active-duty manpower ceiling. Unlike the other combat arms, the Air Service would administer its own research and development, supply, plus weapons procurement programs. The Chief of the Air Service, a major general, remained subject to General Staff direction and could not command tactical flying units. The latter were subordinated to ground forces commanders at the division and corps levels.[21]

Air power radicals were disappointed by the military aviation provisions of the 1920 legislation. Frustrated by his failed efforts to build the case for an air force within official channels, Mitchell turned to the public for support. He continually hammered home his air power message to newspaper reporters, reserve officers and civilian groups. Mitchell also wrote articles and a 1925 book titled *Winged Defense* to promote his views.[22]

Even if circumstances had made them more receptive to the full potential of military aviation, the Army's senior leadership lacked the budgetary resources to embrace it wholeheartedly. War Department expenditures nosedived after the war. During Fiscal Year 1919, the service spent over $9 billion.[23] By Fiscal Year 1926, its expenditures hit their interwar period low of just over $364 million. The Air Service's spending, including appropriated funds and the value of war reserve equipment it used, amounted to approximately 7% of the War Department's total expenditures during Fiscal Years 1920–1926. However, during the next 6 years, aviation's share of the department's spending rose dramatically to 17.7%, soaring from $45.2 million to $76.9 million annually. That dramatic growth driven by a five-year expansion and modernization program approved by Congress in 1926.[24]

The fight for air power was undercut by internal strife and organizational confusion. As had been the case in the AEF during the war, Mitchell was at the center of that storm. Maj Gen Menoher was a non flyer who lacked credibility in the eyes of some of his young aviator subordinates. Recommended for the assignment by his West Point classmate now Gen of the Armies John J. Pershing, Menoher had commanded the 42d *Rainbow Division* in France during World War I. He shared the General Staff's opinion that the air arm's primary job was to support ground forces. That stance outraged some aviation enthusiasts.

Pershing distrusted Mitchell whose lack of management skills and penchant for insubordination troubled him. Moreover, Mitchell was considered an "outsider" because he had been commissioned in a noncombat arm from a source other than the Military Academy. Menoher could not control Mitchell's public outbursts. In 1921, the latter was openly critical of Congress for cutting the Air Service's budget and demanded its separation from the Army. Menoher was so infuriated that he gave President Harding's Secretary of War John W. Weeks an ultimatum: either he or Mitchell had to go. Weeks resolved the issue by transferring Menher to another assignment. But, Mitchell didn't get the top Air Service job he craved. Instead, that assignment went to Maj Gen Mason M. Patrick on 5 October 1921 at Pershing's recommendation. Patrick, a strong disciplinarian and excellent manager who had resolved the personal infighting and mismanagement of the AEF's Air Service during the war, was called upon once more to save Army aviation from itself.[25]

Although the Navy's senior leaders were divided about aviation's future after World War I, their disputes were largely handled internally unlike the Army's. When preparing the Navy's Fiscal Year 1920 budget request, Daniels had turned to the General Board for advice. The Board routinely canvassed the Navy's bureaus and serving officers on various topics ranging from ship designs to strategic issues. During 1919, it had begun an extensive review of naval aviation's future. Contrary to the stereotype of a unified senior officer "gun club"—that promoted traditional surface warfare dominated by battleships—several highly respected senior officers testified in favor of a strong naval aviation program during the board's deliberations. They included Rear Adm Hugh Rodman, commander of the American battleship squadron that had served with Royal Navy during World War I. Adm Williams Sims, the commander of U.S. naval forces in Europe during World War I, was an ex officio board member. For years a strong battleship champion, he pushed for more emphasis on sea-based aviation before the board and Congress. Convinced that battleships were obsolete, he advocated for aircraft and aircraft carriers as integral parts of the fleet while vigorously opposing an amalgamated single air force.[26]

But, the General Board failed to champion the creation of a separate bureau for aviation. It took two more years of hard lobbying within and outside of the sea service (especially on Capitol Hill) to achieve that key organizational initiative. Adm William Benson, the Chief of Naval Operations (CNO), was opposed to the creation of yet another bureau over which he would have no direct control. Those organizations reported directly to the Secretary of the Navy instead of the CNO.[27]

Insiders at the Helm

Ironically, the major responsibility for developing and implementing the early postwar military aviation programs fell on two officers with conventional military backgrounds. Moffett and Patrick were both service academy graduates. Both were strong administrators. They were skilled and respected insiders who knew how to effectively advance their programs within their respective military services as well as with the NACA, Congress, industry and influential civilians outside government. Each had come to appreciate air power's potential. They realized the need to couple it with a strong technological/industrial base as well as commercial aviation heavily subsidized by the federal government and guided by progressive national policies. Each of them was able to place his service's aviation program on a solid footing. Each had a talent for publicity.[28]

Moffett served as BuAer's leader until his death in 1933. He was the most important architect of naval aviation during the interwar period. His distinguished early career followed a conventional naval path. In 1914, Moffett had commanded the scout cruiser USS *Chester* during the occupation of Veracruz, Mexico, by American forces. Moffett became commandant of the Great Lakes Naval Training Station in late 1914.During his tenure, it expanded from a small station into the largest U.S. Navy recruit training depot. While at Great Lakes, Moffett was instrumental in establishing a special Aviation Mechanics School for enlisted recruits. [29] After the war, Moffett was given a prestigious assignment as commander of the battleship USS *Mississippi*. While its skipper, he demonstrated his growing interest in exploring the potential of aviation in the fleet. Next, Moffett was assigned to serve as Director of Naval Aviation on the CNO's staff, a post that enabled him to recommend policy and budget initiatives but not actually mandate and oversee their implementation. According to political scientist Thomas C. Hone, that assignment "...gave Moffett the chance to convince both the active, aggressive fliers and his senior Navy colleagues that he was the right BuAer Chief.... Behind the scenes, he helped draft the General Order which specified the responsibilities and powers of the new Bureau."[30] At age 52, Moffett was too old to train as a pilot but qualified as an aerial observer (i.e., an aircrew member who provided real-time intelligence and other critical information to the pilot but wasn't qualified to fly a plane) in June 1921 (Illustration 4.2).[31]

Moffett was a master of bureaucratic politics. Naval historian Clark G. Reynolds stresses that BuAer's Chief was not "... a meticulous adept manager and left details to subordinates to concentrate on political and policy matters, exuding both charm and toughness in his dealings with congressmen and admirals alike. Moffett rarely explained his decisions to his officers; conversely, he signed virtually anything they put in front of him. He instilled supreme confidence in all who served under him...."[32]

Illustration 4.2 Rear ADM William A. Moffett, Sr. (*Photo* William A. Moffett, Sr., NHHC, history.navy.mil)

His supporters included presidents, members of Congress, industrialists, aviators and journalists.[33]

BuAer focused on embarking aircraft on the fleet's vessels as quickly as possible. Moffett stressed in 1922 that "We are trying to get aviation in the fleet. That has been the first aim of the bureau and the attempt has been to get planes with the fleet and in this way we hope to get the planes on all battleships, destroyers and scout cruisers, and in time we hope on submarines."[34]

Moffett had to fight on three fronts after he took BuAer's helm. Initially, he had to deal with young naval aviators. A few had become "true believers" who were convinced that aviation would revolutionize warfare. They advocated creation of a separate aviation corps within the Navy, like the Marines. Moffett strongly opposed them because he believed that would take them out of the sea service's mainstream and undermine its integration with the fleet. BuAer was a materiel organization that wasn't officially responsible for operational matters. Naval historian William F. Trimble observed that ".. Moffett faced the constant difficulty of maintaining two-way communication with the aviators and ran the risk of insulating himself from their special requirements, demands, and ideas. Moffett overcame the obstacle by establishing and maintaining close, informal contacts with operating personnel and creating a rotation system whereby aviators interspersed tours at the bureau with their flying duties."[35]

Conservative, mostly high-ranking naval officers determined to defend their bureaucratic "turf" and opposed to change constituted Moffett's second front. Because the Washington Naval Conference basically suspended capital ship construction for a decade, American naval officers focused more on aircraft and aircraft carriers than battleships. Although some aviators scorned them as members of the "gun club," Trimble concluded that "Generally they wanted, along with Moffett, to incorporate aviation into the fleet and learn how to maximize the potential of the new technology. Moffett's struggle with conservatives was not so much with a crusty group of battleship admirals... as it was a conflict with senior bureau chiefs in Washington. These men deeply resented Moffett's authority and influence and perceived the new Bureau of Aeronautics as a threat to their long-established prerogatives and a drain on the Navy's limited manpower and resources."[36]

Moffett's third and arguably least dangerous front was typified by Mitchell and other air power radicals who believed that aircraft had revolutionized warfare and made conventional military forces virtually obsolete. Trimble concludes that ".. in many ways Mitchell was easier to deal with than the young Turks or old shellbacks within Moffett's service. Mitchell thrived on controversy and attracted charges like a lightning rod. Moffett knew that all he had to do was wait until Mitchell's intemperance led to his own undoing."[37] Trimble stresses that "... Moffett won the war with Mitchell using an astute combination of carefully worded jabs in the news media and the support of fiscally conservative Republican administrations that did not want to add another military service to the already existing layers of Washington bureaucracy."[38]

As previously noted, Pershing had tapped Patrick in 1921 to head the Army's floundering Air Service. The two were close friends who had graduated together from West Point in 1886. Standing second in his class, Patrick had been allowed to select his branch and chose the prestigious Corps of Engineers.[39] He faced an Air Service in chaos in 1921. Patrick wasted no time in taking charge and working to correct that situation. He later recalled that "There was on hand practically nothing but war-built equipment, and of this vast quantities were scattered all over the country, costing large sums merely to care for it.... no settled policies had been laid down; the men in charge of the various activities were going along without being directed; little or nothing was known of what aircraft and airmen could or should do...."[40]

Patrick was at odds with his deputy, Mitchell, from the outset. When the tough engineering officer arrived at his office, Mitchell lectured him on how to run the Air Service. Mitchell threatened to quit after Patrick suggested that command responsibility was his not Mitchell's. Patrick called his bluff and the younger man stayed on. The Air Service's new head moved vigorously to place its operations on a more professional and economical basis. Concerned that practically all of the Air Service's aircraft were of World War I vintage and rapidly deteriorating, Patrick strongly recommended a program to purchase new aircraft. He concentrated all Air Service storage and maintenance activities in five depots. Flight training was consolidated at two fields. Under his leadership, the Air Service established a new experimental organization at Wright Field, Ohio, and a permanent flight training center at Randolph Field, Texas.

To better understand the needs of his young aviators and earn their respect, Patrick learned to fly, earning a military pilot's rating when he was nearly 60 years old.[41] Unlike Mitchell, Patrick sought to avoid or minimize public controversies. He proved to be adept at building support for his organization with senior War Department civilian officials, the General Staff, Congress, the aircraft industry, the press and civilian organizations. Patrick developed realistic plans to achieve greater Air Service autonomy while placing its operations on a more economical basis (Illustration 4.3). He came to believe in expanding the roles of military aviation. Patrick was willing to compromise and work within the existing system to gradually achieve his objectives. Pershing supported his efforts. In his last report to the Secretary of War before retiring in 1924, the Chief of Staff argued that an increased Air Service and training more noncommissioned officers as pilots were the Army's biggest needs.[42]

Maj Gen Patrick quietly won War Department acceptance of the idea that part of the Air Service (i.e., an "air force") should be set aside for independent offensive operations. Like Mitchell, Patrick came to favor creation of a separate air force but concluded that such a revolutionary change could wait. Instead, he proposed the creation of an aviation organization within the War Department whose status would be similar to that of the Marine Corps in the Navy Department. That initiative led to legislation that created the Air Corps in 1926.[43]

Hanging On By Its Eyelids: The Aircraft Industry

Most of the American aircraft industry was in a deplorable condition after World War I. Scholar Harry Howe Ransom writes that following 1918 it "... was hanging on by its eyelids."[44] The vast majority of participants in the aviation production program had either been created or enormously expanded by wartime demands. Aside from a fledgling airmail program funded by the federal government, there was virtually no commercial aviation in the U.S. after wartime contracts were cancelled.

The impact of those policies was immediate and catastrophic. Although technical expertise and management capabilities remained, approximately 90% of the industry either disappeared or turned to other lines of work in 1919. The great majority of aircraft plants had been converted from other purposes during the war and were expected to return to their previous lines of work. Boeing, influenced by its founder's experience in the lumber industry, began manufacturing furniture until the aircraft market revived. The industry's other survivors included: Wright and Curtiss in the East; Martin, Dayton-Wright, and a few plants organized by veteran pilots in the Midwest; plus Douglas and several very small locally-financed firms in the West.[45]

Military aviation's modernization practices aggravated the aircraft industry's severe problems. The armed forces continued to rely on their own aeronautical organizations to design planes. During 1919–1922, McCook Field's Engineering Division designed and built 27 aircraft. Quantity production orders for such aircraft were let to aircraft makers

Illustration 4.3 VIPS at October Dayton 1924 air races (*Photo* Left to right, Patrick, Mitchell and businessman Frederick Patterson meet during the October 1924 air races held near Dayton, Ohio. Office of History, Air Force Materiel Command, VIRIN: 20086-F-ZS999-003.JPG)

on the basis of competitive bids except for designs that had been developed in cooperation with private firms. When winning designs came from outside the Engineering Division and the designer was the sole source of support, contracts were awarded without competition.

From 1919 through 1921, only some half-dozen private firms received small military aircraft production orders. The American Aviation Mission that traveled overseas and studied European aviation developments strongly recommended in July 1919 that "... competition of the government with industry should be avoided; the only allowable exception being cases where, either on account of expenses or other causes [the armed forces] cannot obtain needed material or design from existing sources."[46] CAPT Henry C. Mustin, the commission's naval representative, dissented regarding aerial materiel stressing the need for an unspecified organization (i.e., the NAF) to conduct experimental research work for naval aviation.[47]

Even though most of the production contracts went to private firms, the aircraft industry, struggling for its very existence, deeply resented the modest aircraft design and production activities of the military services after the Armistice. The manufacturers believed that competition for orders were slanted to favor McCook Field and the NAF and that the whole process inhibited technical progress. Tensions with the military were heightened by the common practice of private firms submitting deliberately low bids to win contracts and then attempting to recover their deficits by sending the government numerous expensive engineering changes. In addition, there were behind-the-scenes industry efforts to grab virtually all the functions of McCook Field and the NAF.[48]

Historian Elsbeth E. Freudenthal writes that "... even with the small amount of government manufacture and large government help in development, Howard E. Coffin and C.M. Keys of the Curtiss company claimed that the government was competing with private industry. Their anger was increased further by the fact that there was a certain amount of competitive bidding on government orders during this period. Keys complained that the government's money 'has been spent only with the thought of getting ships for the service as cheaply as possible no matter whether that was below the cost of production or above it'."[49] Frudenthal stresses that Keys' complaint wasn't entirely true. Maj Gen Patrick testified before Congress that "We have departed somewhat from competitive bidding at many times to keep a core of firms in business."[50]

When the military turned to private firms to design and build new aircraft prototypes, the relatively lucrative follow-on production contracts were awarded after separate competitions. Consequently, aircraft manufacturers sometimes lost the opportunity to produce and profit from their own creations. Winners had to be selected on a lowest-price basis from among competent and reliable bidders. This practice encouraged aircraft firms to reduce their expensive engineering staffs to keep their cost estimates for production costs as low as possible.

That process was based on congressional procurement legislation intended to deal with the purchase of common commercial items like toilet paper. However, it was ill-suited to developing advanced aircraft according to a broad consensus shared by aviation experts. Basically, it discouraged technological innovation, a risky and expensive proposition. Congress, which tried to maintain strong control of the federal government's aircraft procurement practices, insisted that the industry must be preserved as a bastion of free market, laissez-faire capitalism.[51]

BuAer representatives testified on Capitol Hill in December 1925 that the Navy and Army had stopped purchasing the design rights to new aircraft. This had been done under pressure from the aircraft industry, which realized that designs became obsolete after a few years. BuAer's policy was to encourage manufacturers to do design work as much as possible.[52]

Aircraft manufacturers had complained bitterly about the federal procurement policies and practices, especially when they risked losing production contracts on their own designs. Reuben Fleet, the outspoken founder and president of the Consolidated Aircraft Corporation, was especially harsh in his criticism of those procurement practices. Writing in the *New York Times*, he lamented that "Every company employing a staff of designers and engineers knows that under the present system any worthy model it has created may be torn away from its creator and tuned over to an underselling, pirating competitor to manufacture.... and when its product is taken away by the government that it has tried so hard to serve... it is spirit-killing, discouraging beyond words, unethical and unfair."[53]

Furthermore, early on Air Service officers lacked faith in the abilities of American aircraft designers in private industry who hadn't really been given an opportunity to demonstrate their talents during World War I. Their prevailing attitude had been illustrated by an episode at McCook

Field during 1919 involving Claire Egtved, the son of a Norwegian immigrant father and a first generation Norwegian-American mother. Egtved, a Boeing engineer who later became the guiding genius behind Boeing's B-17 "Flying Fortress" of World War II fame, was shown around the installation by one of the Engineering Division's young officers. The latter told Egtved that "We have a hundred different planes here... ninety-seven of them aren't worth a darn."[54] The bewildering array of aircraft included one shaped like a halibut. The officer explained that "The Army paid $75,000 for this design.... The only way you can get it off the ground is with a chain and block. You can see why we would rather design them ourselves."[55]

Col Thurman H. Bane, Chief of the Air Service's Engineering Division at McCook Field, was highly critical of federal procurement policies.[56] He was convinced that the firms which developed winning designs, not others, should be awarded production contracts for their creations.[57] Bane was a strong believer in commercial aviation and linked the potential of the Engineering Division's work to the latter's development. His organization freely shared technical information it had compiled with industry, except for classified material.[58]

Naval aviation's leadership held similar views. In its Fiscal Year 1924 report to the Secretary of the Navy, BuAer complained that "The compulsory award of contracts to the lowest bidder as required by law has not always worked to the advantage of the Government or of the industry. In some cases the lowest bidder has failed to fulfill the contract because of engineering incapacity or because of financial losses sustained. As a result, the Government does not secure the aircraft contracted for on time, and the money has been tied up which should have been available as a potential credit for more reliable contractors to carry on legitimate production."[59]

To ameliorate such problems, leaders of the Army and Navy air arms often obtained administrative rulings from their civilian superiors that categorized production contracts as experimental awards to winning designers which didn't have to be re-competed.[60]

To aid the struggling postwar aircraft industry, Patrick and Moffett moved to minimize competition between it and the armed forces. Patrick ended most of the Engineering Division's practice of designing and building aircraft prototypes. Instead, it participated in developing requirements and then tested new aircraft and engines developed by private firms.[61]

Although Moffett followed a similar policy path, the NAF continued to build a limited number of aircraft for a variety of reasons including the unique requirements of flying while embarked at sea as well as experimental research and establishing yardsticks for gauging private industry costs. While no new aircraft production orders were placed with the NAF after early 1922, Moffett was determined to maintain an expert Navy design staff so that the sea service would never be entirely dependent on private firms to meet its requirements for new aircraft and related equipment.[62]

In their extensive study of relations between government and business, Robert Schlaifer and S. D. Heron conclude that "The development of high-powered aircraft engines in the U.S. from 1919 to 1939 was almost entirely dependent upon the military services, the commercial market made use of engines which were only modifications, usually rather light modifications, of engines originally developed for military use."[63] Both military services tried to develop their own engines but failed. Instead, they provided general technical supervision of engine development and sometimes pushed for projects which they had originated and would probably have been delayed significantly if left to private industry. Nevertheless, firms disliked technical interference and the delays usually involved in development contracts. Consequently, they preferred to finance most of the engineering projects themselves knowing that the military services allowed them to include those expenses plus overhead in engine production contracts.[64]

During the interwar period virtually the entire development of American aircraft engines involved the refinement of existing types rather than revolutionary technological innovations like jets. Schlaifer and Heron conclude that that situation was not accidental. They observe that "... successful engine manufacturers properly gave their primary attention to the progressive improvement of their regular product, and the engineers of the armed services were almost equally obliged to concentrate on problems pressing for immediate solution."[65] At its core, power plant innovation relied on incremental improvements by private firms primarily financed by the armed services in their endless quest for improved aircraft performance.

An Industrial Policy

There was a sharp postwar drop in American aircraft production after the war. To save itself, industry representatives and their political allies pressed the federal government for money, aircraft contracts and protection from foreign competition. They also lobbied for government regulation as well as commercial aviation subsidies. Patrick's Fiscal Year 1923 report was explicit about the industrial policy that the Air Service was promoting. He stressed that "... in accordance with policies approved by the War Department, every effort is made, through the judicious distribution of contracts, to maintain on an active status manufacturers equipped with plant facilities adequate for major aircraft production.... [This is] serving as a measure to nourish an industry which must be capable of rapid expansion in time of war..."[66] Secretary Weeks held similar views. While arguing for increased Army manpower, he informed Congress in 1922 that "Above all, we are short in the development of air power.... The national regulation of air activities and the provision of a system of airways are matters which are essential to the development of air power."[67]

Writing in a January 1924 *New York Times* editorial that American military flyers held 33 of 42 world performance records, Moffett warned of a steady decline in the nation's air power. He stressed that "Lack of federal law regulating and stimulating commercial aviation has retarded the development of this important field of aviation, and the aircraft industry has suffered in that it has been wholly dependent on government orders (Table 4.1)."[68]

By the mid-1920s, Congress had become concerned about the sad state of American aviation. Mitchell's quixotic crusade for air power,

Table 4.1 U.S. Aircaft production, 1918–1926[69]

Year	*Total*	*Military*	*Civil*
1918	14,020	13,991	29
1919	780	682	98
1920	328	256	72
1921	437	389	48
1922	263	226	37
1923	743	687	56
1924	377	317	60
1925	789	447	342
1926	1186	532	654

climaxed by his sensational court-martial conviction in 1925 followed by his resignation from the Army, had helped promote a growing recognition of the need for a comprehensive set of national aviation policies. Other individuals had played equally important roles in that process. Patrick and Moffett had labored tirelessly for legislation designed to achieve that goal in addition to pushing for significantly higher aviation budgets in their respective armed forces. The NACA, industry groups, insurance interests and the Aero Club of America repeatedly pressed for federal legislation to regulate and promote the development of aviation.

In January 1922, representatives of aircraft and engine firms organized the Aeronautical Chamber of Commerce to promote the growth of their industry and the aeronautical art in the U.S. The organization soon attracted additional members from allied industries. In addition to its lobbying activities, the Chamber tried to convince public opinion that commercial aviation was a safe and economical alternative to established surface transportation.[70]

Military aviators continued to press for strong federal government support for commercial aviation as a critical foundation for greatly expanded air power needs in a future war. Maj James A. Mars, head of the Air Service's Industrial War Plans Division, stressed in June 1925 that "Until commercial aviation becomes an established fact, the government service must keep the factories alive.... The Army Air Service believes that commercial aviation will become an established fact and, like the automobile industry, will be a tremendous factor in our civilization and a powerful agent in the event of war."[71] He added that "A healthy industry, capable of immediate expansion to meet wartime needs, is the best war reserve that this country can have."[72]

The first concrete legislative result of growing pressure for a national aviation policy was the Air Mail Act of 1925. Passed in response to President Calvin Coolidge's request for legislation designed to foster commercial aviation, it transferred responsibility for carrying airmail from the Post Office to private firms. That change provided a lucrative financial incentive to design, build, purchase and operate technologically advanced aircraft that could haul mail safely over long distances faster than railroads. The government paid subsidies to the commercial airlines that carried mail, not the firms that built their planes.[73]

Historian Thomas Worth Walterman assesses the changing political dynamics affecting American aviation during the early interwar period.

He observes that "By 1925, concerted efforts by industry and the military had aroused popular interest to the point of making legislation a political imperative (Table 4.2). Consequently, free from the welter of conflicting plans and proposals which had been formulated, the Executive and legislative branches [of the federal government] concurred in a fundamentally conservative program of aid and regulation."[74]

Growing concerns about the state of American aeronautics had also prompted Congress to establish a "House Select Committee of Inquiry into the Operations of the Air Services." Chaired by Representative Florian Lampert, the committee held 11 months of hearings. Although its investigation originally grew out of charges that the poor condition of military aviation materiel was caused by the monopolistic practices of aircraft manufacturers, the committee's proceedings broadened into a general examination of government-industry relations. The hearings were plagued by charges and countercharges about the state of American aviation, especially its military component. Aircraft industry representatives continued to press for increased federal government support, especially aircraft orders, arguing that their firms were on the verge of collapse. In end, the Lampert Committee was receptive to the views of air power radicals. Its report, released on 14 December 1925, concluded that federal government spending on military aviation had not produced commensurate results and that the industry was in a dire condition largely due to official procurement practices.[83]

Table 4.2 Federal military and NACA aircraft appropriations, 1919–1926

Year	*Army*	*NAVY*	*NACA*
1919[75]	$952,304,758[1]	$220,383,119	$205,000
1920[76]	$25,000,000	$25,000,000	$750,000
1921[77]	$33,000,000	$20,000,000	$200,000
1922[78]	$19,200,000	$13,413,431	$200,000
1923[79]	$12,700,000	$14,683,500	$210,000
1924[80]	$12,426,000	$14,647,474	$270,000
1925[81]	$14,113,043[2]	$15,150,000	$470,000
1926[82]	$14,700,000[3]	$14,790,000	$534,000

[1] $485,000,000 of this original figure revoked.
[2] Plus $2,500,000 "contract authorizations" for additional aircraft purchases.
[3] Plus $4,100,000 "contract authorizations" for additional aircraft purchases.

The Lampert Committee recommended that the federal government should stop competing with private industry in designing and building aircraft and engines, a practice that Patrick and Moffett had already largely eliminated. Its other proposals included: creating a single agency to conduct all government aviation procurement; transferring most of the armed forces' aviation research activities to other organizations; creating a separate air force; and establishing a department of national defense. Both the Army and Navy would retain some tactical air organizations to support their own operations.[84]

Meanwhile, Mitchell's crusade for a separate air force had reached its apogee in 1925 before it, along with his military career, crashed and burned. He was reduced to his permanent rank of colonel and exiled from his senior Air Service post in Washington, D.C., to San Antonio, Texas. Then Mitchell lashed out at the Army and Navy following two naval aviation disasters. One was the forced landing at sea of a PN-9 flying boat on a flight to Hawaii from the West Coast when it ran out of fuel due to unpredicted headwinds. The PN-9's crew, led by Cdr John Rodgers, rigged fabric sails on the aircraft and managed to sail the remaining 559 of its 2,400-mile overwater flight to Hawaii. The second was the destruction of the Navy's rigid airship USS *Shenandoah* in a severe squall line over southeastern Ohio on 3 September 1925 killing 14 of the 43 sailors aboard.

Mitchell charged that both of those incidents were caused by ".. incompetence, criminal negligence and almost treasonable of the national defense by the war and navy departments."[85] Regardless of whether or not those accusations were correct, the blatantly insubordinate manner he used clearly violated military law, causing a high-profile court-martial and conviction that terminated his career. His sensational charges also encouraged Coolidge to create the President's Aircraft Review Board. The President, an economy-minded conservative, moved quickly to diffuse support for the revolutionary and expensive policies that Mitchell and the Lampert Committee were promoting. On 12 September 1925, he appointed his friend and former Amherst College classmate banker Dwight W. Morrow to head the board. Coolidge directed it to study American aviation and develop policy recommendations for dealing with its problems.[86]

Representative Carl Vinson was a key member of the Morrow Board. He represented a rural land-locked district in central Georgia. Serving in the House of Representatives from 1914 to 1965, he was a master of congressional politics widely respected for his expertise on military matters and attention to detail. A realist, he focused on what was politically possible through building consensus using persuasion and accommodation. Vinson realized the importance of maintaining a bipartisan approach to issues coming before the Naval Affairs Committee. Although a member of that body and its chairman beginning in 1931, he never set foot on a warship and steadfastly refused to take any of the international junkets that were deemed prerogatives by many of his colleagues. Vinson was determined to "provide for the common defense" of the U.S. by building the world's most powerful navy.[87] Biographer James F. Cook concludes that Vinson's service on the Morrow Board enabled him to understand "... how important the airplane would be in repelling an attack on American shores, bombing naval and fixed targets, and in reconnaissance. In determining America's defense needs, he made building new and improved aircraft carriers and airplanes a high priority."[88]

The Morrow Board issued its report in November 1925. It rejected claims that the U.S. was subject to crippling surprise air attacks and that future wars would be decided by strategic bombardment. Since the board was convinced that there was no radical new air threat to American security, it ruled out creation of a separate air force and a department of national defense. The board acknowledged that aviation could play an independent role in future wars. Noting the numerous international performance records held by American aircraft, it concluded that the federal government's aeronautical research and development programs had fostered substantial aviation progress.[89]

The board recommended modest changes in federal aviation policies. It rejected unification of the military air arms but urged the government aid aircraft manufacturers. It recommended that the Army rename its air arm the "Air Corps" to strengthen the idea that it was an offensive arm rather than an auxiliary service. It also suggested that aviation be given special representation on the General Staff. The board urged the creation of civilian Assistant Secretaries for Aeronautics in both the War and Navy departments to represent aviation in their highest councils.

Recognizing that the aircraft industry would serve as a base for wartime mobilization and technological development, the board urged a five-year program of modern aircraft procurement for the Army and Navy.

Agreeing that many of the procurement problems experienced by military aviation were caused by the failure of Congress to provide an adequate multiyear procurement program, it urged adoption of one and concluded that it would strengthen the aircraft industry. To foster further technological progress, it suggested that aircraft design and product contracts should only be let to firms that had good design staffs and that there should be liberal provisions for experimental procurements to determine standard new military aircraft types. The board also concluded that the government should continue its aeronautical research programs and that its testing facilities ought to be made available to industry. Coolidge fully accepted the board's findings.[90]

While considering the Morrow Board's recommendations, the House Naval Affairs Committee requested that BuAer prepare data regarding the Navy's aviation needs and submit a five-year development program.[91] In a parallel move, the House Military Affairs Committee asked Patrick during January 1926 to prepare a draft bill establishing Army aviation as a separate corps within the War Department and creating a new Assistant Secretary of War for Air among other provisions.[92]

The Morrow Board's findings, coupled with the Air Mail Act, encouraged significant growth in American aviation. It helped inspire both the Air Commerce Act and the Air Corps Act, both passed in 1926. The former enabled the federal government to regulate interstate flying activities as well as inspect and regulate commercial aircraft. It also provided for government development of a national system of airways. Secretary of Commerce, Herbert C. Hoover, was given responsibility for fostering the growth of commercial aviation. To assist him, the law created a new post, the Assistant Secretary of Commerce for Aviation. Relieved of its civil aviation responsibilities, the NACA was free to concentrate its energies on aeronautical research. Like other aviation advocates, Hoover justified his public support of a larger federal role in large part due to national defense concerns.[93]

Historian Irving B. Holley, Jr. concludes that the Air Commerce Act " ... contributed toward the establishment of a stable, healthy, and vigorous aviation industry."[94] Historian Walterman stresses that the legislation's " ... passage, along with three other bills dealing with military aeronautics and government procurement methods, constituted a long-overdue recognition by the Federal authority of the important position achieved in the military and commercial activities of the world. The tardiness of the enactment was accentuated by the fact that none of the bill's provisions had arisen out of recent innovations or developments."[95]

Historian Sean Seyer takes a "wider view" of the legislation and the emergence of U.S. aviation policy. He argues that they " ... developed in response to three interrelated and mutually reinforcing factors: the airplanes unprecedented ability to provide controlled movement within three-dimensional space, shifting notions of constitutional federalism, and the need to maintain compatibility with the provisions of the Convention Relating to the Regulation of Aerial Navigation crafted as part of the Versailles peace Conference in 1919."[96] While the 1919 aerial convention was never ratified by the U.S., American advocates of federal aviation regulation had to take it into account because Canada had. Sever's work stresses " ... how a relatively small group of bureaucrats, military officers, industry leaders, and engineers drew upon previous regulatory schemes and international principles in their struggle to define government's relationship to the airplane."[97]

The Air Corps Act of 1926 made significant changes in Army aviation but fell well short of the fundamental revamping of the national defense policies and organizations advocated by Mitchell. To strengthen aviation's position within the Army, the Air Service was re-designated the "Air Corps" and given a stronger position in the General Staff. The legislation also created a new Assistant Secretary of War for Air. That provision gave the air organization its own political representative at the highest decision-making levels enabling it to circumvent the General Staff. It was the only Army combat branch to have that kind of representation.

The legislation also authorized a five-year expansion program for Army aviation. The Air Corps' goal under the act was to obtain 1,800 serviceable aircraft. When the legislation was passed most of the service's planes and engines were considered obsolescent. Military aircraft were changing rapidly. The Army's yardstick for procurement planning at that point was that its pursuit, training and observation aircraft would be considered obsolescent after about three years. The planned expansion of the Army's air arm under the legislation would double its annual cost to an estimated $56 million.[98]

Historian Vander Meulen observed that the legislation produced continuing "... uncertainty about how its congressional-military client would apply its regulatory powers over their affairs. Congress left it up to contracting officers to risk interpreting the laws in ways that might assist the industry ... ." [99] He added that "For a secretary to allow production contracting to go beyond the simple criteria of lowest price would require 'an extraordinary amount of courage' in the face of close congressional scrutiny. Whether the act intended contracting for military aircraft to be based upon principles of competition or negotiation 'came down to a hopelessly moot question.' This unresolved problem preserved competition by default, as well as resentment and mistrust among all involved in the industry."[100] Nevertheless, Vander Meulen concluded that the legislation "... was one of the most important events in the contemporary American political economy. Supplemented by the Vinson-Trammel Act of 1934, it defined the aircraft industry's business environment until the eve of World War II, and thus the institutional foundation for the nation's vast air power arsenal during that conflict."[101]

Meanwhile, Rear Adm Moffett and his staff drafted an aviation bill and worked to insure its passage on Capitol Hill. President Coolidge signed the Naval Aviation Expansion Act of 1926 on 24 June 1926. It authorized the purchase of 1614 aircraft over five years but didn't specify which kinds. Paralleling the Air Corps Act, the legislation created a new post, the Assistant Secretary of the Navy for Aeronautics. Congress also directed several other key naval aviation reforms to create an attractive career path for flyers and attract some of the service's most capable officers. Command of naval air stations, schools and tactical air units had to be assigned to naval aviators. Likewise, command of aircraft carriers and tenders could only be assigned to naval aviators or naval observers. That provision brought a number of talented and ambitious mid-career officers into aviation including Ernest J. King and William F. Halsey. They were better positioned initially than younger career flyers to rise to senior positions in the Navy. Such officers would play critical roles in integrating aviation into the fleet and leading it during World War II.

However, the armed forces never received large enough congressional appropriations to buy all the aircraft authorized in their five-year expansion programs. In March 1926, Coolidge had signaled his opposition to increased military spending; he was determined to balance the federal budget in 1927 without a tax increase. If the aviation arms of the Army and Navy were to be strengthened, the President favored doing so at the expense of other elements of the armed forces.[102]

The Air Corps Act basically preserved the existing laissez-faire procurement practices. Under the legislation, the federal government would encourage the growth of advanced technical staffs in the aircraft industry by liberal use of experimental orders for aircraft prototypes. The government was prohibited from designing and manufacturing aircraft in its own facilities, a practice that the Army and Navy already had largely abandoned as noted previously. The armed forces cooperated in drafting the contracting provisions of the legislation. For non-experimental aircraft production contracts, the act mandated competitive procurement. The service secretaries were required to award such deals to the lowest bidder capable of doing the work regardless of which firms had designed the new planes. For designs submitted prior to the act, the armed forces could negotiate purchases of experimental equipment and quantity purchases of production goods including aircraft. Experimental aircraft could be purchased without aircraft production competitions.

Regardless of the law's intent, open design competitions were abandoned by the Air Corps. They often resulted in unsatisfactory aircraft and successful designers frequently failed to win the more lucrative production contracts for their creations. Instead, the Air Corps (and to a large extent the Navy) increasingly relied on provisions of the law that enabled it to award negotiated contracts for experimental aircraft. The Air Corps further interpreted the legislation to mean that it could award production contracts for successful experimental aircraft to the firms that designed without additional competition.[103]

The U.S. had finally established a set of policies designed to promote the development of American commercial and military aviation as well as the industry which sustained them. Those policies helped to make commercial flying safe and potentially profitable. Airmail contracts and the aircraft procurement programs of the armed forces, while not fully funded, stimulated significate improvements in aircraft and engine designs by private manufacturers. Meanwhile, Wall Street's great bull market of the 1920s hit its stride, spurring a rapid growth of private investment in American aviation. From 1926 through 1928, that investment soared from \$15 million to \$125 million. Stimulated by profitable military and civilian commercial contracts, the industry was able to devote significantly greater resources to promoting technological progress through innovative designs.[104]

Survival and Rehabilitation

The period immediately following World War I was a difficult one for Army aviation. Its funding and personnel were slashed like the rest of the military's. Mitchell's drive for a separate air force created enormous political controversy and serious tensions within the military establishment. Maj Gen Patrick, as noted earlier, played a key role in rectifying that situation. According to historian Robert P. White "Immediate [Air Service] independence was not the issue for Patrick; the survival of the Air Service in the crises days of World War I was."[105]

Lacking the appropriations and political support needed to build a large modern air force, the Army air arm's leaders largely sacrificed operational readiness and what they saw as an adequate force structure for technological progress in order to maintain the nucleus of an aircraft industrial base for wartime expansion. The Air Service's Fiscal Year 1920 report stressed that it "... must spend its appropriations in experimentation, in limited procurement, and in the test, under field conditions, of experimental productions, in order that it may know what equipment to use when war comes."[106] It added that "the [Engineering] Division must keep ahead of the rest of the world in aviation development if this country is to meet the enemy of the future on equal ground."[107]

Maj Gen Patrick lobbied Congress for enough funds to keep the technologically progressive segment of the postwar American aircraft industry alive and allocated contracts to its strongest members.[108] During 1925, he stated the matter bluntly. Noting that there was not enough money "... to provide aircraft in numbers adequate to equip completely the present tactical units, much less to provide a war reserve," he concluded that the only viable alternative was to "... continue experimentation and research until the best types of planes were ready for quantity production in wartime."[109] His decision to emphasize research and development at the expense of standardization and quantity procurement of advanced designs was realistic considering the budget realities imposed by the federal government in the 1920s. During 1921–1926, 40% of the Air Service's total obligations, amounting to $53.21 million, were spent on research, development and procurement (Table 4.3).[110]

Postwar Army aircraft modernization programs also suffered from a severe shortage of objective studies of World War I tactical operations according to historian Holley.[112] Instead, the Air Service developed a complex formal engineering-oriented system for determining its materiel

Table 4.3 Air Service research and experimentation: appropriations, FISCAL years 1921–1925[111]

FISCAL year	*Appropriation*
1921	$5,000,000
1922	$4,300,000
1923	$3,500,000
1924	$3,000,000
1925	$2,850,000

requirements. Subject to General Staff guidance, Air Service headquarters would formulate its requirements for equipment. Then the Engineering Division would draft general specifications giving the features desired in that equipment. Ordinarily, those specifications incorporated what Air Service tactical units wanted and the latest technical information collected at McCook Field. Sometimes the specifications reflected efforts by the Engineering Division to take advantage of recent aircraft technological developments superior to those currently in use by the Air Service. Or, they could result from the development of superior engines. Finally, they might represent the consolidation of data gathered in testing experimental equipment, especially aircraft.[113]

Once approved the general specifications were sent in circulars to all manufacturers which the Air Service deemed competent to design and produce the needed equipment. Other firms would be sent copies of the circulars upon request. Because of the increasingly large expense involved in constructing prototypes, few manufacturers were actually willing to build them before the Air Service made a contract award.[114] The normal practice was to ask firms for "... a dimensioned three-view drawing showing the general outlines of the proposed design, rough unidimensional sketches showing the types of construction proposed, performance estimates, estimates of weight and center of gravity location, and diagrams showing, as well as possible at such an early stage of the design process, the locations of the elements of the power plant, armament equipment, etc."[115] Those designs would be accompanied by bids estimating what it would cost to build the equipment. Until Patrick ended the practice, the Engineering Division could enter its own designs in those competitions.[116]

The next step in the official process was for the Engineering Division to evaluate the design bids received from the manufacturers as well as its own. Manufacturers intensely disliked that phase because they feared

McCook's engineers would probably favor their own creations. Once his staff had completed its evaluation of competing designs, the head of the Engineering Division would make a recommendation to the Chief of the Air Service. The latter would decide, subject to a final approval by the Secretary of War, which design (if any) would be built. That process often involved intense personal lobbying of the Air Service's top officer and the Assistant Secretary of War by firms competing for contracts. Losing firms, especially those with weaker design and production capabilities, often accused the Air Service of bias or hostility against them. They usually directed their fire against the Engineering Division. The actual procurement contract of a winning design was awarded and supervised by the Office of the Chief of the Air Service. Experimental contracts were an exception. They were administered by the Engineering Division.[117]

Despite its procedural veneer, the Air Service's aircraft procurement decisions were often very personalized ad hoc affairs in the years immediately following the war. "Selection of airplanes..." according to Holley, "... seem to have been dominated by what might have been called the joystick approach of General Mitchell... [He] would leap out of an experimental aircraft, pronounce it a 'hot ship' and urge its immediate procurement."[118] During his turbulent tenure at Air Service headquarters, either Mitchell or one of his close assistants would personally pass judgment on virtually all equipment being considered by the Engineering Division for Air Service procurement. Since those officers did not always fully understand the state of aeronautics and the capabilities of equipment, their personalized approach sometimes led to poor procurement decisions that embarrassed the Air Service while wasting its limited resources.[119]

The Wittemann-Lewis XNBL-1 *Barling Bomber* that Mitchell championed was an embarrassing example of that problem. Built at the firm's plant in Hasbrouck Heights, New Jersey, the giant plane weighed about 20 tons and was broken into eight-foot-long sections for shipment in Pennsylvania Railroad cars to Dayton. It was a six-engine triplane. Published (non-official) accounts rated its top speed at 95 mph. It could carry as much as 5,000 pounds of bombs with a reduced fuel load. But, it couldn't fly high enough to cross the Appalachians.

Maj. Gen. Patrick resisted pressure in January 1922 to reduce the buy from two to one aircraft and offered the bomber's manufacturer another $350,000 to complete the project. But, to no avail. After spending over $1 million for a single prototype, Congress forced the Air Service to abandon the entire project because of its rising costs and poor performance. Due to frequent changes and the fixed-price contract that the Army demanded, only one of the two bombers required in the contract was ever produced. The Wittemann-Lewis firm went bankrupt and was forced out of business. The plane had been designed under a separate contract won by Walter H. Barling the British designer of the RAF's failed *Tarrant Tabor*bomber. It became widely known as "Mitchell's folly" because of his major role in fostering its development.[120]

Patrick viewed the Army aviation's weapons modernization process as a mess and set out to reform it. He established a technical committee at Air Service headquarters that met at least annually to lay down requirements for new aircraft and associated equipment. Patrick also minimized its design activities. Those initiatives were undertaken, with the support of his civilian superiors, once Patrick was convinced by his own careful study of the issues involved that they were the best policies for the Air Service and would help foster a healthy, technologically innovative aircraft manufacturing base capable of massive wartime expansion. Subsequently, McCook Field refocused its energies away from design work and toward interpreting the Air Service's needs to manufacturers and evaluating the latter's work.[121]

Nevertheless, McCook Field's activities remained critical to technical progress during much of the 1920s. Maj Leslie MacDill, the Engineering Division's chief engineer, wrote in 1925 that its development work for "... some engines and airplanes may amount to as much as 50 per cent, yet the article had been and will be referred to as an article produced by the contractor alone and the cost of production named by the contractor as the complete cost of the development of what the article has been and will be placed as the amount paid to the contractor on the contract."[122]

Maj Gen Patrick was determined to tie the Engineering Division's activities more closely to the Air Service's operational needs and put them on a more economical footing. To help formulate policies and programs, he began extensive semi-annual personal inspections of McCook Field along with personnel from other parts of the Air Service. Patrick also directed a drastic curtailment of the number of aircraft development categories. During 1922, the Air Service had 16 separate types of aircraft project categories: 5 pursuit, 1 ground attack, 3 observation, 1 liaison, 3 bombers, 2 training and 1 miscellaneous.[123] They were based upon the aerial experience of America's partners during World War I and implied some confusion about the uses of military aviation. In July

1923, an Air Service board at McCook Field urged a break with that outdated thinking. Instead, it recommended that aircraft should be classified according to construction since, depending upon the circumstances, they could fly during the day or at night and perform several different functions. The board championed an official classification scheme of eight different aircraft types: pursuit, observation to include attack, light bomber, heavy bomber, trainer, transport, ambulance and night observation. By 1925, that list had shrunk to a more manageable five types: pursuit, trainer, observation, bomber and attack.[124]

Postwar Air Service aircraft procurement priorities changed dramatically during the first few years after the Armistice. Patrick wrote Pershing during 1919 that an air service in a war zone should consist of 80% observation and 20% other aircraft, an assessment that mirrored the AEF's wartime experience. Some seven years later, Patrick reversed his stance on these type proportions. His deputy, Mitchell argued in 1926 that observation aviation had had no role in the future.[125] Army historian Raines concludes that "Mitchell's apotheosis as a 'martyr for air power' in 1926 and the passage of the Air Corps Act that same year accelerated observation's decline into orphan status and an intellectual and professional backwater within the Air Corps."[126]

The pursuit plane remained the most highly valued type of aircraft purchased for the Army's air arm during most of the 1920s. That priority reflected a belief among Army fliers that air superiority was necessary before other air operations could be attempted. The Air Service conducted a competition between the Curtiss PW-8 and the Douglas PW-9 in the early 1920s to replace its Boeing MB-3A. The Curtiss entry won that contest. The firm's production model named the PW-8 *Hawk* began operational service in 1923. This single-place open cockpit biplane served as the primary Air Service pursuit aircraft during the mid-1920s. It originated as a private venture by Curtiss based on the designs of the successful racing aircraft the firm had built for the military. The PW-8's fuselage skeleton was made of welded steel with tail surfaces of steel and duralumin. Fabric covered the welded steel tubing in the fuselage behind the cockpit and the tail surfaces. The wings were wooden and covered with spruce planking. An easily removable metal cowling covered the forward fuselage. Powered by a Curtiss D-12 engine generating approximately 400 hp., its top speed at ground level was 159 mph. The aircraft's service ceiling was rated at 21,800 feet with a range of 544 miles.

The *Hawk* represented a relatively modest improvement in pursuit performance over World War I fighters flown by the Allies underscoring the conservative incremental approach to aircraft innovation adopted by the Air Service and its contractors. For example, Britain's Sopworth *Camel*, the top destroyer of German aircraft during the war, had a maximum speed of 113 mph, service ceiling of 19,000 feet and a range of 300 miles.[127]

For fundamental aeronautical research, the Engineering Division relied on the NACA and a few universities. The NACA, whose Langley Memorial Aeronautical Laboratory and its first primitive wind tunnel were formally opened on 11 June 1920, became the primary source of such research for American aviation during the interwar period. A joint body of civilian scientists as well as Army and Navy officers controlled its activities. The NACA made enormous contributions to aeronautics during that era especially in aerodynamics—the study of the properties of moving air and the interaction between the air and solid bodies moving through it. The cost of the Langley Memorial Aeronautical Laboratory's work for the military services would be paid out of NACA's appropriations unless there simply were not enough funds available to cover specific projects. In such cases, the Army and Navy financed such efforts. NACA's aviation research was one of the major reasons why American aircraft were among the most advanced in the world during much of that period.[128]

Dr. Hugh L. Dryden, a distinguished physicist, was one of the most significant scientists who contributed to the progress of American military aviation during and after the interwar period. A child prodigy, he was admitted to Johns Hopkins University at the age of 14 graduating with honors three years later and earning a master's degree in 1918. His work impressed Dr. Joseph S. Ames, the highly respected head of the university's Physics Department and a key NACA official. Serving as Dryden's mentor, Ames recommended him for a position at the National Bureau of Standards. Dryden worked in the bureau's new wind tunnel section while completing his doctorate.

In 1920, he was promoted to head of the bureau's Aerodynamics Section making him responsible for guiding its wind tunnel research and operations. He focused on high-speed aerodynamics, especially problems caused by flying at or near the speed of sound. His research led to the design of low-turbulence wind tunnels, and his data—employed by the NACA—contributed to development of the laminar flow wings used on the P-51 *Mustang* fighter in World War II.[129]

One of the most critical contributors to NACA's preeminence in aeronautical research was a German, Max M. Munk. Holding doctorates in physics and engineering, he came to the U.S. in 1920 and achieved international renown by creating the Langley Variable Density Tunnel (VDT). That achievement was the most significant accomplishment of the embryonic NACA and placed the Langley Memorial Aeronautical Laboratory at the forefront of the world's aeronautical research organizations. The VDT created a number of different conditions which were directly comparable to those in flight. It became operational in 1922. While other nations quickly began building similar tunnels, the NACA led the world in that area for over a decade.[130]

The Army's Engineering Division had also contributed directly or indirectly to many technological developments in aviation prior to 1926. In early 1920, the Air Service purchased three model R-1 air-cooled radial engines from the Lawrence Aeronautical Engine Corporation. The air-cooled radial engine was in part a direct result of work begun at McCook Field in 1919 in the cooling of engine valves and cylinders. The key technological breakthrough was a sodium-filled hollowed out aluminum exhaust valve developed by Englishman Samuel D. Heron, who had been lured away from Great Britain in 1923 to work as a civilian researcher at McCook Field.[131]

In World War I, flying as high as possible while maintaining speed had been critical to victory in aerial dogfighting. To achieve that goal the technical arms of the Allies worked feverishly to supercharge aircraft engines which, without that feature, lost power at higher altitudes because of the reduced atmospheric pressure. From 1917 to 1921, engineers at McCook Field's Power Plant Section worked closely with General Electric's Dr. Sanford A. Moss to develop an engine supercharger. The device's function was to "... 'pep up' the engine, enabling an aircraft to operate effectively at high altitudes by compressing air drawn in from the atmosphere even when the air thinned out."[132] Subsequently, the Engineering Division developed its own supercharger design. By 1924, it was standardized after extensive development work and numerous flight tests by McCook Field personnel. Air Force historian Albert E. Misenko observes that it "...proved effective and was on nearly all turbo-supercharged aircraft through World War II."[133]

Feet Wet: Aviation Goes to Sea

The overriding goal of the Navy's postwar aviation programs was to enable aircraft to go to sea with the fleet. During December 1918, the CNO's office requested that BuC&R renew efforts to develop aircraft that could operate from warships. In March 1919, Adm Benson issued a preliminary program for naval aircraft development. It emphasized fighters, torpedo carriers and bombers that could be embarked on warships. Land bases would support single and twin-engine long distance patrol and bomber planes. The Marines would need a combination land and seaplane.

In June 1919, the General Board issued the last of several reports to Secretary Daniels recommending policies for developing naval aviation. Stressing that aviation had become vital to the fleet, it advocated a large peacetime program that would create an air arm able to operate at sea around the world. Secretary Daniels approved the report's recommendations on July 24th.[134] The BuC&R let production orders with the Loeing, Martin and Vought firms for aircraft that were suitable or could be adapted to go to sea with the fleet. During Fiscal Year 1920, eight battleships were equipped with catapults for launching aircraft.[135]

Technical issues including reliability problems with Liberty engines plagued postwar naval aviation programs. Alterations in the engines directed by BuAer greatly increased the reliability of those engines during the early 1920s.[136] But, that proved to be only an interim fix. Meanwhile, in January 1920, the Bureau of Steam Engineering[137] had allocated $100,000 for the development and purchase of 200 hp. air-cooled radial engines from the Lawrance Aeronautical Engine Corporation. Those engines evolved into efficient and reliable sources of propulsion. The firm's founder and president, engineer Charles L. Lawrance, had been assigned to work on engine development while a naval officer during World War I.

Air-cooled radial engines soon became standard in naval aircraft and, with few exceptions, powered commercial airlines until the advent of jet power plants. The Navy had pressed their development because they eliminated the weight and technical complications (and maintenance workload) involved in the water, pipe fittings and accessories used in their water-cooled counterparts. Since no other firm was interested at that time in air-cooled radial engines, the Navy turned to Lawrance. Although enthusiastic about those engines, the Navy became concerned that his firm couldn't produce enough of them to meet its burgeoning requirements. Consequently, it told Wright Aeronautical to purchase the Lawrance corporation. If it did not, the sea service threatened not to buy any more engines or spares from Wright. Since the Navy was a major customer, Wright was forced to comply. It acquired Lawrance's firm in May 1923, and he was installed as Wright's vice president.[138]

Another major technological innovation sponsored by the Navy during the interwar period involved a highly classified piece of equipment usually associated with the strategic bombing campaigns of the Army Air Forces (AAF) during World War II. During September 1920, Carl L. Norden, a Dutch-born engineer educated in Switzerland who had emigrated to the U.S., was loaned a naval aircraft to begin testing his new bombsight. In August 1921, naval aviators tested a modified World War I high-altitude bombsight mounted on a gyroscopically stabilized base. It marked the end of the initial phase of the development of the Norden bombsight for the Bureau of Ordnance.[139]

Beginning in 1915, the Navy's General Board had recommended developing an aircraft carrier several times to the Secretary of the Navy.[140] That initiative was finally approved by Congress in the Fiscal Year 1920 Naval Appropriations Act on 11 July 1919. It provided for the conversion of the former coal collier USS *Jupiter* into the Navy's first aircraft carrier, later named USS *Langley*.[141]

Secretary Daniels stressed after the Armistice that the NAF, which had been built as a temporary wartime expedient, was being retained as a permanent part of the Navy. It was needed to continue developing and producing experimental aircraft as well as to provide overhaul and repair services. The NAF would also give the sea service a mechanism to check the costs of developing and producing aircraft from private sources.

The service's Aeronautical Engine Laboratory was transferred from the Washington Navy Yard to the NAF in 1923 underscoring the latter's role as the center of naval aviation research and development. The laboratory's primary role was to test engines and their accessories built by private firms and to determine whether they were acceptable for naval service.[142]

The Navy's Fiscal Year 1922 annual report noted that it was working with several manufacturers and the NAF to explore metal construction of aircraft because of the rapid deterioration in service of wood and fabric planes, especially at sea. On 25 April 1922, the first all-metal airplane developed for the Navy, the ST-1 twin-engine, mid-wing monoplane torpedo plane built by the Stout Engineering Laboratory, made its maiden flight. It was tested, along with several other competitors at Naval Air Station (NAS) Anacostia in Washington, D.C. Although it lacked adequate longitudinal stability and never went into bulk production, the ST-1 marked a significant step forward in the transition from wood to metal aircraft. During the following month, Navy Secretary Edwin Denby announced that the Glenn L. Martin Company had been authorized to develop a seaplane made of duralumin, a strong, lightweight alloy of aluminum developed in Germany in 1906 by metallurgist Alfred Wilm.

BuAer's experts reported during November 1922 that tests of an all-metal aluminum alloy monoplane designed by them and built by Martin, the MO-1, had shown promise for directing battleship gunfire by radio from its planned maximum altitude of approximately 10,000 feet. The MO-1 aircraft had an all-metal structure with a fabric covering. The Navy purchased 36 of the aircraft for about $1 million. They were convertible to either sea or land planes.

During February 1921, the General Board had discussed converting one or more of the Navy's battle cruisers under construction since 1916 into an aircraft carrier but rejected the idea. However, a number of newspapers, public officials and aviation advocates urged the service to reconsider. On 22 April 1922, the BuC&R submitted an outline sketch for the conversion of two unfinished battlecruisers to aircraft carriers. The department approved them during the following May. Congress authorized that conversion on 1 July 1922 as allowed by the Washington Naval Treaty.[143]

USS *Langley* was commissioned on 20 March 1922 and began operational experiments with land planes modified with strengthened landing gear and arresting gear to operate from the ship. The first aircraft to fly from the carrier was the VE-7 *Bluebird*, originally designed for the Army as a trainer by Chance M. Vought, co-founder of the Lewis and Vought Corporation. Vought was a trained engineer who had been taught to fly by the Wright brothers. Patterned after some European designs, his aircraft was originally designed as a two-place biplane trainer for the Army.

During 1923, the Vought company began delivering its UO-1 aircraft to the fleet, a catapult-launched biplane that was placed on battleships and cruisers. It was a redesigned version of the VE-7 and VE-9 equipped with floats and new engines. Vought built 141 of the aircraft for the sea service beginning in 1922. The unarmed aircraft performed observation missions. It was the first big military order for the firm which became a mainstay of naval aviation.

While USS *Langley* was finding its sea legs, two other American carriers were under construction on unfinished battlecruiser hulls. On 17 November 1924, USS *Langley* reported for duty with the fleet, ending its over two years as an experimental vessel. It embarked 16 aircraft earlier that year. During March 1925, BuAer asked manufactures to develop new aircraft designs specifically for use on aircraft carriers. That initiative resulted in Martin's all-metal version of its SC, several F6C types by Curtiss, Vought's O2U and the Douglas T2B.

Long-range naval patrol aircraft had also been developed and began performing scouting functions. Aircraft were beginning to be assigned to battleships. The decade saw impressive technological progress by BuAer, the aircraft industry, NACA and other federal government agencies even though funding was tight (Illustration 4.4). Technological highlights included air-cooled radial engines that evolved into efficient and reliable sources of propulsion, better instruments and an accurate bombsight. The Navy actively promoted the transition from wooden to all-metal aircraft. Every year its planes flew faster, higher and longer while naval aviators set their share of world aircraft performance records. During December 1924, Navy Secretary Curtis D. Wilbur and Rear Adm Moffett testified before a congressional committee that the sea service's aircraft were the world's best; the problem was there just weren't enough of them.[144]

Illustration 4.4 USS *Langley* (*Photo* USS *Langley* photographed in 1922–1924. Catalog #: NH 79,703. Copyright Owner: NHHC, history.navy.mil)

Notes

1. Foulois and Glines, *From the Wright Brothers to the Astronauts: The Memoirs of Major General Benjamin Foulois*, 199; Jeffrey S. Underwood, *The Wings of Democracy: The Influence of Air Power on the Roosevelt Administration, 1933–1941* (College Station, TX: Texas A&M University Press, 1 December 1991), 11.
2. Foulois and Glines, *From the Wright Brothers to the Astronauts: The Memoirs of Major General Benjamin Foulois*, 199.
3. "Annual Report of the Chief of the Bureau of Construction and Repair to the Secretary of the Navy for the Fiscal Year 1919," 13; "Development of the NC Seaplanes," *Aviation and Aeronautical Engineering*, 1 June 1919, 468–470, Hathi Trust Digital Library; "Annual Reports of the Department of the Navy for the Fiscal Year 1919," (Washington, D.C.: GPO, 1920), 27–32, U.S. Congress, House of Representatives, 66th Cong., 2nd sess., Document No. 729, ProQuest Congressional Documents, Pentagon Library; Turnbull and Lord, *History of United States Naval Aviation,* 164–169; Van Fleet, and Armstrong, *United States Naval Aviation, 1910–1980.NAV AIR 00-80P-1,* 37.
4. Jacob A. Vander Meulen, *The Politics of Aircraft: Building an American Military Industry* (Lawrence, K.S.: University Press of Kansas, 1991), 41–42.
5. Office of Naval Intelligence, "The U.S. Navy as an Industrial Asset: What the Navy Has Done for Industry and Commerce," October 1922, revised to 1 July 1924 (Washington, D.C.: GPO, 1924), 155, Hathi Trust Digital Library.
6. Draft Memorandum, Maj Gen Charles T. Menoher, "Air Service Program U.S. Army 1919," 15 April 1919, 1, AAF Records, RG 18, Series 211 (U-Stencils), Box 1, National Archives; J. Mayhew Wainwright, Assistant Secretary of War, "The Situation in the Army Air Service," *U.S. Air Service*, April 1923, 19, Hathi Trust Digital Library; U.S. Congress, House, Committee on Appropriations, *Army Appropriation Bill [Fiscal Year] 1922. Hearings before a subcommittee of the House Committee on Appropriations. Part 1.* 66th Cong., 3rd sess., 20 December 1920, 331 (Washington, D.C.: GPO, 1921), Hathi Trust Digital Library; U.S. Congress, Senate, *Navy Department Appropriation Bill [Fiscal Year] 1925, Hearings Before the Subcommittee of the Committee*

on Appropriations, U.S. Senate, 68th Cong., 1st. sess., 29 March 1924, 204, 209, Hathi Trust Digital Library; Weigley, *History of the United States Army*, 395–396; "Annual Reports of the Navy Department for the Fiscal Year 1919," 38–39; "Annual Reports of the Navy Department for the Fiscal Year 1925 (Including Operations to November 15, 1925," (Washington, D.C.: GPO, 1926), 618, Hathi Trust Digital Library; Maj Gen Charles T. Menoher, Director, Air Service, U.S. Army, "Peace Problems of U.S. Air Service," address to the Southern Aeronautic Conference at Macon, Georgia on 5 May 1919, reprinted in the *Army and Navy Journal*, 17 May 1919, 1284, Nimitz Library; "Report of the Director Air Service to the Secretary of War [Fiscal Year] 1920" (Washington, D.C.: GPO, 1920), 9, Hathi Trust Digital Library; Adrian O. Van Wyen, *The Aeronautical Board, 1916–1947* (Washington, D.C.: History Unit, Deputy Chief of Naval Operations (Air), USGPO, 1947), 28, Hathi Trust Digital Library; "Aeronautics: Eighth Annual Report of the National Advisory Committee for Aeronautics [Fiscal Year] 1922," (Washington, D.C.: GPO, 1923), 5, Hathi Trust Digital Library; "Aeronautics: Ninth Annual Report of the National Advisory Committee for Aeronautics [Fiscal Year] 1923 (Washington, D.C.: GPO, 1924), 5, Hathi Trust Digital Library; "Aeronautics: Twelfth Annual Report of the National Advisory Committee for Aeronautics [Fiscal Year] 1926," (Washington, D.C.: USGPO, 1927), 13, Hathi Trust Digital Library; "Twenty-Third Annual Report of the National Advisory Committee for Aeronautics [Fiscal Year] 1937," (Washington, D.C.: USGPO, 1937), 40, 47, Hathi Trust Digital Library; Jerome C. Hunsaker, "Naval Architecture in Aeronautics," *Aviation and Aeronautical Engineering*, 1 August 1920, 10–14, reprint of a paper read by Hunsaker on 22 June 1920 at the annual Wilbur Wright Lecture of the Royal Aeronautical Society, Hathi Trust Digital Library.

7. Millett and Maslowski, For the Common Defense: A Military History of the U.S. of America, 364–365; Millis, Arms and Men: A Study in American Military History, 244–246; Russell F. Weigley, The American Way of War: A History of U.S. Military Strategy and Policy (New York, NY: Macmillan, 1973), 244–245; Hagan, This People's Navy: The Making of American Sea Power,

266–267; Hone and Utz, History of the Office of the Chief of Naval Operations, 1915–2015, 67, NHHC, history.navy.mil.

8. "Limitation of Naval Armament (Five-Power Treaty or Washington Treaty)," signed at Washington, 22 February 1926, terminated 31 December 1936, *Multilateral Agreements 1918–1930*, 353, 354, aits.utexas.edu.
9. U.S. Office of Naval Intelligence, "Information on the U.S. Navy and Other Navies," (Washington, D.C.: GPO, May 1925), 148, Hathi Trust Digital Library; Weigley, *The American Way of War: A History of U.S. Military Strategy and Policy*, 244–245; Millis, *Arms and Men: A Study in American Military History*, 246–248; Ransom, "The Air Corps Act of 1926; A Study of the Legislative Process," 89; BuAer, *Naval Aviation: The Eyes of the Fleet* (Washington, D.C.: GPO, 1923), 4, Hathi Trust Digital Library.
10. Millett and Maslowski, *For the Common Defense: A Military History of the U.S. of America*, 365.
11. "Annual Reports of the Navy Department for the Fiscal Year 1920 (Including Operations and Recommendations to December 1, 1920)," (Washington, D.C.: GPO, 1921), 44, 53, 54, Hathi Trust Digital Library; "Annual Reports of the Navy Department for the Fiscal Year 1922 (Including Operations to November 15, 1922)," (Washington, D.C.: GPO, 1923), 382–383, Hathi Trust Digital Library; U.S. Congress, House, *Naval Appropriation Bill. Naval Aviation*, Hearings the Committee on Naval Affairs, 66th Cong., 2d sess., 5 February 1920, 1476, 1477, Hathi Trust Digital Library; "Annual Report of the Bureau of Construction and Repair for the Fiscal Year Ending June 30, 1922," (Washington, D.C.: GPO, 1922), 15 October 1922, 1, Hathi Trust Digital Library; Davis, *The Admirals Lobby*, 76; Roger E. Bilstein, *Flight in America 1900–1983: From the Wrights to the Astronauts* (Baltimore, MD: Johns Hopkins University Press, 1984), 45; Turnbull and Lord, *History of U.S. Naval Aviation*, 186; Thomas C. Hone, "Navy Leadership: RADM W. A. Moffett As Chief of the Bureau of Aeronautics," final draft, prepared for the April 1984 meeting of the American Military Institute, Air Force Historical Foundation, and Military Classics Seminar, 10, 11; Theodore D. Roosevelt, Jr., Assistant Secretary of the Navy, to Herbert C. Hoover, Secretary of Commerce, Subj.: Aviation, 8 April 1922, Commerce Papers, Box 122,

File "Commerce Department, Aeronautics, Bureau of, Legislation, 1922," Herbert Hoover Presidential Library and Museum (HHPL&M); U.S. Congress, Hearings Before a Subcommittee of the House Subcommittee on Aeronautics of the Committee on Naval Affairs on H.R. 273 (Later amended and reintroduced as H.R. 5219) *To Create a Bureau of Aeronautics in the Department of the Navy*, 25 April 1921, passim, ProQuest Congressional Documents, Pentagon Library; "Annual Reports of the Navy Department for Fiscal Year (Including Operations to November 15, 1921) 1921," (Washington, D.C.: GPO, 1921), 62, U.S. Congress, House of Representatives, 67th Cong., 2d sess., House Document No. 224, ProQuest Congressional Documents, Pentagon Library; Navy Department General Order No. 65, Subj.: BuAer's Duties and its Place in the Navy Department, 10 August 1921 in "Aircraft: Hearing Before the President's Aircraft Board," Volume 1, 22 September 1925, 187 (Washington, D.C.: GPO, 1925), ProQuest Congressional Documents, Pentagon Library; Howard M. Humphray, "Budget Formulation and Execution in the Bureau of Aeronautics," Navy Graduate Comptrollership Program, Business Administration 268A (Washington, D.C.: George Washington University, January 1956), 14, Internet Archive.

12. Davis, *The Admirals Lobby*, 77.
13. Ibid.
14. Ibid.
15. Weigley, *History of the United States Army*, 396–397; U.S. Congress, Senate, *Navy Department Appropriation Bill [Fiscal Year] 1925, Hearings Before the Subcommittee of the Committee on Appropriations, U.S. Senate*, 68th Cong., 1st. sess., 29 March 1924, 201, Hathi Trust Digital Library; "New Program for Air Service," *Aviation And Aeronautical Engineering*, 15 April 1919, 338, Hathi Trust Digital Library.
16. Alfred F. Hurley, *Billy Mitchell: Crusader for Air Power* (New York, NY: Franklin Watts, 1964), 40, 42–45.
17. Rondall R. Rice, *The Politics of Air Power: From Confrontation to Cooperation in Army Aviation Civil–Military Relations* (Lincoln, NE: University of Nebraska Press, 2004), 13, 15; "American Air Commission Returns from Tour," *Aircraft Journal*, 26 July 1919, 5, Hathi Trust Digital Library; U.S. Congress, House,

United Air Service, Hearings Before a Subcommittee of the Committee on Military Affairs, 66th Cong., 2d sess., December 1919–February 1920, passim; Henning, *Harnessing the Airplane: American and British Cavalry Responses to a New Technology, 1903–1939*, 75–78.

18. Millett and Maslowski, *For the Common Defense: A Military History of the United States of America*, 363–364.
19. Ibid.
20. Ibid., 363, 364, 366; Weigley, *History of the United States Army*, 400–401, 568–569.
21. Maj J.A. Mars, Air Service, "Mass Procurement of Air Service Material[:] A Production Study," Office of the Chief of Air Service, War Department, 15 February 1923, 1, RG 107, Records of the Office of the Secretary of War, Entry 144, Box 1, National Archives; Memorandum, Mason M. Patrick, Maj Gen, Chief of Air Service, to the Adjutant General, Subj.: "Reorganization of air forces for national defense," 19 September 1924, 4, William Mitchell Papers, Box 10, Manuscript Division, LOC; Hurley, *Billy Mitchell: Crusader for Air Power*, 49; Weigley, *History of the United States Army*, 400; John F. Shiner, *Foulois and the U.S. Army Air Corps, 1931–1935* (Washington, D.C.: Office of Air Force History, 1983), 17; "War Department Annual Reports [Fiscal Year] 1920. Volume I," (Washington, D.C.: GPO, 1921), 13, U.S. Congress, House of Representatives, 66th Cong., 3rd sess., Document No. 863, ProQuest Congressional Documents, Pentagon Library; Roger E. Bilstein, *Flight Patterns: Trends of Aeronautical Development in the U.S.* (Athens, GA: University of Georgia Press, 1983), 9, 10.
22. Hurley, *Billy Mitchell: Crusader for Air Power*, 40, 60–63; Shiner, *Foulois and the U.S. Army Air Corps, 1931–1935*, 17; Arnold, *Global Mission*, 91–94; "The Neglect of Aviation," *New York Times*, 21 August 1919, 10, ProQuest Historical Newspapers, Pentagon Library; William Mitchell, *Winged Defense: The Development and Possibilities of Modern Air Power* (New York, NY: G.P. Putnam's Sons, 1925).
23. Wesley Frank Craven and James Lea Cates, eds., *The Army Air Forces in World War II. Vol. One. Plans and Early Operations. January 1939 to August 1942* (Chicago, IL: University

of Chicago Press, 1948), 34–35; Weigley, *History of the United States Army*, 561.

24. *Bicentennial Edition: Historical Statistics of the United States Colonial Times to 1970. Part 2* (Washington, D.C.: Bureau of the Census, Department of Commerce, September 1975), 1114, census.gov; Maurer Maurer, *Aviation in the U.S. Army, 1919–1939* (Washington, D.C.: Office of Air Force History and USAF Historical Research Center, 1987), 467–477.
25. White, *Mason Patrick and the Fight for Air Service Independence*, 54–57; Thomas Wildenberg, *Billy Mitchell's War with the Navy: The Interwar Rivalry Over Air Power* (Annapolis, MD: Naval Institute Press, 2013), 3; Mason M. Patrick, *The U.S. in the Air* (Garden City, NY: Doubleday, Doran and Co., 1928), 89; Ransom, "The Air Corps Act of 1926; A Study of the Legislative Process," 218; Hurley, *Billy Mitchell: Crusader for Air Power*, 68–69; Weigley, *History of the United States Army*, 558; Rice, *The Politics of Air Power: From Confrontation to Cooperation in Army Aviation Civil–Military Relations*, 23; "Menoher to Quit Army Air Service," *New York Times*, 17 September 1921, 3, ProQuest Historical Newspapers, Pentagon Library; "General Patrick Appointed C. [Chief] of A.S. [Air Service]," *Aviation and Aircraft Journal*, 3 October 1921, 395, Hathi Trust Digital Library; Coffman, *The Hilt of the Sword: The Career of Peyton C. March*, 204–205; Henry Blaine Davis, Jr., *Generals in Khaki* (Raleigh, NC: Pentland Press, Inc., 1 April 1998), 267–68; D. Clayton, James, *The Years of MacArthur, Volume 1, 1880–1941* (1st ed.) (Boston, MA: Houghton Mifflin Company, 1970), 148.
26. CAPT T. T. Craven, USN, Director of Naval Aviation, "Naval Air Service Independence," *Aviation and Aeronautical Engineering*, 7 March 1921, 305, Hathi Trust Digital Library; Hone and Utz, *History of the Office of the Chief of Naval Operations, 1915–2015*, 59–60, NHHC, history.navy.mil; "Admiral Sims on Aviation," *Aviation and Aircraft Journal*, 5 December 1921, 659, Hathi Trust Digital Library.
27. Hone, "Navy Leadership: RADM W. A. Moffett as Chief of the Bureau of Aeronautics," 5, 10.
28. Gross, *American Military Aviation: The Indispensable Arm*, 55–56; White, *Mason Patrick and the Fight for Air Service Independence*, 102–104; "Annual Reports of the Navy Department for

Fiscal Year (Including Operations to November 15, 1924) 1924," (Washington, D.C.: GPO, 1925), 611, U.S. Congress, House of Representatives, 68th Cong., 2d sess., House Document No. 555, ProQuest Congressional Documents, Pentagon Library.

29. *"William Adger Moffett Papers, 1904–1948 (Bulk, 1918–1933): Finding Aid,"* Special Collections & Archives, Nimitz Library, published 1989; *"Chester 1 (Cruiser No. 1)," NHHC, published 22 February 2016,* history.navy.mil; William F. Trimble, *Admiral William A. Moffett: Architect of Naval Aviation (*Annapolis, MD: Bluejacket Books, Naval Institute Press, 1994), 23–24, 28*;* Hone, "Navy Leadership: RADM W. A. Moffett as Chief of the Bureau of Aeronautics," 2, 3.
30. Hone, "Navy Leadership: RADM W. A. Moffett as Chief of the Bureau of Aeronautics," 3–4; Trimble, *Admiral William A. Moffett: Architect of Naval Aviation*, 59–63.
31. "William Adger Moffett, Rear Admiral, U.S. Navy," Arlington National Cemetery, last updated 21 November 2010, www.arlingtoncemetery.net.
32. Reynolds, *Admiral John H. Towers: The Struggle for Naval Air Supremacy*, 220–221.
33. Ibid., 221.
34. U.S. Congress, House, *Navy Department Appropriation Bill [Fiscal Year] 1924, Hearing Before Subcommittee of [the] House Committee on Appropriations*, 67th Cong., 4th sess., 23 November 1922, 574, Hathi Trust Digital Library.
35. Trimble, *Admiral William A. Moffett: Architect of Naval Aviation*, 7.
36. Ibid., 8.
37. Ibid., 8–9.
38. Ibid., 9.
39. Maj Robert P. White, USAF, "Air Power Engineer: MGen. Mason M. Patrick and the Air Force Road to Independence," Office of Air Force History, Headquarters USAF, May 1997, 8.
40. Patrick, *The U.S. in the Air*, 89–90; "General of the Armies of the United States John Joseph Pershing," The Army Historical Foundation, undated, armyhistory.org.
41. Patrick, *The U.S. In the Air*, 89–93; White, *Mason Patrick and the Fight for Air Service Independence*, 60–61; Peake, "Major General Mason M. Patrick: Military Aviation Pioneer," 203; Mason M.

Patrick, "[Office] Diary of Mason Patrick, Chief of Air Service, 1921–1924," passim, papers of Benjamin Foulois, Box 6, Folders 8 and 9, USAFAL; James Phillip Tate, "The Army and Its Air Corps: A Study of the Evolution of Army Policy Towards Aviation, 1919–1941," Ph.D. dissertation, Department of History, Indiana University, 1976, 27; "Annual Report of the Chief of Air Service for the Fiscal Year Ending June 30, 1922," (Washington, D.C.: GPO, 1922), 48, 49, Pentagon Library; Maj H.A. Dargue, "Training a Major General to Fly," *U.S. Air Service*, September 1923, 18–19, 21–22; Hathi Trust Digital Library; Maj Gen Mason M. Patrick, "[Office] Diary of Major General Mason M. Patrick, Chief of Air Service, 1922," 22 March 1922, 151, 28 March 1922, 157, 3 April 1922, 161, 17 May 1922, 200–201, 8 June 1922, 213, 17 August 1022, 233, 3 October 1922, 238, MS-17, Benjamin D. Foulois Papers, Box 6, Folder 9, USAFAL; Maj Gen Mason M. Patrick, "[Office] Diary of Major General Mason M. Patrick, Chief of Air Service, 1923," 5 December 1923, 256, MS-17, Benjamin D. Foulois Papers, Box 6, Folder 9, USAFAL; Maj Gen Mason M. Patrick, "[Office] Diary of Major General Mason M. Patrick, Chief of Air Service, 1924," 26 February 1924, 269–270, MS-17, Benjamin D. Foulois Papers, Box 6, Folder 9, USAFAL.

42. White, *Mason Patrick and the Fight for Air Service Independence*, 72–76; Ransom, "The Air Corps Act of 1926; A study of the Legislative Process," 218; "Annual Report of the Chief of the Air Service, Fiscal Year 1922," (Washington, D.C.: GPO, 1922), 41–46, Pentagon Library; "Bigger Air Service Urged By Pershing," *New York Times*, 30 November 1924, E1, AACPL, www.aacpl.net.
43. Thomas H. Greer, *The Development of Air Doctrine in the Army Air Arm, 1917–1941, Air Force Historical Study No. 89* (Maxwell AFB, AL: USAF Historical Division, Air University, September 1953), 25–27; White, *Mason Patrick and the Fight for Air Service Independence*, 115; "Air Inquiry By Davis Stirs House Attack," *New York Times*, 11 February 1926, 6, ProQuest Historical Newspapers, Pentagon Library.
44. Ransom," The Air Corps Act of 1926; A Study of the Legislative Process," 230.

45. Patrick, *The U.S. in the Air*, 97; "Annual Report of the Director of the Air Service, Fiscal Year 1919," (Washington, D.C.: Department of War, 1919), 15, 55, Pentagon Library; Earl Remington, "The Future of American Aeronautics," *The North American Review*, January 1919, 58, JSTOR, Pentagon Library; Freudenthal, *The Aviation Business: From Kitty Hawk to Wall Street,* 62–63; Harold Mansfield, *Vision: A Saga of the Sky* (New York, NY: Duell, Sloan and Pearce, 1956), 30–31; Howard Mingos, "Birth of an Industry," *The History of the American Industry: An Anthology*, G. R. Simanson, Ed. (Cambridge, MA: MIT Press, 1968), 53.
46. U.S. Congress, House, "Report of American Aviation Mission," "United Air Service," Hearing Before a Subcommittee of the Committee on Military Affairs, House of Representatives, 66th Cong., 2nd sess., 4 December 1919, 16–31 (Washington, D.C.: GPO, 1921), ProQuest Congressional Documents, Pentagon Library; "McCook Field Builds 1,000 Hp. Engine," *Aviation and Aeronautical Engineering*, 16 May 1921, 640, Hathi Trust Digital Library; Patrick, *The U.S. in the Air*, 98; Rae, *Climb to Greatness. The American Aircraft Industry, 1920–1960*, 7; *Mansfield, Vision: A Saga of the Sky*, 42; Vander Meulen, *The Politics of Aircraft: Building an American Military Industry*, 2–5, 57–58; *The Aircraft Year Book for 1934,* 113–115, AIA, aia-aerospace.org; "Annual Report of the Chief of the Air Service for the Fiscal Year Ending June 30, 1923," (Washington, D.C.: GPO, 1923) 36, 40, Pentagon Library.
47. CAPT Henry C. Mustin, USN, Memorandum No. 2 [Dissent], "Report of American Aviation Mission," "United Air Service," Hearing Before a Subcommittee of the Committee on Military Affairs, House of Representatives, 66th Cong., 2nd sess., 4 December 1919, 4, 30–31 (Washington, D.C.: GPO, 1921), Pentagon Library.
48. Boyne, "The Treasure Trove of McCook Field," pt.1, *Airpower*, July 1975, 12; Purtee, *History of the Army Air Service, 1907–1926, Air Force Logistics Command (AFLC) Historical Study No. 279*, 128, 154–155; Air Service, "Report on the Engineering Division," November 1924, 3, 1221; Patrick, *The U.S. in the Air*, 102–103; Maj Gen Mason M. Patrick, "Problems of Air Service Procurement of Material," for delivery at the Army Industrial

College, 20 May 1925, 3–5, RG 18, AAF Records, Entry 229, Box 1, National Archives; Maj Gen Mason M. Patrick, "[Office] Diary of Major General Mason M. Patrick, Chief of Air Service, 1921," 12 October 1921, 10, 17 October 1921, 15, MS-17, Benjamin D. Foulois Papers, Box 6, Folder 8, USAFAL; Trimble, "The Naval Aircraft Factory, the American Aviation Industry, and Government Competition, 1919–1928," 181, reprinted from the *Business History Review*, Summer 1986; Trimble, *Admiral William A. Moffett: Architect of Naval Aviation*, 112–113; Army Air Service, "The Engineering Division: A Consideration of Its Status With Respect to the Air Service and the Aeronautical Industry in the U.S. and Abroad [Report for the Investigation of the Lampert Committee]," revised 20 November 1924, 27, RG 18, AAF Records, File 333.5, Entry 91, Box 1, National Archives.

49. Elsbeth E. Freudenthal, *The Aviation Business: From Kitty Hawk to Wall Street* (New York, NY: Vanguard Press, 1940), 74, Hathi Trust Digital Library.
50. Ibid.
51. Patrick, *The U.S. in the Air*, 98; Rae, *Climb to Greatness. The American Aircraft Industry, 1920–1960*, 7; "Annual Report of the Chief of the Air Service for the Fiscal Year Ending June 30, 1921," (Washington, D.C.: GPO, 1921), 31, Pentagon Library; Mansfield, *Vision: A Saga of the Sky*, 42; Vander Meulen, *The Politics of Aircraft: Building an American Military Industry*, 2–5, 57–58; *The Aircraft Year Book For 1934*, 113–115, AIA, aia-aerospace.org; Gardner W. Carr, "Production Methods and the Aircraft Industry," *U.S. Air Services*, July 1925, 37, Hathi Trust Digital Library.
52. U.S. Congress, House, *Department of the Navy, Appropriation Bill For [Fiscal Year] 1927, Hearing before Subcommittee of House Committee on Appropriations,* 69th Cong., 1st sess., 21 December 1925, 600–601, Hathi Trust Digital Library.
53. Jacob Goodwin, *Brotherhood of Arms: General Dynamics and the Business of Defending America* (New York, NY: Times Books, 1985), 48–52, 58–59.
54. "The Engineering Division: A Consideration of Its Status With Respect to the Air Service and the Aeronautical Industry in the United States and Abroad [Report for the Investigation of the Lampert Committee]," revised 20 November 1924, 27, RG 18,

Records of the Army Air Forces, File 333.5, Entry 91, Box 1, National Archives; Mansfield, *Vision: A Saga of the Sky*, 34; 1st Lt Charles N. Monteith, Air Corps, Engineering Division, McCook Field, Dayton, Ohio, "Simple Aerodynamics and the Airplane: Prepared by Direction of the Chief of the Air Corps – U.S. Army. Reprint of the Third Edition, November 1925," (Washington, DC: USGPO, 1927), 268, Hathi Digital Trust Library.

55. Mansfield, *Vision: A Saga of the Sky*, 34.
56. "Biographical Sketch of Colonel Thurman H. Bane, 1884–1932," USAF Historical Division, Research Studies Institute, August 1957, passim; Col Bane quoted in Albert Misenko, "Aero Propulsion and Power Directorate: The McCook Field Years (1917–1927)," AFRL-PR-WP-TR-1998-2020, 18, ASD/HO.
57. Misenko, "Aero Propulsion and Power Directorate: The McCook Field Years (1917–1927)," 19.
58. Thurman H. Bane, "Recent Advances in Aviation," *SAE Transactions, 1920, Vol. 15, PART II,* 1920, 84, JSTOR, Pentagon Library; "Annual Report of Chief of Air Service for the Fiscal Year Ending 30 June 1921," Hathi Trust Digital Library, 30.
59. "Annual Reports of the Navy Department for the Fiscal Year 1922 (Including Operations to November 15, 1922)," 389.
60. Gross, *American Military Aviation: The Indispensable Arm*, 63–64.
61. Patrick, "Problems of Air Service Procurement of Material," for delivery at the Army Industrial College, 20 May 1925, 5; U.S. Congress, House, Committee on Appropriations, *War Department Appropriation Bill, 1925, Military Activities. Hearings Before a Subcommittee of the House Committee on Appropriations. Part 1.* 68th Cong., 1st sess., 6 February 1924, 911–912 (Washington, D.C.: 1924), ProQuest Congressional Documents, Pentagon Library.
62. "Annual Report of the Bureau of Aeronautics to the Secretary of the Navy, Fiscal Year 1923," (Washington, D.C.: GPO, 1923), 15, 39, Hathi Trust Digital Library; Trimble, "The Naval Aircraft Factory, The American Aviation Industry, and Government Competition, 1919–1928," 183–185; U.S. Congress, House, *Department of the Navy, Appropriation Bill for [Fiscal Year] 1927, Hearing Before Subcommittee of House Committee on Appropriations,* 69th Cong., 1st sess., 21 December 1925, 598; U.S.

Congress, Senate, *Navy Department Appropriation Bill [Fiscal Year] 1927, Hearings Before a Subcommittee of the Committee on Appropriations, U.S. Senate*, 69th Cong., 1st sess., 5 February 1926, 121–122, Hathi Trust Digital Library; "Aircraft: Hearing Before the President's Aircraft Board," Volume 1, 22 September 1925, 206 (Washington, D.C.: GPO, 1925), ProQuest Congressional Documents, Pentagon Library; "Annual Reports of the Navy Department for the Fiscal Year 1920 (Including Operations and Recommendations to December 1, 1920)," (Washington, D.C.: GPO, 1921), 51, Hathi Trust Digital Library; "Annual Reports of the Navy Department for the Fiscal Year 1923 (Including Operations and Recommendations to November 15, 1923)," (Washington, D.C.: GPO, 1924), 541, Hathi Trust Digital Library; [Admiral] R. E. Coontz, CNO to all Bureaus, Offices and Stations Concerned, Subj.: "Naval Aeronautic Organization for Fiscal Year 1923," 17 June 1922, 16, Hathi Trust Digital Library; U.S. Congress, House, *Department of the Navy, Appropriation Bill for [Fiscal Year] 1927, Hearing Before Subcommittee of House Committee on Appropriations,* 69th Cong., 1st sess., 21 December 1925, 607.
63. Robert Schlaifer and S. D. Heron, *Development of Aircraft Engines and Aviation Fuels. Two Studies of Relations Between Government and Industry* (Boston, MA: Harvard School of Business Administration, 1950), 7.
64. Ibid., 8–9.
65. Ibid., 12–13.
66. "Annual Report of the Chief of the Air Service for the Fiscal Year Ending June 30, 1923, 35.
67. "Army Inadequate, Weeks Declares," *New York Times*, 2 December 1922, 10, AACPL, www.aacpl.net.
68. Rear Adm William A. Moffett, Chief, BuAER, "Aviation," *New York Times*, 1 January 1924, 29, ACCPL, www.aacpl.net.
69. McAllister, ed., *Aerospace Facts and Figures 1964, 24*, AIA, aia-aerospace.org.
70. *Aircraft Year Book 1922* (New York, NY: Aeronautical Chamber of Commerce of America, Inc., 1922), 2, AIA, aia-aerospace.org.
71. "Annual Report of the Chief of the Air Service for the Fiscal Year Ending June 30, 1925," 37, AFHRA Microfilm # 167-4011; Maj. James A. Mars, Chief, Industrial War Plans Division, Army

Air Service, "Government Must Encourage Commercial Aviation," *U.S. Air Services*, June 1925, 27, 28, Hathi Trust Digital Library.

72. Mars, "Government Must Encourage Commercial Aviation," *U.S. Air Services*, June 1925, 29.
73. Donald R. McCoy, *Coolidge: The Quiet President* (New York, NY: Macmillan, 1967), 304; Freudenthal, 75; Rae, *Climb to Greatness. The American Aircraft Industry, 1920–1960*, 34; *Aircraft Year Book, 1922*, 161–166; "Aircraft Industry Forms Corporation," *New York Times*, 1 January 1922, 6, ProQuest Historical Newspapers, Pentagon Library; "An Act to Encourage Commercial Aviation and to Authorize the (H. R. 7064.) Postmaster General to contract for air mail service," 68th Cong., 2nd sess., Chapter 128, 2 February 1925, H.R. 7064, PL68-359, 805, 43 Stat. 805; 68 H.R. 7064, ProQuest Congressional Documents, Pentagon Library; Louis De Gouyon Matigon, "The Airmail Act of 1925," 13 October 1920, NASA, nasa.gov; *Aircraft Year Book, 1926*, 1, 292, AIA, aia-aerospace.org.
74. Thomas Worth Waterman, "Airpower and Private Enterprise: Federal-Industrial Relations in the Aeronautics Field, 1918–1926," Ph.D. dissertation, Department of History, Washington, University, June 1970, 3–4.
75. *Aircraft Year Book 1922* (New York, NY: Aeronautical Chamber of Commerce of America, Inc., 1922), 179, AIA, aia-aerosp ace.org.
76. Ibid.
77. Ibid.
78. Ibid.
79. *Aircraft Year Book 1923* (New York, NY: Aeronautical Chamber of Commerce of America Inc., 1923) 310, AIA, aia-aerosp ace.org.
80. Ibid.
81. *Aircraft Year Book 1926* (New York, NY: Aeronautical Chamber of Commerce of America, Inc., 1926), 301, AIA, aia-aerosp ace.org.
82. Ibid; Charles D. Walcott, NACA Chairman, to Herbert Hoover, Secretary of Commerce, Subj.: National Aviation Policy, 23 March 1921, with attached "Digest of National Aviation Policy Contained in the Sixth Annual Report of the National Advisory

Committee for Aeronautics [Fiscal Year] 1920," 23 March 1921, passim, HHPL&M, Commerce, Box 39, File "Aviation, 1920–21;" Notes, Subj.: "Meeting of the Air Craft Men Monday," Department of Commerce, 16 July 1921, with attached list of attendees, "Aviation--Conference in Office of Secretary Of Commerce, " 18 July 1921, HHPL&M, Commerce, Box 39, File "Aviation, 1920–21."

83. U.S. Congress, House, *Report of the Select Committee of Inquiry Into the Operations of the U.S. Air Service*, House Report No. 1653, 68th Cong., 2nd sess., 1925, passim, Pentagon Library; Ransom, "The Air Corps Act of 1926; A Study of the Legislative Process," 233–243; Mooney and Layman, *Organization of Military Aeronautics, 1907–1935 (Congressional and War Department Action). U.S. Air Force Historical Study No. 25,* 63–64; Trimble, "The Naval Aircraft Factory, the American Aviation Industry, and Government Competition, 1919–1928," 188–192; "Air Industry Asks Nation to Save It," *New York Times*, 16 January 1925, 8, ProQuest Historical Newspapers, Pentagon Library.
84. U.S. Congress, House, *Report of the Select Committee of Inquiry Into the Operations of the U.S. Air Services*, House Report No. 1653, 68th Cong., 2nd sess., 1925, passim, Pentagon Library.
85. Jason Ryan, "Ten Days Lost at Sea: The First Flight (and Voyage) to Hawaii," *Naval History Magazine*, February 2019, U.S. Naval Institute, usni.org; Grossnick, *U.S. Naval Aviation, 1910–1995*, 63; Hone, "Navy Leadership: RADM W. A. Moffett as Chief of the Bureau of Aeronautics," 22; Turnbull and Lord, *History of U.S. Naval Aviation,* 250–251; Wildenberg, *Billy Mitchell's War With the Navy: The Interwar Rivalry Over Air Power,* 129–139; Hurley, *Billy Mitchell: Crusader for Air Power*, 98, 101; "Annual Reports of the Navy Department for Fiscal Year (Including Operations to November 15, 1925)," (Washington, D.C.: GPO, 1926), 11, U.S. Congress, House of Representatives, 69th Cong., 1st sess., House Document No. 200, ProQuest Congressional Documents, Pentagon Library; Van Fleet, and Armstrong, *U.S. Naval Aviation, 1910–1980,* 60; "Statement of Rear Admiral W.A. Moffett, Chief of the Bureau of Aeronautics, Navy Department, on Rigid Airships, Before the Committee on Naval Affairs of the House of Representatives, Thursday January 28, 1926 (H.R. 7375)," (No. 103), 455–464, U.S. Congress,

House, *Sundry Legislation Affecting the Naval Establishment, Hearings Before the Committee on Naval Affairs of the House of Representatives, 1925–1926*, 69th Cong., 1st sess., ca 1926, Hathi Trust Digital Library; "Annual Report of the Chief of the Bureau of Aeronautics to the Secretary of the Navy, Fiscal Year 1926," (Washington, D.C.: GPO, 1926), 41, Hathi Trust Digital Library.

86. Ransom, "The Air Corps Act of 1926; A Study of the Legislative Process," 264; Mooney and Layman, *Organization of Military Aeronautics, 1907–1935 (Congressional and War Department Action). U.S. Air Force Historical Study No. 25*, 71–72; Trimble, *Admiral William A. Moffett: Architect of Naval Aviation*, 162; Turnbull and Lord, *History of U.S. Naval Aviation*, 251; McCoy, *Coolidge: The Quiet* President, 305; *Aircraft Year Book, 1926*, 6, 8, AIA, aia-aerospace.org; "Personnel of Board to Make Air Inquiry," *Boston Globe*, 13 September 1925, B6, ProQuest Historical Newspapers, Pentagon Library.
87. James F. Cook, *Carl Vinson: Patriarch of the Armed Forces* (Macon, G.A.: Mercer University Press, 2004), ix–xi, 51–53, 62, 63, 64; "VINSON, Carl: 1883–1981," *Biographical Directory of the U.S. Congress*, undated, bioguide.congress.gov; Grossnick, *U.S. Naval Aviation, 1910–1995*, 83.
88. Cook, *Carl Vinson: Patriarch of the Armed Forces*, 57.
89. "Aircraft in National Defense: Message From the President of the U.S.," U.S. Senate, Document No. 18, 69th Cong. 1st sess., November 1925, passim, ProQuest Congressional Documents, Pentagon Library.
90. Ibid; Draft Ltr., unsigned, Secretary of War to Representative Carl Vinson, Subj.: War Department Views on Aircraft Procurement, 20 May 1926, 1–3, William P. MacCracken, Jr. Papers, Box 51, Subject Series, Aviation, Folder "Air Law, Federal 1926, Aircraft and Equipment Procurement, HR 9690, HR 9953," HHPL&M; Rae, *Climb to Greatness. The American Aircraft Industry, 1920–1960*, 20–22; Arthur Sears Henning, "Coolidge Backs Aircraft Board in Full," *Chicago Daily Tribune*, 5 December 1925, 7, ProQuest, Historical Newspapers, Pentagon Library.
91. "Annual Reports of the Navy Department for Fiscal Year (Including Operations to November 15, 1926) 1926," (Washington, D.C.: USGPO, 1927), 597–598, U.S. Congress, House

of Representatives, 79th Cong., 2nd sess., House Document No. 625, ProQuest Congressional Documents, Pentagon Library.

92. "Air Corps Favored By House Probers," *Boston Daily Globe*, 27 January 1926, 5, ProQuest Historical Newspapers, Pentagon Library; "Patrick Will Draft Bill for Air Corps," *New York Times*, 27 January 1926, 25, ProQuest Historical Newspapers, Pentagon Library; "Army Air Corps Bill Offered in House," *New York Times*, 29 January 1926, 23, ProQuest Historical Newspapers, Pentagon Library.
93. Freudenthal, *The Aviation Business: From Kitty Hawk to Wall Street*, 77; Hunsaker, "Forty Years of Aeronautical Research," *Annual Report of the Smithsonian Institution, 1955*, 254–255; "Annual Report of the Chief of the Air Corps [Fiscal Year] 1926," (Washington, D.C.: GPO, 1926), 158, Pentagon Library; "Statement of Secretary Hoover on Commercial Aviation," 24 September 1925, passim, Herbert C. Hoover, Secretary of Commerce, Commerce Papers, Box 40, File "Aviation 1925," HHPL&M; "Aeronautics: Twelfth Annual Report of the National Advisory Committee for Aeronautics [Fiscal Year] 1926," 9–10; "Annual Report of the Director of Aeronautics to the Secretary of Commerce for the Fiscal Year Ended June 30, 1927," (Washington, D.C.: USGPO, 1927), 1, Internet Archive; James E. Fetchet, *Flying* (New York, NY and London, UK: The Century Book Co., A Williams & Wilkins Book, 1933), 53.
94. Holley, *Buying Aircraft: Materiel Procurement for the Army Air Forces. Special Studies. U.S. Army in World War II*, 14.
95. Waterman, "Airpower and Private Enterprise: Federal-Industrial Relations in the Aeronautics Field, 1918–1926," 1–2.
96. Sean Seyer, *Sovereign Skies: The Origins of America Civil Aviation Policy* (Baltimore, MD: Johns Hopkins University Press, 2021), 5–8.
97. Ibid.
98. U.S. Congress, Senate, *The Army Air Service, Hearing Before the Committee on Military Affairs*, 69th Cong., 1st sess., 10 May 1926, 29, Hathi Trust Digital Library; Ransom, "The Air Corps Act of 1926; A Study of the Legislative Process," 338–340; Rice, *The Politics of Air Power: From Confrontation to Cooperation in Army Aviation Civil–Military Relations*, 75; Goldberg, *A History*

of the U.S. Air Force, 1907–1957, 36; "1926—The U.S. Army Air Corps Act," AFHSD, 4 February 2011, www.afhistory.af.mil.

99. Vander Meulen, *The Politics of Aircraft: Building An American Military Industry*, 82.
100. Ibid.
101. Ibid.
102. "Annual Reports of the Navy Department for Fiscal Year (Including Operations to November 15,1926) 1926," 1, 597, 598, 601, 602, 618; Shanahan, *Procurement of Naval Aircraft, 1907–1939*, 337; Lt. Col. Craig F. Morris, USAF, *The Origins of American Strategic Bombing Theory* (Annapolis, MD: Naval Institute Press, 2017), 164; Trimble, *Admiral William A. Moffett: Architect of Naval Aviation*, 174–179; Van Fleet, and Armstrong, *U.S. Naval Aviation, 1910–1980,* 62; Ernest J. King and Walter Muir Fairchild, *Fleet Admiral King: A Naval Record* (New York, NY: W.W. Norton, Inc., 1952), 187.
103. Danhoff *Government and Technological Change*, 26–27; Futrell, *Ideas, Concepts, Doctrine: A History of Basic Thinking in the U.S. Air Force, 1907–1964*, 31–32; Goldberg, *A History of the U.S. Air Force, 1907–1957*, 36–37; Ransom, "The Air Corps Act of 1926; A Study of the Legislative Process," 338–340; *The Aircraft Year Book for 1943* (New York, NY: The Aeronautical Chamber of America, Inc., 1943), 122–125, AIA, aia-aerospace.org; Holley, *Buying Aircraft: Materiel Procurement for the Army Air Forces. Special Studies. U.S. Army in World War II,* 89–93; Trimble, *Admiral William A. Moffett: Architect of Naval Aviation*, 182.
104. Rae, *Climb to Greatness. The American Aircraft Industry, 1920–1960*, 23, 34–35; Mingos, "Birth of an Industry," *The History of the American Industry: An Anthology*, 63.
105. White, *Mason Patrick and the Fight for Air Service Independence*, 133.
106. "Report of the Director of the Air Service [Fiscal Year] 1920," 17, 18.
107. Ibid., 41.
108. Maj Gen Mason M. Patrick, "[Office] Diary of Mason M. Patrick, Chief of Air Service, 1921," MS-17, 4 November 1921, 34, 9 December 1921, 70, Benjamin D. Foulois Papers, Box 6, Folder 8, USAFAL; Maj Gen Mason M. Patrick, "[Office] Diary of Mason M. Patrick, Chief of Air Service, 1924," 15 February

1934, 264, 20 February 1924, 266, Benjamin D. Foulois Papers, Box 6, Folder 8, USAFAL; U.S. Congress, House, Committee on Appropriations, *Army Appropriation Bill [Fiscal Year] 1922. Hearings Before a Subcommittee of the House Committee on Appropriations. Part 1.* 66th Cong., 3rd sess., 18 December 1920, 260 (Washington, D.C.: GPO, 1921), Hathi Trust Digital Library.

109. Goldberg, *A History of the U.S. Air Force, 1907–1957*, 32.
110. "Annual Report of Chief of Air Service for the Fiscal Year Ending 30 June 1921," 4 October 1921, 31; "Annual Report of Chief of Air Service for the Fiscal Year Ending 30 June 1922," 9 September 1922, 23, Hathi Trust Digital Library; Foulois and Glines, *From the Wright Brothers to the Astronauts: The Memoirs of Major General Benjamin Foulois*, 208; Holley, *Ideas and Weapons: Exploitation of the Aerial Weapon by the U.S. in World War I: A Study in the Relationship of Technological Advance, Military Doctrine, and the Development of Weapons*, 177.
111. "Report of the Secretary of War to the President [Fiscal Year] 1924," (Washington, DC: GPO, 1924), 23, Hathi Trust Digital Library.
112. Irving B. Holley, Jr., *Evolution of the Liaison Type Airplane, 1917–1944. Army Air Forces Historical Studies: No. 44* (Washington, D.C.: AAF Historical Office, Headquarters, AAF, April 1946), 113, Numbered USAF Historical Studies, Microfilm Reel IRIS Number K1006 467635, AFHRA, afhra.af.mil.
113. W.V. Andrews, "The Purchase of Airplanes, Supplies, and Equipment," *U.S. Air Services*, November 1923, 42, Hathi Trust Digital Library; "Suggested Speech for Major General Mason M. Patrick Before the Army War College, Washington, D.C.," March 1922, 4, RG 18, AAF Records, Entry 229, Box 1, National Archives; Purtee, *History of the Army Air Service, 1907–1926, Air Force Logistics Command (AFLC) Historical Study No. 279*, 118; Alfred S. Niles, "Development of the New Types of Airplanes and Accessories," 4 November 1927, 3–4, RG 18, AAF Records, Entry 211, Box 9, Vol XVII, National Archives.
114. Niles, "Development of the New Types of Airplanes and Accessories," 4 November 1927, 3–4; Andrews, "The Purchase of Airplanes, Supplies, and Equipment," *U.S. Air Service*, November 1923, 42–43; Purtee, *History of the Army Air Service, 1907–1926, Air Force Logistics Command (AFLC) Historical Study No. 279*,

118; "Suggested Speech for Major General Mason M. Patrick Before the Army War College, Washington, D.C.," March 1922, 4; Carr, "Production Methods and the Aircraft Industry," *U.S. Air Services*, July 1925, 37, Hathi Trust Digital Library.

115. Niles, "Development of the New Types of Airplanes and Accessories," 4 November 1927, 4.
116. Patrick, *The U.S. in the Air,* 102–103; Purtee, *History of the Army Air Service, 1907–1926,* 154–155.
117. Andrews, "The Purchase of Airplanes, Supplies, and Equipment," *U.S. Air Service*, November 1923, 42; "Suggested Speech for Major General Mason M. Patrick before the Army War College, Washington, D.C.," March 1922, 118; Patrick, "1921–1924 Office Diaries," passim.
118. Holley, *Buying Aircraft: Materiel Procurement for the Army Air Forces. Special Studies. U.S. Army in World War II*, 112.
119. U.S. Congress, House of Representatives, "United Air Service," Hearing Before a Subcommittee of the Committee on Military Affairs, House of Representatives, 66th Cong., 2nd sess., 8 December 1919, 106 (Washington, D.C.: GPO, 1921), ProQuest Congressional Documents, Pentagon Library; Hurley, *Billy Mitchell: Crusader for Air Power*, 53–54; Brig Gen William Mitchell, Assistant Chief of Air Service, for Chief of Air Service, Subj.; "Report on Inspection of Aeronautical Equipment and Organization at McCook Field," 4 November 1921, passim, General William Mitchell Papers, 1888–1946, Box 41, Manuscript Division, LOC; Brig Gen William Mitchell, Assistant Chief of Air Service, for Chief of Air Service, Subj.; Inspection Visit to McCook Field, 15 May 1922, 5, 8–9, General William Mitchell Papers, 1888–1946, Box 41, Manuscript Division, LOC; Brig Gen William Mitchell, Assistant Chief of Air Service, for Chief of Air Service, Subj.; "Report of Inspection of Air Service Engineering Division, McCook Field, by Brigadier General William Mitchell, April 1923," April 1923, 1–3, General William Mitchell Papers, 1888–1946, Box 41, Manuscript Division, LOC.
120. H. V. Pat Riley, "Aviation Pioneers: The Wittemann Brothers," Aviation Hall of Fame and Museum of New Jersey, 18 July 2005, njahof.org; Diana G. Cornelisse, *Splendid Vision, Unswerving Purpose: Developing Air Power for the United States Air Force*

During the First Century of Powered Flight (Wright-Patterson Air Force Base, OH: ASD/HO, ASD, AFSC, U.S. Air Force Publications, 2004), 70; Haulman, *One Hundred Years of Flight: USAF Chronology of Significant Air and Space Events 1903–2002,* 23; Maj. Gen. Mason M. Patrick, " [Office] Diary of Major General Mason M. Patrick, Chief of Air Service, 1921," 19 October 1921, 18, MS-17, Benjamin D. Foulois Papers, Box 6, Folder 8, USAFAL; Maj. Gen. Mason M. Patrick, " [Office] Diary of Major General Mason M. Patrick, Chief of Air Service, 1922," 4 January 1922, 93, 6 January 1922, 97, MS-17, Benjamin D. Foulois Papers, Box 6, Folder 9, USAFAL; "Barling Bomber – Army's Super Airplane," *U.S. Air Service*, August 1923, 15, Hathi Digital Trust Library; "Largest Bomber Soon to Appear," ***[Baltimore] Sun,*** 30 September 1922, 8, ProQuest Historical Newspapers, AACPL, www.aacpl.net; "Biggest Airplane's Test Flight July 15," *[Baltimore] Sun*, 18 June 1923, 2, ProQuest Historical Newspapers, AACPL, www.aacpl.net; Monteith, Air Corps, Engineering Division, McCook Field, Dayton, Ohio, "Simple Aerodynamics and The Airplane: Prepared by Direction of the Chief of the Air Corps – U.S. Army. Reprint of the Third Edition, November 1925," 287, Hathi Digital Trust Library; Coffee, *HAP: The Story of the U.S. Air Force and the Man Who Built It, General Henry H. "Hap" Arnold,* 138–139.

121. White, *Mason Patrick and the Fight for Air Service Independence*, 94–99; "Suggested Speech for Major General Mason M. Patrick Before the Army War College, Washington, D.C.," March 1922, 2–3; Speech, Patrick, "Problems of Air Service Procurement of Material, for delivery at the Army Industrial College," 5–6; Patrick, *The U.S. in the Air*, 98, 101–103; Maj Gen Mason M. Patrick, "[Office] Diary of Mason M. Patrick, Chief of Air Service 1921," 17 October 1921, 15, MS-17, Benjamin D. Foulois Papers, Box 6, Folder 8, USAFAL; U.S. Congress, House, *Report of the Select Committee of Inquiry Into the Operations of the U.S. Air Services*, House Report No. 1653, 68th Cong., 2nd sess., 1925, 129–133, 144–145, 166, Pentagon Library; U.S. Congress, House of Representatives, *Appropriation for Purchase of Airplanes. Hearing Before a Subcommittee of the Committee on Military Affairs, House of Representatives,* 66th Cong., 2nd sess.,

on H.R. 11809, 2 February 1920, 6, ProQuest Congressional Documents, Pentagon Library.

122. Freudenthal, *The Aviation Business: From Kitty Hawk to Wall Street*, 73–74.

123. Gardner W. Carr, Assistant Chief of Production, Engineering Division, Air Service, "Organization and Activities of Engineering Division of the Army Air Service (continued from January 1922 issue)," *U.S. Air Service*, February 1922, 22, Hathi Trust Digital Library; "General Patrick Inspects McCook Field," *Army and Navy Journal*, 26 November 1921, 296, Nimitz Library; Edward O. Purtee, *The Development of Light and Medium Bombers, AFLC Historical Study No. 196* (Wright Field, OH.: Historical Section, Air Materiel Command, December 1946), 6–11.

124. "Suggested Speech for Major General Mason M. Patrick Before the Army War College, Washington, D.C.," March 1922, 2–3; Patrick, "Problems of Air Service Procurement of Material, for delivery at the Army Industrial College," 6; U.S. Congress, House, *Report of the Select Committee of Inquiry Into the Operations of the U.S. Air Services*, House Report No. 1653, 68th Cong., 2nd sess., 1925, 129–133, Pentagon Library; Patrick, *The U.S. in the Air*, 98, 101–103.

125. Raines, *Eyes of the Artillery: The Origins of Modern U.S. Army Aviation in World War II,* 14; Holley, *Evolution of the Liaison Type Airplane, 1917–1944,* 3, 7.

126. Raines, *Eyes of the Artillery: The Origins of Modern U.S. Army Aviation in World War II,* 14.

127. Nalty, *Winged Shield, Winged Sword: A History of the U.S. Air Force, Volume I, 1907*–1950, 85, 87, 90; Bryan R. Swopes, "Curtiss PW-8 Hawk Archives – This Day in Aviation, Tag Archives: Curtiss PW-8 Hawk: 23 June 1924," 2017, www.thisdayinaviation.com; "New [Curtiss] Pursuit Plane for the Army Air Service," *Aerial Age*, July 1923, 338, Hathi Trust Digital Library; Robert L. Cavanagh, "Boeing PW-9's in the U.S. Army," *American Aviation Historical Society Journal*, Summer 1962, 111, 113; 1st Lt Charles N. Monteith, Air Corps, Engineering Division, "Simple Aerodynamics and the Airplane: Prepared by Direction of the Chief of the Air Corps – U.S. Army. Reprint of the Third Edition, November 1925," 271, Hathi Digital Trust Library; "Sopworth F.1 Camel," SNASM, undated, airandspace.si.edu.

128. "The Engineering Division: A Consideration of Its Status With Respect to the Air Service and the Aeronautical Industry in the U.S. and Abroad [Report of the Investigation of the Lampert Committee]," revised 20 November 1924, 66–67, RG 18, AAF Records, File 333.5, Entry 91, Box 1, National Archives; Purtee, *History of the Army Air Service, 1907–1926*, 128–129; 26; *Aeronautics: Seventh Annual Report of the National Committee for Aeronautics [Fiscal Year] 1921* (Washington, D.C.: GPO, 1923), 26 Hathi Trust Digital Library; Hunsaker, "Forty Years of Aeronautical Research," *Annual Report of the Smithsonian Institution, 1955*, 251–260; Martin P. Claussen, *Comparative History of Research and Development Policies Affecting Air Materiel, 1915–1944. U.S. Air Force Historical Study No. 20* (Washington, D.C.: AAF Historical Office, Headquarters, AAF, June 1945), 29–30; Henry H. Arnold and Ira C. Eaker, *Winged Warfare* (New York, NY: Harper and Brothers, 1941), 217–221; Claussen, *Materiel Research and Development in the Army Air Arm, 1914–1945*, 20–23; *Fifth Annual Report of the National Advisory Committee for Aeronautics [Fiscal Year 1919]* (Washington, D.C.: GPO, 2 December 1919), 6, U.S. Congress, Senate, 66th Cong., 2nd sess., Document No. 166, ProQuest Congressional Documents, Pentagon Library; "*Aeronautics: Sixth Annual Report of the National Advisory Committee for Aeronautics [Fiscal Year] 1920* (Washington, D.C.: GPO, 1920), 8, Hathi Trust Digital Library.
129. Dr. Michael H. Gorn, "Hugh L. Dryden's Career in Aviation and Space," Monographs in Aerospace History, Number 5 (Washington, D.C.: NASA History Office, 1996), 1–2, 4, nasa.gov; Laura Newton, page editor, "Dr. Hugh L. Dryden," NASA, 8 September 2021, nasa.gov.
130. "Max M. Munk," NASA, last updated 7 August 2017, nasa.gov; John David Anderson, *A History of Aerodynamics and Its Impact on Flying Machines* (New York, NY: Cambridge University Press, 1997), 290, Internet Archive.
131. H.C. Dinger, "The Development of the Wright Air-Cooled Aviation Engine," *Journal of the American Society of Naval Engineers*, November 1926, 858, Hathi Trust Digital Library; Spenser, *The Airplane: How Ideas Gave Us Wings*, 218; "The Engineering Division: A Consideration of Its Status With Respect to the Air Service and the Aeronautical Industry in the U.S. and Abroad [Report

of the Investigation of the Lampert Committee]," 20 November 1924, 69, 81–83, RG 18, AAF Records, File 333.5, Entry 91, Box 1, National Archives; "History of the Air Corps Materiel Division," v-8110, A.C., Office of Air Force History (AFCHO) Microform No. 200-1, undated, no page numbers; J. F. Curry, "What McCook Field Means to Aviation," *U.S. Air Services*, August 1926, 44, Hathi Trust Digital Library; "Annual Report of the Chief of the Air Service for the Fiscal Year Ending June 30, 1925," 28–31; Gardner W. Carr, Assistant Chief of Production, Engineering Division, Air Service, "Organization and Activities of Engineering Division of the Army Air Service," *U.S. Air Service*, March 1922 (continued from February 1922)," 25–26, Hathi, Digital Trust Library; Claussen, *Comparative History of Research and Development Policies Affecting Air Materiel, 1915–1944, 20,* 30; "Minutes of Monday Meeting of [Air Service Headquarters] Division Chiefs," 6 March 1922, 17, Benjamin D. Foulois Papers, USAFAL Box 7, Folder 5.

132. Alexander Klemin, "Latest Developments in Technical Aeronautics: New Supercharging Device to Enable Military Airplanes to Achieve the Highest Possible Ceiling – Gluing Up Propeller Laminations," *Aircraft Journal*, 5 April 1919, 6, Hathi Trust Digital Library; Dr. Sanford A. Moss, General Electric Company, "The General Electric Turbo Supercharger for Airplanes," *Journal of the American Society of Naval Engineers*, August 1920, 586–588, Hathi Trust Digital Library; David Gregg, Supercharger Research Department, McCook Field, "The Story of the Supercharger," *The Slipstream*, March 1924, 17, Internet Archive.
133. Misenko, "Aero Propulsion and Power Directorate: The McCook Field Years (1917–1927)," 106, 107.
134. "Annual Report of the Chief the Bureau of Construction and Repair for the Secretary of the Navy for the Fiscal Year 1920," (Washington, D.C.: GPO, 1920), 12, Hathi Trust Digital Library; "Annual Report of the Bureau of Aeronautics to the Secretary of the Navy, Fiscal Year 1923," 16; "Annual Reports of the Navy Department for Fiscal Year 1919," 5, 36–37, 39; Grossnick, *U.S. Naval Aviation, 1910–1995*, 37, 38, 40; "Annual Reports of the Navy Department for Fiscal Year (Including Operations to November 15, 1921) 1921," (Washington, D.C.: GPO, 1921), 3, U.S. Congress, House of Representatives, 67th Cong.,

2nd sess., House Document No. 224, ProQuest Congressional Documents, Pentagon Library.

135. "Annual Report of the Chief the Bureau of Construction and Repair for the Secretary of the Navy for the Fiscal Year 1920," 12.

136. "Now Admit Defects of Liberty Motor: U. S. Experts Concede First Engines Failed to Meet War Demands," *[Baltimore] Sun*, 31 July 1922, 2, ProQuest Historical Newspapers, AACPL, www.aacpl.net.

137. The Bureau of Steam Engineering was renamed the Bureau of Engineering by the naval appropriation act approved on 4 June 1920.

138. Dinger, "The Development of the Wright Air-Cooled Aviation Engine," *Journal of the American Society of Naval Engineers*, November 1926, 858; Taylor, "Aircraft Propulsion: A Review of the Evolution of Aircraft Powerplants," 269–270, 273; Spenser, *The Airplane: How Ideas Gave Us Wings*, 214–216; Crouch, *Wings: A History of Aviation from Kites to the Space Age*, 242; Van Fleet, and Armstrong, *U.S. Naval Aviation, 1910–1980,* 48, 54; B. G. Leighton, "Recent Developments in Air-Craft Engines in the Navy," *SAE Transactions*, Vol. 18, Part I, 1923, 862, JSTOR, Pentagon Library; E. E. Wilson, "The Trend of Aircraft Engine Development," *The Journal of the American Society of Naval Engineers*, February 1926, 130–143, Hathi Trust Digital Library; "Annual Report of the Bureau of Aeronautics to the Secretary of the Navy, Fiscal Year 1923," 26; "Annual Reports of the Navy Department for the Fiscal Year (Including Operations to November 15, 1922) 1922," (Washington, D.C.: GPO, 1923), 395, 397, U.S. Congress, House of Representatives, 67th Cong., 4th sess., House Document No. 611, ProQuest Congressional Documents, Pentagon Library; "Annual Reports of the Navy Department for Fiscal Year (Including Operations and Recommendations to November 15, 1923) 1923," (Washington, D.C.: GPO, 1924), 552, U.S. Congress, House of Representatives, 68th Cong., 1st sess., House of Representatives Document No. 150, ProQuest Congressional Documents, Pentagon Library; "Annual Reports of the Navy Department for the Fiscal Year 1920 (Including Operations and Recommendations to December

1, 1920)," (Washington, D.C.: GPO, 1921), 699, Hathi Trust Digital Library.

139. Grossnick, *U.S. Naval Aviation, 1910–1995*, 50; "Navy Aircraft Equipment Tests," *Aircraft Journal*, 27 September 1920, 13, Hathi Trust Digital Library; Timothy Moy, *War Machines: Transforming Technologies in the U.S. Military, 1920–1940* (College Station, TX: Texas A&M University Press, 2001), 84, 85, 86; Don Sherman, "The Secret Weapon," *Air & Space Magazine*, February/March 1995 [no page numbers] smithsonianmag.com.
140. H. R. Keller, "U.S.S. Langley, Our First Aircraft Carrier," *Journal of the American Society of Naval Engineers*, August 1923, 500, Hathi Trust Digital Library.
141. "Annual Report of the Chief the Bureau of Construction and Repair for the Secretary of the Navy for the Fiscal Year 1920," 13–14; "U.S. Congress, House, *Naval Appropriation Bill. Naval Aviation*, Hearings the Committee on Naval Affairs, 66th Cong., 2d sess., 5 February 1920, 1539, Hathi Trust Digital Library; "Annual Reports of the Navy Department for Fiscal Year (Including Operations and Recommendations to December 1, 1920) 1920)," (Washington, D.C.: GPO, 1921), 6, 44, 704, U.S. Congress, House of Representatives, 66th Cong., 3rd sess., House of Representatives Document No. 994, ProQuest Congressional Documents, Pentagon Library.
142. "Annual Reports of the Navy Department for Fiscal Year (Including Operations to November 15, 1921) 1921," (Washington, D.C.: GPO, 1921), 51, U.S. Congress, House of Representatives, 67th Cong., 2nd sess., House Document No. 224, ProQuest Congressional Documents, Pentagon Library; "Annual Reports of the Navy Department for the Fiscal Year (Including Operations to November 15, 1922) 1922," (Washington, D.C.: GPO, 1923), 365–369, U.S. Congress, House of Representatives, 67th Cong., 4th sess., House Document No. 611, ProQuest Congressional Documents, Pentagon Library; Trimble, "The Naval Aircraft Factory, The American Aviation Industry, and Government Competition, 1919–1928," 186–188; "Annual Reports of the Department of the Navy for the Fiscal Year (Including Operations to November 15, 1924) 1924," (Washington, D.C.: GPO, 1925), 630, U.S. Congress, House of Representatives, 68th Cong., 2nd sess., House Document No. 555,

ProQuest Congressional Documents, Pentagon Library; "Annual Report of the Chief of the Bureau of Aeronautics to the Secretary of the Navy, Fiscal Year 1926," (Washington, D.C.: GPO, 1926), 41, Hathi Trust Digital Library; Charles G. McCord, "The Aeronautical Engine Laboratory, Naval Aircraft Factory, Philadelphia, Penna.," *Journal of the American Society of Naval Engineers*, May 1925, 275, 276, Hathi Trust Digital Library.

143. "Annual Reports of the Navy Department for Fiscal Year (Including Operations to November 15, 1921) 1921," (Washington, D.C.: GPO, 1921), 51, 56, U.S. Congress, House of Representatives, 67th Cong., 2nd sess., House Document No. 224, ProQuest Congressional Documents, Pentagon Library; "Annual Reports of the Navy Department for the Fiscal Year 1922 (Including Operations to November 15, 1922)," 384; "Annual Report of the Bureau of Construction and Repair for the Fiscal Year Ending June 30, 1922)," (Washington, D.C.: GPO, 1922), 3, Hathi Trust Digital Library; Trimble, "The Naval Aircraft Factory, The American Aviation Industry, and Government Competition, 1919–1928," 178–180; "Navy to use Metal Seaplanes," *U.S. Naval Institute Proceedings,*" July 1922, 1218, Hathi Trust Digital Library; William A. Moffett, "Naval Aviation," *Aviation and Aircraft Journal*, 19 December 1921, 715, Hathi Trust Digital Library; "Glenn L. Martin Company [to use duralumin in Navy seaplanes]," *U.S. Air Service*, May 1922, 31, Hathi Trust Digital Library; "New Naval Plane Spots Ship Firing," *New York Times*, 27 November 1922, 12, ProQuest Historical Newspapers, Pentagon Library; "Battleship Mast Ten Thousand Feet High," *U.S. Air Service*, May 1923, 26; Hathi Trust Digital Library; "Spotting Planes to be Placed in Service," *Aviation*, 19 November 1923, 638, Hathi Digital Trust Library; "The Navy Spotting Plane," *Aerial Age*, January 1923, 31, Hathi Trust Digital Library; Van Fleet, and Armstrong, *U.S. Naval Aviation, 1910–1980,* 51, 52.

144. "Annual Report of the Bureau of Aeronautics to the Secretary of the Navy, Fiscal Year 1924," undated, 1 (Washington, D.C.: GPO, 1924), Hathi Trust Digital Library; U.S. Congress, Senate, *Navy Department Appropriation Bill [Fiscal Year] 1925, Hearings Before the Subcommittee of the Committee on Appropriations, U.S. Senate*, 68th Cong., 1st. sess., 29 March 1924,

200, Hathi Trust Digital Library; Keller, "U.S.S. Langley, Our First Aircraft Carrier," *Journal of the American Society of Naval Engineers*, August 1923, 510–511, Hathi Trust Digital Library; Turnbull and Lord, *History of U.S. Naval Aviation,* 206, 207; "1920–1929: Reynolds, *Admiral John H. Towers: The Struggle for Naval Air Supremacy*, 198, 199; Richardson, "Development in Naval Aeronautics," *SAE Transactions*, Vol. 19, PART II, 1924, 674, JSTOR, Pentagon Library; Van Fleet, and Armstrong, *U.S. Naval Aviation, 1910–1980. NAV AIR 00-80P-1,* 39, 51, 59, 66; AP, "U.S. Airplanes Best: Secretary Wilbur and Rear-Admiral Moffett Appear Before Special Committee," *[Baltimore]* ***Sun***, 23 December 1924, 2, ProQuest Historical Newspapers. AACPL, www.aacpl.net.

CHAPTER 5

A Golden Age of Innovation, 1926–1938

Air Force historian Richard P. Hallion emphasizes the enormous technological progress in aviation that had taken place in the interwar years. He observes that it "... was a particularly fruitful one for aircraft design, and major developments occurred in the evolution of aerodynamics, structures, and propulsion.... Generally speaking, thanks to the airframe, structures, and propulsion advances of the 1920s and 1930s, these new aircraft were increasingly able to fly farther and faster, were stronger than their predecessors (and thus more tolerant of battle damage)...."[1] He added that military aircraft during that period "... had higher useful weights (and thus had greater payload—fuel and armament—potential)."[2]

The American military played a critical role in shaping that golden age of aeronautical innovation. In 1928, the National Advisory Committee for Aeronautics (NACA) observed that "Much of the fundamental research of the committee has grown out of requests received from the Army and Navy for the study by the committee of particular problems encountered in the services...."[3] During 1929, the Aeronautical Chamber of Commerce of America lauded the key role of the nation's military in promoting technological innovation in aviation. It emphasized that "The present success of American aircraft engines, new designs of planes, improved instruments and other auxiliaries must be attributed to the

C. J. Gross, *Technological Innovation and the Rise of Aviation, 1903–1941*, Palgrave Studies in the History of Science and Technology,
https://doi.org/10.1007/978-3-031-71108-4_5

effort put forth by the Army and Navy in cooperation with industry in recent years."[4]

For much of the interwar period, the quality of American military aircraft equaled or surpassed that of other major powers. On 26 June 1927, Dr. Joseph S. Ames, chairman of the NACA's main committee, wrote that "American military and naval aircraft as a whole is not excelled by European types. In certain characteristics the latter excel and in others we excel."[5]

There were enormous technological advances in American aviation technology that came to fruition during the 1926–1938 period. Demands for larger, faster and longer-range aircraft which could carry greater payloads produced new generations of sleek and powerful planes. The airframe revolution saw sleek all-metal monoplanes replace box-like wood, wire and cloth biplanes. The innovative ideas behind those technical advances came from a variety of individuals and firms in the U.S. and Europe.[6]

But, claims to U.S. aeronautical superiority became increasingly problematic when other major powers embarked on a destabilizing arms race as the post-World War I international security order began unraveling. In March 1936, the *New York Times* warned about the growing challenge to American technical superiority posed by the rapidly expanding aeronautical research programs of European powers, especially Nazi Germany and Great Britain. That May, the NACA warned that American aerial supremacy was being threatened by recently completed superior research facilities in Germany, France, Italy and Great Britain. It requested $1.367 million from Congress to improve and expand its experimental laboratories at Langley, Virginia.

During that period, senior Air Corps generals pinned their plans for the future on new long-range heavy bombers and unproven theories of strategic bombardment as the pathway to a separate air force. Changing aircraft procurement patterns underscored the growing emphasis on bombers after the Air Corps Act of 1926. Its Fiscal Year 1927 procurement program requested 139 pursuit aircraft and 65 bombers. Some 10 years later, it asked for 78 pursuits and 184 bombardment aircraft.

Meanwhile, BuAer was increasingly focused on developing carrier aviation as a striking arm of the fleet and building long-range patrol aircraft. Both military services' aviation programs continued to incrementally modernize their equipment at a moderate pace because of congressional funding constraints.[7]

Gathering Storms

Buoyed by a booming economy and a relatively tranquil international scene during the 1920s, the U.S. allowed much of its small military establishment to stagnate. Instead, the nation returned to traditional unilateral approaches to advancing the nation's extensive overseas interests and providing for the common defense. But, the global Great Depression that began in 1929, expansionist Japanese militarism and the rise of aggressive totalitarian regimes in Italy and Germany undermined the assumptions upon which American national security had rested.

Despite the gathering storms abroad, most Americans were determined that they should never again risk their nation's blood and treasure in foreign quarrels. That historic pattern was driven by growing disenchantment with the disappointing results of the nation's unprecedented military intervention in Europe during World War I and economic woes at home. Congress passed a series of neutrality acts during the 1930s designed to prevent America from being embroiled in another foreign war. Those popular American attitudes did not change significantly until the Nazis conquered France in 1940.[8]

The defense policies of the U.S. were hardly any better attuned to the emerging dangerous realities of the world situation than public opinion. Until the late 1930s, military planners repeatedly complained about the lack of concrete policy guidance from top civilian officials. Preoccupied with the crisis posed by the Great Depression, Presidents Herbert C. Hoover and Franklin D. Roosevelt (initially) focused their energies on domestic matters. Military planners reverted to traditional American patterns. They stressed the Monroe Doctrine, hemispheric defense, protection of overseas territories, preservation of China's Open Door and defending the Panama Canal. But, the funds required to carry out those missions were simply not available until the U.S. began a modest rearmament program in 1934 which focused on building aircraft and warships. Historians Allan R. Millett and Peter Maslowski observe that "For twenty years the United States relied again on its military potential rather than standing forces."[9]

The Airframe Revolution

The armed forces and the NACA played critical roles in stimulating the airframe revolution which culminated in the development of iconic American aircraft including the legendary Douglas DC-3 airliner and Boeing's B-17 bomber. Replacing the box-like wooden frame, wire and cloth-covered biplanes with open cockpits and external-brace wings, the radically new aircraft designs were sleek low-wing monoplanes with all-metal monocoque airframes constructed with aluminum alloys. Many of their power plants were radial air-cooled designs that burned high-octane aviation gasoline. Their innovative features included low-drag NACA engine cowlings, retractable landing gear, enclosed pressurized cockpits and cabins (on a limited basis) as well as controllable-pitch metal propellers. The airframe revolution rendered the old-fashioned biplanes obsolete. It presaged the huge fleets of deadly aircraft that dominated the skies during World War II.[10]

Despite the public's isolationist mood and meager budgets, the middle years of the interwar period were hardly the dark ages of military aviation portrayed by some pioneering flying officers. The armed forces devoted significant shares of their tight budgets to aviation. While the aircraft and aircraft engine industries continued to rely on the armed forces as their primary customers, commercial aviation began to flourish. Military aviation programs shifted their primary emphasis in the late 1920s from research and development of essentially experimental aircraft to advanced development and procurement of small numbers of standardized aircraft for operational service. That change was initially driven by significantly increased funding for aircraft modernization and expansion authorized by Congress in 1926 (but never fully funded for the Army). Military demand was limited to an estimated 10 firms which received approximately 90% of the business. The pace of modernization and expansion picked up in the mid-1930s in response to the deteriorating international security situation as well as efforts by President Roosevelt to stimulate increased employment during the Great Depression and strengthen the nation's defenses.[11]

The airframe revolution in the U.S. was largely a product of the symbiotic relationships between aircraft manufacturers, the military services, NACA and the emerging airlines. It was aided by the economic boom of the 1920s, newly adopted federal aviation regulations and federal airmail contracts to commercial air carriers. The great expanse of the continental U.S. meant that modern commercial aircraft could carry passengers long distances with significant savings in time. For the first time, civilian transports were profitable and relatively safe for passengers.[12] However, Historian David Courtwright stresses that even with the Douglas DC-3 airliner, many commercial flights were restricted to about 12,000–14,000 feet due to lack of pressurization and could be terribly uncomfortable due to near-constant turbulence encountered while flying through (rather than over) storms.[13]

Development of military and civilian aircraft could be closely interconnected in the work of some manufacturers. Boeing's XP-9 pursuit was a single-seat open cockpit all-metal design begun with company funds in 1928 in response to an Air Corps requirement for a monoplane pursuit aircraft. Although a highly innovative design, only one prototype was built primarily because of poor handling and bad pilot visibility. However, its design spurred Boeing's interest in the civilian market. Begun the same year as the XP-9, the firm's *Model 200 Monomail* mail plane marked a dramatic technological advance. First flown in 1930, it was an all-metal cantilevered (self-supporting with the structure buried under the aerodynamic skin providing a clean appearance with low drag) low-wing monoplane with a streamlined fuselage and retractable landing gear. It began service in 1931 with mail, passenger and express service. Neither aircraft was ever produced in quantity as the era of small, single-engine transports was over, replaced by larger twin-engine types. Boeing used the same basic designs for its military and commercial planes.[14]

Historian John B. Rae concludes that "... the culminating synthesis of the advances in aeronautical technology was achieved by... Douglas Aircraft."[15] The DC-3 was a streamlined twin-engined, low-wing, all-metal monoplane with conventional retractable landing gear, controllable-pitch propellers, engine cowlings and wing flaps coupled with an enclosed cabin and cockpit. Powered by two air-cooled Wright R-1820 *Cyclone* radial engines, generating 1100 hp. apiece, the DC-3A had a top speed of 230 mph and a cruising speed of 207 mph with a range of 1500 miles. Over 16,000 commercial and military copies of the DC-3 were built.[16] Rae stresses that the DC-3 "... was made possible because engine development was keeping pace with aircraft design and by 1935 radial air-cooled engines of 1,000 hp. were available.... The DC-3 was described as the first airplane capable of supporting itself economically as well as aerodynamically."[17]

Military orders subsidized, to a significant extent, the talented and expensive engineering design staffs that were responsible for Douglas and Boeing commercial planes. Those transports also benefitted from the technologies developed by and for the military services and the NACA. Rear Adm Ernest J. King, BuAer's chief from 1933 to 1936, highlighted the critical roles the military had played in fostering those innovative design features. King stressed that such a connection had "... been particularly true since the time that is marked by the appearance of the Douglas transport plane. The preparation for that plane had been going on for years. But in that particular plane were embodied the results of research and development in engines, aerodynamics, and other aeronautical equipment, and they combined all those things in one ship, and made the rest of the transport business obsolete. We made a great any of those developments in the Army and the Navy."[18]

The Air Corps' Materiel Division, working with the oil industry and other government agencies, was a key catalyst in developing high octane, anti-knock aviation fuels during those years. Historian Stephen L. McFarland observes that "It [the Air Corps] was the largest single purchaser of aviation gasoline and of aviation engines. Navy policy was to procure engines and try to match them to the best fuel then available. The Air Corps' approach was to encourage the development of the best fuels possible and then convince manufacturers to develop engines to take advantage of the improved fuels."[19]

French-born and educated mechanical engineer Eugene Jules Houdry provided another partial solution. Following combat service in World War I, he developed processes to convert coal to gasoline. Houdry demonstrated his processes in the U.S. during 1930. By 1942, 14 Houdry fixed-bed catalytic units were producing high-octane aviation fuel for the armed forces. The increased performance meant that Allied planes were better than similar Axis planes by a factor of 15% to 30% in engine power for takeoff and climbing; 25% in payload; 10% in maximum speed; and 12% in operational altitude.[20]

In another key innovative development, the Navy had begun a rapid shift toward all-metal aircraft in the mid-1920s. Rear Adm Moffett stressed in 1926 that "To date the most important application of metal construction in the construction of planes has been the extensive adoption of metal tubular construction to succeed the old standard wood and wire truss for the framing of fuselages."[21] By 1930, the Navy had eliminated the last of its aircraft with wooden framed fuselages. Metal designs replaced wooden wing ribs and tail structures. Experiments with monocoque fuselages and corrosion-resisting steels continued.[22]

Historian Glenn E. Bugos stresses the astonishing speed of the change from wood to metal construction writing that "The airframe revolution came swiftly in the early 1930s, as aircraft took on their sleek, metallic, modern look. In 1927 all major aircraft were made mostly of wood. By 1932 every major new aircraft was built mostly of metal."[23]

The ideas behind those modern designs had come from many different sources in the U.S. and Europe. For example, the first all-metal aircraft had been designed in Europe before World War I. During the late 1920s, NACA engineer Fred E. Weick led the development of engine cowlings that streamlined the airflow over air-cooled radial power plants, reducing drag. That innovation added over 13% to the effective horsepower of radial air-cooled engines. At that time, NACA considered the cowling to be the single most important contribution to aviation technology since World War I.[24]

Daniel Guggenheim, the son of Jewish immigrants from Switzerland, was another key contributor to American aeronautical progress. He became a wealthy mining magnate and philanthropist. Working with then Secretary of Commerce Herbert C. Hoover, he donated $2.5 million in 1926 to launch the "Guggenheim Foundation Fund for the Promotion of Aeronautics." Initially, the fund made gifts to the California Institute of Technology, Stanford University, the University of Michigan and MIT to promote aeronautical education. It also sponsored the creation of aeronautical engineering degree programs at various other universities.[25]

Hungarian-born Theodore von Karman, a world-renowned physicist, mathematician and engineer, was recruited to join the California Institute of Technology by Nobel Prize-winning physicist Robert A. Millikan. The Hungarian's primary task was to advise the school on designing its wind tunnel which was completed in 1928. Von Karman later worked at the school's Guggenheim Aeronautical Laboratory (GALCIT). After he was promoted to its directorship in 1930, GALCIT gained prominence by helping numerous West Coast aviation firms test their aircraft models.[26] Historian Michael H. Gorn observes that, in a few years, von Karman "...turned the fledgling GALCIT aeronautics program into one of the finest in America and a contender for international prominence."[27]

The armed forces remained the primary market for the American aircraft industry throughout the interwar period. While only about 33% of the industry's annual production in the early 1930s were earmarked for the military, it accounted for some 66% of its dollar value. It would soon become almost impossible to bring an increasingly costly new major aircraft design to fruition without direct or indirect financial support from the military. The stock market crash of 1929 and the subsequent Great Depression had coincided with the revolutionary leap in aircraft technologies that had sent aircraft costs soaring. The shift from wood to metal tubular construction had increased engineering costs alone by 48% while the transition to monocoque and semi-monocoque fuselages[28] added another 50%. To remain in business, many companies were forced to obtain huge loans as well as rely on military contracts for increasingly high performance and more expensive planes. Those facts drove the industry to consolidate into fewer, larger firms with costly innovative design staffs.[29] Table 5.1 shows the relative importance of the military and civilian aircraft markets in the U.S. during the 1926–1936 period as measured by unit funding of planes.

Historian Irving B. Holley, Jr. concludes that "Clearly, military sales whether for domestic or exports destinations, constituted the most important element of the [American] aircraft market even during the years of peace from 1928 to 1938."[31] Holley also observes that research and development played a key role in the progress of American military aviation. He stresses that the "... insistent requirement for ever better performance in military aircraft made the military market the most difficult to enter in the technical sense. The need for research and development along with the growing complexity and size of military aircraft made the military market a costly business, a veritable bottomless pit for

Table 5.1 American military and civilian aircraft markets, 1926–1936[30]

Year	*Military units sold*	*Value in thousands*	*Civilian units sold*	*Value in thousands*
1926	552	$6,154	604	$,2716
1928	1,219	$19,066	3,582	$17,194
1930	747	$10,723	1,937	$10,746
1932	593	$10,389	549	$,2,338
1934	437	$8,837	772	$9,958
1936	1,141	$27,836	1,159	$12,380

Table 5.2 American aircraft production, 1926–1938[33]

Year	*Total*	*Military*	*Civil*
1926	1,186	532	654
1927	1,995	621	1,374
1928	4346	1,219	3,127
1929	6,193	677	5,516
1930	3,437	747	2,690
1931	2,800	812	1,988
1932	1,396	593	803
1933	1,324	466	858
1934	1,615	437	1,178
1935	1710	459	1,251
1936	3,010	1,141	1,869
1937	3,773	949	2,824
1938	3,623	1,800	1,823

funds."[32] Table 5.2 shows the distribution of American aircraft production between military and civilian customers during the 1926–1938 period.

Military expenditures accounted for the bulk of the aviation industry's progress during the interwar period. The engineering and scientific staffs which major firms maintained in order to remain technologically competitive were extremely expensive. As noted earlier, the shift from wood to metal construction required a major reimagining of the way firms did business. In addition, competitive economic pressures to win federally-funded airmail contracts aided the emergence of revolutionary commercial aircraft designs.[34] Table 5.3 shows the distribution of military and NACA aviation appropriations during the 1926 through 1938 period.

Exports had become an important element of the American aviation industry's economic performance by the late 1920s. In 1928, the industry sold 162 aircraft overseas valued at $1,759,653, a substantial increase over the 63 planes exported for $848,658 the previous year. Sales of American military aircraft had increased during the next three years and accounted for about 25% of American exports abroad by May 1931. By May 1934, American aircraft and engine firms were heavily involved in selling their wares abroad, especially in South America, Mexico, Turkey, China and the Soviet Union. During 1925–1934, Germany was the seventh largest purchaser of American aviation exports, buying $3,382,000 worth of

Table 5.3 Military and NACA aircraft appropriations, FISCAL years 1926–1938[35]

FISCAL year	*ARMY*	*NAVY*	*NACA*
1926	$18,061,191	$15,150,000	$534,000
1927	$18,256,294	$19,065,288	$513,000
1928	$25,612,494	$20,100,000	$550,000
1929	$33,911,431	$32,189,000	$628,770
1930	$34,910,059	$31,430,000	$1,508,000
1931	$38,892,968	$32,033,211	$1,321,000
1932	$31,859,892	$31,145,000	$1,051,070
1933	$25,673,236	$32,745,420	$920,000
1934	$34,037,769	$21,957,459	$695,000
1935	$30,917,702	$34,842,253	$726,492
1936	$50,287,197	$40,732,310	$1,177,550
1937	$62,607,727	$38,588,270	$2,544,550
1938	$67,308,374	$49,500,000	$1,733,850

them. Violating terms of the Versailles Peace Treaty (which the U.S. never ratified), American aircraft firms supplied Germany with products that their executives acknowledged could be used for military purposes.[36]

Stretching Its Wings: Treaty Era Naval Aviation

Notwithstanding the airframe revolution and its enormous impact on American aviation, the fleet, especially its proud battleships, remained the nation's first line of defense. Since World War I, the Navy had navigated a treacherous course between treaty limits on its strength and inadequate budgets. It also weathered the internal storms created by integrating two revolutionary weapons, submarines and aircraft, into the fleet. Aircraft played an increasingly important role in going to sea with the fleet on its giant new aircraft carriers USS *Saratoga* and USS *Lexington*. They were commissioned respectively on 16 November 1927 and 14 December 1927. By late 1928, all of the Navy's battleships and its most modern cruisers were equipped with catapults for launching aircraft (Illustration 5.1).

The U.S. launched modest aircraft and shipbuilding programs during 1933–1934 to begin bridging the gap between its naval forces on hand and those actually required under the "War Plan Orange" scenario for

Illustration 5.1 USS *Lexington* launching aircraft (*Photo* USS *Lexington* launching Martin T4M-1 torpedo planes, circa 1929. Collection of LT Cdr Abraham DeSomer, donated by Myles DeSomer, 1975, NHHC, Catalog #: NH 82117, history.navy.mil)

a war with Japan. They weren't fully funded by Congress until 1938. A major problem facing the sea service was the lack of a defensible base in the Western Pacific. The Marine Corps suggested a solution to that problem by developing plans and techniques for amphibious assaults by "leathernecks" against fortified islands controlled by the Japanese.

Marine Corps aviation units played major roles in plans to seize and control those islands for use as advanced fleet bases. Under the leadership of Col Thomas C. Turner, who led the Marine Corps program during 1921–1925 and again from 1929 until his death in an accident during 1931, the leathernecks built a flying force of over 100 pilots and 1,000 enlisted men operating 12 squadrons early in the 1930s.[37]

The growing personnel demands of aviation in the fleet had exacerbated its manning problems. Vessels assigned to aviation had to be fully manned at the expense of other types of ships.[38] During that challenging period, naval aviation received some important political top cover. MIT professor, Dr. Edward P. Warner, became the first Assistant Secretary of the Navy for Aeronautics in July 1926.[39] Historian Roger D. Launius writes that "In this position, Warner shaped the course of naval aviation, bringing to fruition plans that had been gestating for several years for the development of ship-based air power. During his tenure, the aircraft carriers *Lexington* and *Saratoga* entered service. Warner was tireless in his quest for additional funding for naval aviation, expanding facilities, personnel, and operations."[40]

Although reporting to Warner, Rear Adm Moffett continued to be the dominant force in the development of naval aviation. His BuAer staff remained in close contact with the design staffs of industry, the Air Corps, NACA and other government agencies involved in aviation's advancement. BuAer relied on NACA for almost all of its basic research by the early 1930s, especially aerodynamics.

Normally, naval aircraft were procured through competitive bidding instead of negotiation because of congressional legislation. Plane construction contracts were allocated in the late 1920s to enable a few key manufacturers like Boeing to maintain reasonably active and continuous production levels. That policy was made possible by the five-year modernization program which enabled private firms to retain skilled and innovated engineering staffs on their payrolls. BuAer believed that the policy also insured that the Navy could obtain the quality products at a reasonable cost.

The NAF also conducted experimental, repair and overhaul work for the Navy. While it didn't build production scale quantities of aircraft, the organization continued to design and construct a few experimental prototypes while carrying on unspecified confidential experiments and serving as a general storehouse for aviation materiel. In April 1934, Representative Vinson endorsed a proposal to expand the NAF's activities to enable it to establish yardsticks for determining the cost of aircraft and accessories.[41]

BuAer aided the transition to metal aircraft by financing fundamental research in the properties of materials and the science of strength of materials. Those projects advanced the general field of thin sheet metal use in aircraft. The Bureau also pioneered advanced development work on monocoque fuselage structures. Experimental deliveries of aircraft featuring them began arriving at the sea service during 1929. In 1930, the Navy reported the phasing out of the last wooden aircraft fuselages stressing that they had been replaced by metal construction in new designs.[42]

During 1932, BuAer issued contracts for nine experimental designs in various classes. It also contracted for significant modifications of six experimental aircraft that had already been delivered to the Navy. Following successful trials of experimental models, the Navy ordered its first planes designed to serve as dive bombers. On 18 November 1934, Northrop received a contract from the Navy to develop the XBT-1 dive bombing and torpedo aircraft. Featuring innovative perforated flaps on its wings that eliminated tail buffeting during dives, it was the initial prototype of a series of aircraft that led to the SBD *Dauntless* series of dive bombers which served in the fleet starting in 1938 through World War II.[43]

The armed forces and industry continued to experience their share of aircraft failures. For example, Boeing delivered its Model 205 fighter, the first monoplane designed for carrier use, to NAS Anacostia for testing during February 1930. Later designated the XF5B-1 fighter by the Navy, the single-seat plane was built around a semi-monocoque fuselage and parasol monoplane metal wing. It was equipped with a Pratt & Whitney SR-1340C engine. The aircraft was purchased for $50,000. It was tested for three years as an experimental plane but never accepted for production possibly due to its poor rate of climb and high landing speed. A similar model was rejected by the Air Corps. Begun in 1929 as a private venture, the aircraft nevertheless marked an important milestone in the technological evolution of carrier borne fighters.[44]

Change in American naval aviation wasn't limited to aircraft. Congress authorized the construction of another aircraft carrier on 13 February 1929. Because of its extremely limited experience with those vessels, preliminary design work on USS *Ranger* was largely based on war games at the Naval War College. That began to change as the Navy gained more experience with its first three carriers. The other major factor influencing its design was the Washington Naval Treaty which limited America's carrier displacement to 70,000 tons after USS *Lexington* and USS *Saratoga* were taken into account. Since war game modeling had suggested that there would be severe losses of aircraft and hulls in wartime, the sea service selected a design option that could enable it to build up to five 13,500 ton vessels. Engineers designed it in such a way that it could embark 76 aircraft, nearly equal to those on the USS *Lexington* but with only 50% of the latter's displacement.[45]

Moffett won approval for several new 20,000-ton aircraft carriers after studies demonstrated that the 13,800-ton USS *Ranger* was too small to provide the optimum mix of speed, armor protection, hangar and flight deck space as well as an adequate aircraft complement. Because of improvements in dive bombing aircraft, torpedo-armed aircraft were omitted from the newest carrier in favor of additional fighter-bomber types.[46]

Although Congress initially refused to appropriate money for the proposed new flattops, President Roosevelt, a former assistant Navy Secretary with a soft spot in his heart for the sea service, allocated $238 million from the National Industrial Recovery Act (NIRA) for several warships including two aircraft carriers on 16 June 1933 after Navy Secretary Claude A. Swanson requested them. The flattops became USS *Yorktown* and USS *Enterprise*. Those aircraft carriers were built as part of Roosevelt's public works program to help alleviate unemployment in the Great Depression. That episode also underscored Roosevelt's adroit political maneuvering to start rearming the U.S. in the face of stiff pacifist and anti-interventionist opposition.[47]

Meanwhile, American naval aviation had been able to achieve the goal established in the Naval Expansion Act of 1926 of acquiring 1,000 new aircraft by the end of 1931 due in no small part to Moffett's finely honed bureaucratic and political skills. Unlike the Army, Moffett and the Navy were able to win the congressional appropriations needed to achieve their ambitious aircraft procurement goals during a period of constrained budgets. In June 1931, Moffett announced that the Navy had completed its five-year aircraft expansion and modernization program in four years for $64,771,600 instead of its original $85,078,750 estimate.[48] Historian Craig F. Morris stresses that "Not only did the Navy's program compete for research and acquisition [i.e., development and procurement] funds, but it had an operational side effect that threatened long-range bomber production for the Army."[49] Table 5.4 shows the amounts authorized and appropriated as well as funding reductions for naval aircraft purchases for Fiscal Years 1927 through 1931.

Following the tragic death of Moffett during the destruction of the rigid airship USS *Akron* in a thunderstorm off the New Jersey coast on 4 April 1933, CAPT Earnest J. King was picked to replace him. King, a tough, talented and hard driving careerist, had worked diligently to win

Table 5.4 Naval aircraft procurement funding, FISCAL Years 1927–1931[50]

Fiscal year	*Authorized*	*Appropriated*	*Reduction*
1927	$12,285,000	$9,062,500	$3,222,500
1928	$16,223,750	$13,557,000	$2,666,750
1929	$17,582,700	$16,385,000	$1,197,500
1930	$18,941,250	$13,441,000	$5,500,250
1931	$20,046,250	$12,326,100	$7,720,150
Totals	$85,078,750	$64,771,600	$20,307,150

that key assignment. President Roosevelt, not wishing to ignore the naval high command's wishes despite the strong support within the naval aviation community for pioneer flyer CAPT John H Towers who had been groomed by Moffett to be his successor, approved that choice as did the U.S. Senate.

King was an extremely bright and efficient officer as well as a strict disciplinarian. He also had a reputation for being haughty and arrogant. A 1901 Naval Academy graduate, King had pursued a conventional naval career before switching to aviation and winning his gold wings in 1927. But, he lacked political support and did not have a strong following in the Navy. Unlike Moffett and Towers, he had little experience in lobbying Congress and industrial leaders to win appropriations. The accession of the pro-Navy Roosevelt to the presidency with his allocations of funds from the NIRA helped compensate for King's shortcomings.

Amid the governmental spending reductions initially imposed by the Roosevelt administration because of the Great Depression, King's most formidable early challenges included finding money to purchase planes for the USS *Ranger*, then under construction, while funding rigorous aeronautical research and development programs. The existing congressionally authorized 1,000 aircraft requirement was based on the needs of the Navy in 1926; however, it didn't take into account the requirements of the new vessels added or authorized for the fleet by 1934. The latter were commissioned as USS *Yorktown* in September 1937 and USS *Enterprise* in the following May. They represented a shift from the "minimal" carrier concept embodied in USS *Ranger* toward larger vessels equipped with more powerful engines, larger aircraft complements and a revival of the torpedo-plane carrying capability. In November 1933, the Roosevelt administration allotted the Navy $7.5 million in NIRA funds to purchase naval aircraft and equipment enabling BuAer to maintain its revised aircraft program (Illustration 5.2).[51]

King implemented several key policy changes after he took BuAer's helm. He accelerated the development of land-based long-range fixed wing aircraft. BuAer's Chief also encouraged the transition to more agile monoplane fighters and bombers while persuading Congress to adopt a major expansion program for training additional naval aviators to provide aircrews for the new carriers. King continued to support torpedo bombers despite their slowness and lack of maneuverability. He persuaded his superiors to award Douglas Aircraft a contract in June 1934 for the all-metal XTBD-1 prototype which became the TBD-1 *Devastator* torpedo bomber when it joined the fleet in 1937. Due to the marginal benefits of

Illustration 5.2 *Photo* Rear Adm King, BuAer's Chief, arrives on board USS *Lexington* in his new Curtiss SOC-1 aircraft, 2 June 1936. Official U.S. Navy Photograph, now in the collections of the National Archives. Catalog #: 80-G-457421, NHHC, history.navy.mil

so many new aeronautical technologies arriving so rapidly, King arranged to purchase smaller numbers of each new plane except Consolidated's PBY *Catalina*, a twin-engine long-range patrol flying boat.[52]

King acknowledged that airplanes embarked on naval vessels were relatively more expensive than their comparable land-based ones. The Navy had to equip its carrier aircraft with costly landing and arresting hook gear, flotation gear and emergency life rafts. Their fuselages had to be designed to take the shocks of carrier landings and catapult launchings. Flying boat hulls were much more expensive than those of comparable large landplanes. Controlling salt water corrosion involved costly fabrication methods and materials. The cost of experimental airplanes was high. They were handcrafted and required an enormous amount of expensive engineering work. The Navy was paying around $100,000 in the early 1930s for representative experimental one- and two-seat airplanes. Despite that high price tag, contractors were losing money on them.[53]

King explained to Congress in February 1934 that "With respect to cost to the Government and profit to the contractor, an analysis of the audits of the aircraft industry indicates an average profit of 8% on their naval aircraft production and a loss of 34% on their experimental aircraft. This is an average operating profit of 4.3% on total Navy business; that is, before deducting State and Federal income taxes."[54] He noted that those figures were based upon audits of Navy airplane contractors from 1927 to 1933. King added that the data did not include the cost of engines, radios and gunnery equipment. The Navy furnished those items to contractors. He added that there were only two engine companies in the U.S. that produced engines of the size and power required for military aircraft. One was the Wright Aeronautical Company of Paterson New Jersey. The other was the Pratt & Whitney Aircraft Company of Hartford, Connecticut.[55]

The armed forces deemed it necessary to have at least two sources of supply for each type and class of engine. BuAer worked hard to realize that goal and coordinated its efforts with the Army. The engine contracts of both services were written so that each received the benefit of price reductions obtained by the other. Starting in 1931, Army and Navy engine awards contained a provision that, if either service let a contract for a specific power plant needed by the other and got it at a lower price, then the latter could get it at that same price.[56] Both services purchased and provided engines used by their aircraft manufacturers.[57]

Naval aviation leaders had cooperated with the Air Corps on intelligence, training, experimentation, procurement and maintenance matters since at least the mid-1920s. They exchanged intelligence information from foreign sources. Technical reports produced by one service were furnished to the other. Representatives of both armed forces' aviation organizations met twice a year to standardize airplane fittings and parts and material specifications to help ease procurement problems. When identical material was required by both the Army and the Navy, the two services negotiated who would make a joint purchase for both. Generally, the service with the most expensive requirement took the lead. The Army and Navy had also taken mutual advantage of each other's aviation industry inspection forces. Sometimes, they conducted joint inspections.[58]

An Unbalanced Force: The Army

Compared to the Navy, the Army's ground forces were in terrible shape during the interwar period. The 1920 Army Reorganization Act called for a force of 290,000 but it was never funded enough to support anything close to that number. Between 1926 and 1934, its average active-duty personnel strength hovered between 134,938 and 138,464. War Department expenditures climbed from $364.09 million to $408.59 million during those years. Given those modest resources, the Army was never able to create the massive ground force envisioned in the 1920 legislation. Instead, it tried to maintain as many skeleton units and individual mobilization fillers as possible.[59]

When Gen Douglas MacArthur became Army Chief of Staff in November 1931, he was determined to end the deplorably low readiness of his service's ground forces. He directed the Army to focus its limited resources on training and modernizing the equipment of the most critical portions of the regular establishment and its reserve components. The General Staff began to formulate war plans to deal with the conflicts which it judged would be most probable. It also sought ways to increase the President's mobilization authority free of congressional interference. MacArthur's goal was to replace the World War I-style mass mobilization on paper force with an "Immediate Readiness Force" of 400,000 soldiers that could deploy quickly in a crisis, especially in a war with Japan. He wanted a balanced distribution of resources throughout the Army and strongly opposed Air Corps' pressures for additional funds.[60]

Meanwhile, Maj Gen Patrick's leadership style and strategy had been largely ignored by his first two successors as they fought for a separate air force after he retired in 1927. Air Force historian Roger Miller observes that Mitchell's tactics were adopted by "... James Fechet and, especially Benjamin D. Foulois, whose intemperate claims, constant complaints, and willingness to appeal to Congress and the public outside the chain-of-command alienated the War Department between 1928 and 1935. Initial distrust became outright hostility and Army leaders ceased listening to their own airmen."[61] Historian Rondall R. Rice stresses that Foulois "... began behaving more like Billy Mitchell, even though the two hated each other. Foulois derided the Navy, wrote and submitted legislation for air-friendly congressmen (circumventing the War Department), openly chided the General staff and the War Department, and pursued programs unsupported by the president."[62]

Historian Miller emphasizes that "Between 1928 and 1933, Air Corps leaders consistently sought a disproportionate slice of the budget pie to fund that [rapidly developing aviation] technology.... Between 1920 and 1934, the Air Corps spent between 13.1 and 22.7% of the Army's annual budget. On the average, this branch – which comprised about 11%, or slightly over one-tenth, of the Army – spent 18.2%, almost one-fifth of the annual budget each year."[63] Moreover, the Air Corps' manpower complement prospered compared to the rest of the Army's. Miller added that the "... Five-Year Program required the Army to man the Air Corps fully even at the expense of its other branches. Under the Five-Year Program some 6,240 men transferred to the Air Corps.... As of 1929, these transfers had forced the army to inactivate five battalions of infantry and most of a field artillery regiment."[64] In addition, the Army was forced to transfer enlisted men from several other branches as well as cut its operations and reduce troops levels even more while funding items including research, housing and technical construction that were not specified in the 1926 Air Corps Act.[65]

Scholar Thomas Alexander Hughes points out that, while all that was happening "... airmen had continually carped about lack of support, other branches had actually suffered outsized reductions in the name of Air Corps growth.... the Cavalry had received exactly forty-seven tanks, while more than half of its officers were still on horseback. The Infantry had not purchased a single rifle.... line officers had grown weary of pilot complaints. For many of them, the air arm was terribly spoiled by the Air Corps Act of 1926, and pilots had little conception of the severe economic cuts that had been endured by other branches to sustain their growth."[66]

Along with expanded funding and its upgraded status as the Air Corps, Army aviation acquired a new organization to modernize its weapons inventory. The Materiel Division was a modern research and development center established during October 1926 at Wright Field near Dayton, Ohio. Much of its land was purchased and donated to the federal government by local citizens after they learned that the Engineering Division at McCook Field was going to be closed and its functions moved elsewhere. The new organization's principal duties were engineering, procurement as well as storage, supply, repair and maintenance, salvage and industrial war planning.[67]

The intricate official process for modernizing the Army's aerial weapons after the 1926 legislation featured several, sometimes conflicting, objectives. It sought to remove the capriciousness of command influence associated with the Mitchell era and protect the Air Corps from external challenges (mostly congressional) to its procurement decisions. More importantly, it was determined to assure that the Army's air arm fielded technologically superior and reliable aircraft at a reasonable cost. Lacking either effective national policy or official War Department doctrinal guidance and continuing to rely on small but growing equipment purchases, the Air Corps' basic approach remained to adapt and gradually push forward existing technologies. That incremental approach, which both stimulated and exploited the airframe revolution as well as other technical innovations developed by NACA and the aircraft industry, eventually produced enormous improvements in military aircraft performance. Like BuAer, the Air Corps' air weapons modernization process sought to build a strong technical and industrial base for the nation's military aviation.[68]

Coupled with the accelerated modernization process, the Materiel Division reported a pronounced increase in the performance of new experimental military aircraft because of their better aerodynamic and structural efficiency as well as their improved engines. Those improvements included monoplane designs, retractable landing gears, enclosed cockpits and streamlined monocoque fuselages and a trend toward all-metal aircraft. Examples of the promising new all-metal monoplane designs included Boeing' P-26 pursuit aircraft and Martin's XB-10 bomber prototype. The Materiel Division reported in 1932 that "This general advance in design and operating efficiency has led to the promulgation of new type requirements which exceed by a substantial margin the best performances of present tactical types."[69] A year later, it reported that almost all of the new aircraft research and experimentation sponsored by the Air Corps involved metal monoplanes, a radical departure from its existing tactical aircraft which were all biplanes.[70]

The creation of the Materiel Division was accompanied by a significant shift in priorities. From 1919 to 1926, the Engineering Division had devoted the lion's share of its efforts to research and development. The emphasis now shifted to the procurement, production and testing of standardized aircraft. The major drivers of that switch were increased funding under the five-year program and the beginnings of the airframe revolution. With the shift of the Air Corps' procurement staff from Washington, D.C. to Wright Field, demands for testing standardized production items increasingly absorbed the energies of the Materiel Division's technical staff to ensure that they conformed to established specifications and contract provisions. That diverted scarce engineering talent from research and development work.

Since experimental funds were extremely limited, the Air Corps rarely awarded research and development contracts that defrayed all of a manufacturer's costs. Instead, contracting officers allowed firms to include such expenses as overhead if and when production contracts were awarded, a practice that further encouraged an incremental approach to technological change. Since revolutionary innovations like jet propulsion ordinarily took years and huge amounts of money to perfect, American companies had little incentive to pursue them. Senior military and NACA leaders ignored the potential of jet propulsion suggested by theoretical and practical research in Europe and Great Britain. Meanwhile, the significant progress by the Army and Navy air arms in cooperation with industry in promoting the airframe revolution and much more powerful engines created a false sense of American technological superiority in the early 1930s.[71]

During 1927, Maj Gen James E. Fetchet, Patrick's immediate successor, decided that bomber development was now the Air Corps' highest modernization priority because of the need to replace the obsolete Curtiss-Martin NBS-1 twin-engine bomber designed in the early 1920s. That decision was apparently partially inspired by the growing interest of several major aircraft firms in developing long-range commercial aircraft during the emerging airframe revolution. Their designs were similar to those of the requirements for new long-range bombers. Another factor possibly involved in that decision was the distinctive theories of air warfare that were being developed by influential Air Corps officers in flying units and the Air Corps Tactical School (ACTS) during the 1920s. Previously, pursuits had been the top modernization priority.

Those theories developed at the ACTS were not based on official doctrine, established national defense policy, hard military requirements or the existing technical capabilities of operational weapons. Nor were they moored in practical realities and systematic experimentation as was the Navy's aviation program. Instead, they were grounded in the air power theories of such thinkers as Italian general Giulio Douhet and Marshal of the Royal Air Force Hugh M. Trenchard. While differing widely in detail, pioneer Army air power thinkers agreed that aviation offered a way to overcome the stalemated slaughter experienced on the western front during World War I. Properly employed air power, they reasoned, could bypass surface-bound armies and navies and directly attack the heartlands of enemy nations. The distinctive contribution of Army airmen was the idea that a long-range bombing force could decide a war quickly and relatively cheaply by mounting precision bombing attacks upon an enemy's industrial capacity to wage war.[72]

The shift away from pursuits was gradual. For example, the Fiscal Year 1930 Army procurement program called for 88 pursuits, 142 observation planes, 37 bombers, 8 amphibians, 120 trainers and 15 other aircraft. The price tag for the aircraft, engines and support equipment was $12,531,724.[73]

During the early 1930s, a larger portion of Air Corps research and development funding began to flow into bomber development at the expense of other types of planes. While the emerging strategic bombing doctrine was usually cited as the reason, other factors may have been more important. The service's pilots and engineers doubted the ability of the new breed of sleek all-metal aircraft designs associated with the airframe revolution to withstand the enormous stresses of high-speed combat maneuvering by pursuit aircraft. Bombers didn't require those capabilities. Instead, like the sleek new commercial airliners, their designs focused on long-range and a heavy load carrying capacity. The aviation industry was reluctant to invest consistently in developing costly high performance designs which advanced all-metal pursuit monoplanes required. In their intensely competitive industry with its low-profit margins, bombers and commercial airliners were safer bets. Technological progress in pursuit of aviation was also initially inhibited by the resistance of pilots to innovations like enclosed cockpits.[74]

The B-10 bomber integrated technologies and design innovations that were being applied to commercial airliners. It was an all-metal, cantilever mid-wing monoplane with retractable landing gear and enclosed cockpits (Illustration 5.3). Its top speed of 207 mph and 2,080 mile range represented a dramatic improvement in performance over the Air Corp's Keystone 6A bomber's maximum speed of 121 mph and 825 mile range.[75]

The bomber's prototype made its first flight in January 1932. The results were so poor that the project was nearly canceled. But, the Air Corps was very interested in the plane's excellent speed characteristics. Its additional design breakthroughs included retractable landing gear and a rotating gun turret. Coupled with its advanced streamlined design, internal bomb bay and full engine cowlings, the B-10's twin 775-hp Wright R-1820 *Cyclone* air-cooled engines made the plane 50% faster than contemporary biplane bombers and on par with most pursuit aircraft of the day. The bomber's strange appearance due to its deep belly and

Illustration 5.3 Martin B-10 bomber (*Photo* Martin B-10B bomber in flight. U.S. Air Force photo, Air Force Historical Research Agency (AFHRA), undated, afhra.af.mil)

blunt nose gave rise to its nickname, the "Flying Whale." Its performance convinced Air Corps planners that the bomber could attack strategic targets at long range without fighter escorts. On 27 November 1933, the Army accepted delivery of its initial production model B-10. It was America's first all-metal monoplane bomber produced in quantity. The Air Corps ordered 121 B-10s during 1933–1936. The contract saved Martin from bankruptcy. The Air Corps bought 32 more of them, powered by Pratt & Whitney's 700-hp R-1690 *Hornet* engines, designating them B-12s.[76] The bomber's reported unit cost was $52,083.[77]

During 1931, the Air Corps finally temporarily acquired the mission that justified its costly long-range bomber programs, coastal defense. Responding to presidential pressure, Gen MacArthur and his Navy counterpart, Adm William V. Pratt, temporarily resolved that sensitive interservice issue. They agreed that all land-based air attacks against invading enemy fleets would be the Air Corps' responsibility. While the mission was real enough, the Air Corps used it to justify continued development of long-range bombers that could perform strategic bombing missions.

Coastal defense was a godsend to Air Corps officers who wanted a mission that would help pave the way toward their goal of an independent air force. This mission, coupled with the engineering successes of the B-10, encouraged the Air Corps to offer contracts in 1934 for development of a bomber with a 2,000 mile range. That initiative resulted in the July 1935 debut of Boeing's four-engine Model 299 bomber prototype which evolved into the famed World War II B-17. After viewing the bomber's rollout, *Seattle Times* reporter Dick Williams dubbed it the "*Flying Fortress*" because of its extensive defensive armament. The name stuck. Air Corps officers liked the nickname because it suggested (falsely) that the bomber's real mission was to defend America's coasts. Meanwhile, after Adm Pratt retired in June 1933, his successor, Adm William H. Standley had promptly repudiated the coastal defense agreement on the grounds that it had been a purely personal arrangement.[78]

The Air Corps continued to develop pursuit aircraft during the 1930s as well as bombers. During early 1931, much of the discussion about new pursuit aircraft under development by aircraft manufacturers focused on their quest to equal or surpass 200 mph in top speed while equipping such planes with more powerful engines and heavier armament. Some engineers at the Materiel Division were convinced that future pursuit performance could equal or surpass that of bombers. However, pursuit development continued to lag behind that of bombers until the late 1930s primarily because of funding shortfalls and inadequate engines. Nevertheless, the Air Corps purchased 1,070 pursuit aircraft between 1927 and 1938.[79]

Following a pursuit aircraft flying competition in 1936 involving Seversky, Curtiss, Consolidated and Vought, the last two firms' entrants were eliminated after they failed to meet Air Corps requirements. After review by the Air Corps Pursuit Evaluation Board, Seversky was judged the winner and received a contract on June 16th of that year for 77 aircraft: 76 P-35s and an XP-41. The company had been founded by Russian immigrant and trained engineer Alexander de Seversky in 1923.

With a price tag of $34,900 per aircraft, the contract's total value was $1,636,250. It was the Air Corps' first streamlined production model, single-seat, low-wing, all-metal pursuit aircraft with retractable landing gear and an enclosed cockpit. The aircraft was powered by a Pratt &Whitney R-1830 engine that generated 850 hp. and a maximum speed of 280 mph. The P-35, whose prototype had first flown on 15 August 1935, was an outgrowth of the design of the firm's unsuccessful SEV-3 amphibian. It had a range of 625 miles with an operational ceiling of 30,600 feet. Its manually controlled, multi-position constant speed propeller provided maximum efficiency during takeoff, climb and high-speed flight.

In 1937, the Air Corps awarded Seversky another contract for 210 P-35s priced at $29,000 each. The first production model P-35 was delivered to the service during May of that year. Its innovative design foreshadowed many of the key aircraft features Americans pilots would fly combat missions during World War II.[80]

In addition to aircraft modernization incorporating the airframe revolution, Secretary of War George H. Dern had approved the creation of a General Headquarters (GHQ) Air Force on 11 October 1933. Its mission was to develop and operate an independent aerial strike force. Air Corps officers saw this as a major step toward their cherished goal of a separate air force. Dern had established a board chaired by former Secretary of War Newton Baker to review the condition of Army aviation and make recommendations to improve it. The Baker Board strongly endorsed the GHQ Air Force initiative. Although it had existed on paper for some time, GHQ Air Force wasn't activated until March 1935. Its first commander, veteran Army flyer Col. Frank M. Andrews, was promoted to the temporary rank of brigadier general.[81]

Because Army doctrine continued to emphasize aviation's role in supporting ground troops, there was a continuing, albeit slow modernization of observation and attack aircraft in the Air Corps' inventory from 1926 to 1938. Its leaders downplayed those missions despite the importance placed on them by the General Staff. Furthermore, there remained unresolved disputes within the Army's air arm concerning the desired characteristics of such aircraft. The basic split was between those who advocated battlefield survival through speedy and highly maneuverable single-engine planes and others who advocated larger, heavily armored twin-engine aircraft. The absence of suitable engines and clear employment concepts hindered attack warplane development. Modifying pursuit ships for the attack mission was considered but rejected. Instead, the Air Corps relied on modified observation aircraft to fulfill that mission. An early example was a two-place, open cockpit biplane developed from a modified O-1B observation aircraft by Curtiss at the Air Service's request following a competition with 11 other prototypes in 1924.[82]

Observation and attack aircraft remained the poor relations of the Air Corps' funding and development priorities during the interwar period. Its Materiel Division gave equal emphasis to their key design characteristics including speed, maneuverability, weight-carrying capacity and durability. That approach resulted in planes that exhibited mediocre performance in every one of those areas according to Army historian Edgar F. Raines, Jr. He stresses that "The exclusion of ground officers from preparing military characteristics for such craft for much of the period and the absence of effective large-scale maneuvers before 1939 combined to produce planes ill fitted for the role ground officers envisioned for them. As a consequence, the evolution of the standard type of observation aircraft helped separate observation from its natural constituency, the ground arms, at the same time that institutional and doctrinal developments increasingly isolated observation aviation from the rest of the Air Corps."[83]

During January 1936, the Army approved a requirement for three types of new planes to fulfill the observation mission: Army Reconnaissance, Corps and Division Observation and Observation—Short Range. But, due to funding limitations, it had to concentrate on a single or, at the most, two types of observation aircraft. Based on the recommendations of an Observation Board, the Chief of the Air Corps and the General Staff decided that North American's O-47 aircraft was best suited to perform the first two of the original three mission categories. It was an all-metal cantilevered low-wing monoplane with retractable landing gear, a three-bladed propeller and windows in its deep belly to facilitate visual observation.

In late 1936, Secretary of War Henry H. Woodring announced the award of over $3.429 million to North American for the delivery of O-47 aircraft to the Air Corps that had been developed to replace the existing fleet of Thomas Morse O-19 and Douglas O-38 biplanes. North American had spent $125,000 of its own money to build a prototype that was completed in July 1935. The Air Corps liked the demonstrator and purchased it in January 1936 as the XO-47.

Planners originally assumed that its machine guns would enable the aircraft to survive over the battlefield even if the enemy enjoyed air superiority. But, the aircraft was large, heavy and difficult to maintain. It needed a 1,200-foot takeoff roll on a sod airfield. Once airborne, it maneuvered sluggishly. The aircraft's top speed increased its aircrew's difficulty in conducting effective observation missions. At some 100 mph slower than the latest American pursuits, Air Corps flyers were convinced that the O-47 was extremely vulnerable to enemy fire.

Training exercises in 1941 highlighted the aircraft's glaring deficiencies. It was relegated to target towing or coastal patrol and anti-submarine operations during World War II. Light aircraft, pursuit planes and twin-engine bombers proved to be more capable of performing its original proposed observation missions.[84] Historian Holley concludes that "On the eve of World War II... strategic plans called for 37% of the available [aircraft] productive capacity to be reserved for observation aircraft. When the war finally arrived, this whole class of aircraft proved unusable and was abandoned as a separate type."[85] In retrospect, that should have been unsurprising given the very limited attention devoted to the proper development of observation aircraft and tactics throughout the interwar period. Table 5.5 documents the annual numbers of pursuit, bomber and observation aircraft accepted by the Army's air arm from 1920 through 1937.

Observation was a cavalry mission that the rise of Army aviation significantly impacted during the interwar years. Historian Lori Henning stresses that "Cavalrymen accepted aviation's assumption of the cavalry's more burdensome duties and worked on creating the best possible cooperative relationship with air forces during the 1930s. Even as the cavalry tried to fully integrate aviation and create organic aerial capabilities, the 1930s proved to be a decade with new aviation challenges and

Table 5.5 Army aircraft acceptances, 1920–1937[86]

Year	*Pursuit*	*Bomber*	*Observation*
1920	112	20	1000
1921	200	85	270
1922	60	25	200
1923	–	–	180
1924	35	–	127
1925	18	2	126
1926	83	1	176
1927	49	9	56
1928	69	34	151
1929	78	22	185
1930	103	30	122
1931	143	73	233
1932	207	68	136
1933	6	13	44
1934	113	38	27
1935	7	10	24
1936	23	89	3
1937	3	32	88

ground threats that targeted the branch's foundation: its identity as horse-mounted soldiers."[87] She adds that "The cavalry continued to express a desire to work with aviation, a desire increasingly refused by more and more politically powerful aviation supporters. When their requests failed to produce the desired results, cavalrymen attempted to gain control of their own aircraft units and aircraft. The United States... saw cavalry trying, albeit unsuccessfully, to establish organic air arms and promote the new technology of the Autogiro."[88]

Horse soldiers weren't the only Army professionals who were interested in autogiros. Despite the growing vulnerabilities of aircraft like the O-47, the service's General Staff hadn't given up on using specialized aircraft to perform the short-range observation mission. For that purpose, ground officers turned to autogiros despite Air Corps warnings that they were technologically problematic. During the interwar period, aeronautical engineers attempted to design and build aircraft able to make short takeoffs and landings. Those efforts eventually produced useful helicopters, but too late for wide American employment during World War II. Like helicopters, autogiros used a rotary wing to produce lift. But, their engines didn't power the autogiro's rotor. Instead, aerodynamic forces made the autogiro rotor spin, while its engine propelled the plane. The development of rotary-wing aircraft appeared to provide one solution to the observation problem. Pressured by Army ground arms officers and reform-minded congressmen, the Air Corps agreed to experiment with autogiros.[89]

Invented by Spanish engineer Don Juan de la Cierva and first flown in 1923, the autogiro was built around a conventional aircraft fuselage powered by a tractor engine, with rotary wings substituting for traditional fixed wings. The flow of air past the blades caused them to rotate generating lift. American entrepreneur Harold Pitcairn, impressed by autogiro flight demonstrations in Europe, partnered with Cievra to build the first American versions of the ship in 1929, the PC-1 and PCA-1. In 1931, President Hoover presented the Pitcairn-Cierva Autogiro Company with the National Aeronautic Association's prestigious Collier Trophy[90] for aeronautical achievement after the PCA-1 landed on the White House lawn to help demonstrate its capabilities.

The Navy ordered the first of three autogiros from Pitcairn-Cierva on 22 January 1931. Delivered to the Flight Test Section at NAS Anacostia, D.C., the autogiro soon was put through its paces by veteran Navy test pilots, the goal being to evaluate the possibilities of their use for reconnaissance work over terrain where adequate landing fields didn't exist. BuAer announced in August 1931 that the autogiro was finally nearing the end of its test status after a very large amount of flying compared to a normal program. On 23 September 1931, LT Alfred M. Pride piloted an autogiro in three successful landings and takeoffs on USS *Langley* while the aircraft carrier was underway. After that, there was little mention of the aircraft in Navy sources, though the Marines field tested the type during operations in Nicaragua.

Meanwhile, W. Wallace Kellett established his own autogiro firm to compete with Pitcairn-Cierva. Kellett manufactured 12 K-2 autogiros during 1931. Improving on the existing Pitcairn-Cierva design, his K-2 had a much larger blade area, a simplified landing gear and a wider fuselage to accommodate side-by-side seating. Powered by a 165 hp Continental A-70 engine, the machine could carry a useful load of 609 pounds at a top speed of 100 mph.

In 1932, the Kellett's firm produced an improved model, the K-3. Powered by a 210-hp Kinner C-5 engine, it had a top speed of 110 mph. Kellett produced six of those aircraft. The Air Corps tested both versions of the Kellett autogiros at Wright Field, but they failed to meet its desired performance characteristics.

During the summer of 1934, Maj Gen Foulois, now the Air Corps Chief, was informed of President Roosevelt's interest in autogiros and Secretary of War Dern's approval of testing them for observation missions at the Army's combat arms schools. Senior cavalry officers had continued to actively campaign for those planes. Finally, the Cavalry School at Fort Riley, Kansas received an autogiro for testing in May 1937. In 1938, the House Committee on Military Affairs approved $2 million for autogiro for research, development and experimental procurement at the behest of the Office of the Chief of Cavalry. The ensuing tests during 1938–1939 produced mixed results. During late 1939, an Army board convened at Fort Knox, Kentucky at Roosevelt's request to test and compare the YG-1B autogiro and the O-47 observation plane strongly argued against the former. Instead, it recommended developing light, slow aircraft for such missions.

Historian Raines writes that "... the almost decade-long tests revealed what Air Corps engineers had suspected all along: the autogiro at its current stage of development lacked sufficient lift to carry aloft a pilot, an observer, and a service radio, which made it impracticable for field use...."[91] He adds that "By demonstrating the feasibility of short takeoffs and landings on unprepared fields close to the front, autogiros increased the constituency in the ground arms for an aircraft of more conventional design but with similar capabilities. Second, they stimulated Congress to appropriate funds for further development of rotary-wing aircraft. The legislative language was broad enough to encompass helicopters as well as autogiros."[92]

Historian Henning concludes that "Despite its potential, the autogiro was a failed technology in the United States because it was never widely adopted by civilian or military organizations. It did not become a permanent part of the America cavalry, or the army in general, because of financial and operational considerations. Another hindrance to autogiro development was the outbreak of World War II, because design developments had to focus on aircraft immediately available for action."[93]

Mr. Vinson's Navy

Primarily driven through Congress by Representative Vinson, the powerful chairman of the House Naval Affairs Committee, the Vinson-Trammel Act, authorized the U.S. to build up its naval forces to the limitations imposed by the 1922 Washington as well as the 1930 and 1935 London treaties. The legislation, drafted by the Navy Department and pushed through the upper chamber of Congress by Senator Alexander Trammel, was strongly supported by President Roosevelt and signed by him on 27 March 1934. It authorized construction of a new aircraft carrier to replace the venerable USS *Langley* without limiting the number of aircraft the Navy could obtain to bring it up to "treaty" strength. At Rear Adm Moffett's request, Vinson's committee amended the bill to authorize the purchase of aircraft numbers commensurate with the needs of the treaty sea service navy. While the act authorized the Navy to build 70 new warships and 730 more naval aircraft, Congress did not rush to fund them.

Although its Capitol Hill opponents were unable to scuttle the legislation, they added two provisions that significantly curbed naval aircraft procurement. First, aircraft manufacturers were limited to 10% profits on the planes, engines and aircraft accessories they produced. A firm was not allowed to make up for losses on a sample plane contract with excess profits on its production contract for that same type of aircraft. Manufacturers found those provisions unacceptable. The Aeronautical Chamber of Commerce formally protested the provision to the House Naval Affairs Committee on the grounds that it weakened the industry, discouraged innovation, increased costs and was difficult to administer but to no avail. Earlier, the profit policy had been 15%. Boeing stopped designing and producing naval aircraft in favor of Army awards which had no profit limits. The act transferred responsibility for determining profits on naval aircraft contracts to the Commissioner of Internal Revenue (Illustration 5.4).

Congress took until April 1923 to pass a similar law for the Army, establishing a 12% profits limit. That same limit was also applied to naval aircraft. Unlike the Vinson-Trammel Act, firms were allowed to average their profits and losses over five years instead of a single one.

The other restrictive legislative clause in the 1934 legislation had required that 10% of all naval aircraft and engines be built in government factories. The NAF in Philadelphia was the only government facility

Illustration 5.4 President Roosevelt signs 1934 Vinson-Trammel ACT (*Photo* President Franklin D. Roosevelt (seated) signs the Vinson-Trammel Act at the White House on 27 March 1934. Standing behind him (left to right) are Representative Carl Vinson, Assistant Secretary of the Navy Henry L. Roosevelt and Representative Fred Britton. Catalog #: NH 973, copyright owner, NHHC, his tory.navy.mil)

capable of building aircraft; however, it had stopped doing that type of work years earlier with the exception of a few experimental planes.

The NAF had not produced engines previously and had to gear up to perform that highly specialized function. It began preparing for that responsibility by building proven engines under license to the Wright Aeronautical Corporation for the latter's seven and nine-cylinder "Whirlwind" power plants.

While the legislation permitted the President to suspend production limits in emergencies, the practical effect was to curtail the number of private firms that could produce naval aircraft in peacetime. Roosevelt allotted nearly $3 million to the Navy in August 1934 to expand the NAF's capability to construct the 10% of the new naval aircraft and engines mandated by the legislation. The NAF began manufacturing engines during Fiscal Year 1937. By January 1940, it was building approximately 10% of the Navy's new aircraft and engines annually. Its manufacturing portfolio had expanded to include scouting-observation aircraft as well as training planes.[94]

The Navy was able to pursue an ambitious aviation modernization program because of technological progress and the Vinson-Trammel Act. The legislation authorized the composition of the fleet at the limits prescribed by the Washington and London naval treaties including one new aircraft carrier of about 15,000 tons. Under that authorization, USS *Wasp* was laid down in 1936 and commissioned on 25 April 1940. Historian Curtis A. Utz stresses that "Improvements in aircraft technology proved to be the most important changes which occurred in the carrier aviation forces during the mid and the late 1930s. The development of monoplane airframes with retractable landing gear led to more streamlined design[s]."[95] More importantly, the modernization of the internal combustion aircraft engine, including the radial engines used in carrier planes, greatly increased the capabilities of all planes.[96] The greater lifting capabilities of those aircraft enabled them to carry more ordnance, greater fuel loads, improved radios and better armor.[97]

From Fiasco to Team Player: Army Aviation

Army aviation, propelled by the airframe revolution, more powerful engines and increased (but inadequate) congressional appropriations made significant gains in aircraft performance during the late 1920s and early 1930s. But, primarily because of budget constraints, it did not achieve the goals established for it in the five-year expansion program approved by Congress during 1926. Nevertheless, it had played a significant role in developing aircraft accessories including navigation instruments, engine superchargers, autopilots and controllable-pitch propellers. All-metal monocoque fuselage monoplanes replaced wood, cloth and wire biplanes. During the early 1930s, the Air Corps speeded up its method of acquiring aircraft by approximately a year which previously, depending on the type of plane being developed, might take three to five years. That was accomplished by placing more money in the initial stage and virtually eliminating the experimental test stage from what had been a three-stage process (experimental, service testing and production).

The Air Corps launched another modest modernization and expansion program in the mid-1930s backed by President Roosevelt. However, first its leaders had to deal with a serious fiasco that damaged its reputation, called into question its capabilities and ended the long military career of pioneer flyer Maj Gen Benjamin Foulois, then the Air Corps' chief. Based upon companion Senate and Post Office Department probes, Roosevelt concluded that federal airmail contracts had been arranged through collusion and fraud. Growing public and congressional indignation about that situation forced him to cancel existing contracts. Maj Gen Foulois readily agreed to have the Air Corps fly the mails instead of commercial airlines. That mission had been first suggested for it by Roosevelt. During February 1934, the Air Corps was officially tasked, with 10 days notice, to carry the mail. Foulois had neglected to consult either with his military superior, Gen MacArthur, or his own principal operational deputy, Col Carl A. Spaatz, before accepting the mission. The result was a bloody fiasco. Air Corps pilots were ill-trained and poorly equipped for night instrument flying. Most of the line unit flyers were young, inexperienced lieutenants who had recently completed flight school and were getting several years of flying time before returning to civilian life as reservists. Complicating matters, the winter of 1934 was the worst on record. After numerous crashes and 12 deaths, the President relieved the Air Corps of its airmail assignment and moved to negotiate new airmail contacts with commercial airlines.[98]

Congress was outraged. After hearings on the issue, the House Military Affairs Committee erroneously concluded that Foulois had been personally responsible for the non-competitive nature of Air Corps' procurement practices and that had caused the airmail debacle. Actually, a shortage of procurement funds, not poor management or a lack of personal integrity by Foulois, was largely responsible for the Air Corps' poor airmail performance.

Nevertheless, Foulois should have known that his equipment was inadequate for the task. Moreover, he apparently failed to realize (or ignored) the fact that his pilots were unprepared for the dangerous challenges they would face flying the mails at night and in poor weather. Amid unfounded rumors of excessive profits by aircraft manufacturers, a congressional subcommittee began an investigation of the War Department's aircraft contracting practices. The subcommittee ignored audited evidence provided by the Air Corps that its aircraft manufacturers' profits were generally low while reiterating the standard and mistaken congressional explanations of troubles in military aircraft production programs: conspiracy, collusion, corruption, profiteering and the violation of free competition.

Apparently inspired by the Air Corps' airmail fiasco, a subcommittee of the House Naval Affairs Committee reached an opposite conclusion for BuAer's aircraft procurement practices. It found that manufacturers' profits on naval aircraft had been "moderate and reasonable." In its estimation, the NAF's operations had provided a brake on excessive private profits and should be employed in the future to fill service needs and provide experimental aircraft for the Navy. Rather than collusion to award naval aircraft contracts to members of a small group of select firms in an "Aircraft Trust," the subcommittee concluded that there was intense competition between rival manufacturers to win such awards. It rejected the idea that naval officers had acted improperly in awarding aircraft contracts.

In some respects, Foulois had been his own worst enemy even before the airmail crisis. His constant complaints, intemperate charges and willingness to go outside official channels to appeal to Congress had already undermined his standing in much of the federal government. As a result of the airmail crisis, he was branded incompetent and forced to retire in December 1935.[99]

Like Mitchell before him, Foulois had misread the political environment. Historian Rice stresses that each of them "... failed to temper his desire for air independence with an understanding of the political and public landscape and, more importantly, failed to work within the programs of the presidents who drove those policies. Roosevelt, like Harding and Coolidge before him, did not want an independent air service and did not like high-ranking air officers pushing that end against presidential wishes."[100]

Under a new group of senior Air Corps leaders including Henry H. Arnold, Army aviators learned to become team players who worked within established institutional norms to develop air power by fostering technological innovation and displaying military aviation's growing utility without abandoning their ultimate goal of establishing a separate air force. President Roosevelt and Secretary of War Dern had selected Brig Gen Oscar Westover to be Foulois' replacement largely because he was considered to be a competent leader who wasn't a pro-air independence firebrand. Westover had opposed air independence since 1919. He was determined to cooperate with the General Staff and the Roosevelt administration.

Meanwhile, two other investigative committees within the executive branch disagreed with the conclusions of the House Military Affairs Committee. As noted earlier, Roosevelt had appointed a special panel in April 1934 headed by Newton D. Baker to investigate the Air Corps. That June, the President established a Federal Aviation Commission led by newspaper editor Clark Howell to examine all aspects of American aviation policy. Both bodies endorsed negotiated military aircraft procurement contracts. But, Roosevelt was well-attuned politically to congressional insistence on price competition in military aircraft contracting. That consideration decided the issue for him, at least temporarily.[101]

In accordance with Roosevelt's decision and to forestall further congressional intervention, Assistant Secretary of War Henry H. Woodring instituted competitive bidding policies and procedures for all Air Corps aircraft procurement contracts. Although Congress had mandated competitive bidding on aircraft contracts in most cases in the 1926 Air Corps Act, it had permitted negotiated contracts for experimental planes. Air Corps leaders had largely ignored congressional intent and continued to negotiate contracts. To circumvent Capitol Hill, they relied on a legal opinion of the Army's Judge Advocate General (JAG) that the Air Corps did not have to rely entirely on the 1926 legislation for such actions. Instead, it applied provisions of Army regulations when it needed to justify negotiated deals. Through December 1933, the Secretary of War and the service's JAG had approved such negotiated purchases.

Woodring eliminated that basic aircraft procurement legal workaround. But, to avoid wasting time and money on empty paper promises from firms whose designs were either unsound or probably would never be constructed, he required that all bids for production contracts be accompanied by aircraft prototypes. Experimental aircraft contracts could either be let on a competitive basis or negotiated under provisions of the 1926 Air Corps Act. The new procurement policies were accompanied by a set of official standardized evaluation procedures developed by the Materiel Division intended to foster more objective procurement decisions. Essentially, they centered upon a formula for scoring the technical merits of sample aircraft. However, the new procedures failed to come to grips with the critical problem of balancing cost against performance.[102]

Despite growing congressional appropriations during the 1930s, the Air Corps' inventory of serviceable aircraft continued to shrink. Maj Gen Westover explained the reasons for that counterintuitive situation to a skeptical congressional audience in 1937. Although tremendous strides in aeronautical technology in recent years had sharply increased aircraft performance, that development had been accompanied by an enormous increase in their costs.[103] Moreover, a growing percentage of the aircraft being built for the Air Corps were large, complicated and very expensive long-range bombers. Westover emphasized that "... even though it might be possible to procure a larger number of airframes for the money expended, it is believed it would be a greater mistake to sacrifice the increased factors in national defense for the sake of merely providing more planes of less effectiveness."[104]

The Air Corps continued its unrelenting push to exploit and accelerate rapid progress in aircraft and engines. In July 1933, the Materiel Division had been tasked to conduct an engineering study of the maximum practical range of a new bomber that could carry a 2,000-pound bomb load. Later that year, the Air Corps launched a bomber competition to replace the Martin B-10. It published a design circular that called for a multi-engine bomber with a top speed of 250 mph and a range of 1,020 miles. The winning firm was expected to obtain a lucrative production contract for up to 220 aircraft. The new bomber had to be capable of reinforcing American bases in Hawaii, Panama and Alaska.

The winning product of that requirement was the Douglas B-18 *Bolo*. It was a twin-engine bomber based on the firm's DC-2 airliner. The B-18 had competed in a "fly off" in 1935 to determine the next standard Air Corps bomber. It flew against Martin's entry, an improved version of its B-10, and Boeing's revolutionary Model 299 bomber, forerunner of the B-17 *Flying Fortress*. Initially, it had appeared that Boeing's Model 299 would win the contest. The aircraft's prototype was a gamble by Boeing that its board of directors had allocated the firm's own money to design and build with no guarantee that the government would purchase it. The plane went from drawing board to flight test in under 12 months. During August 1935, the Model 299 clearly outperformed its rivals (Illustration 5.5). Based in part on its Model 247 commercial airliner, it reflected Boeing's practice of using the same basic design for its civilian and military aircraft. But, the firm's future looked dim after the aircraft crashed during a takeoff at Wright Field. Rather than a structural failure or engine malfunction, the accident was caused by human error when no one remembered prior to takeoff to disengage a locking mechanism designed to prevent the rudder and elevator to be damaged by the wind when the plane was parked on the ground. At its own expense, Boeing had produced the bomber for approximately $425,000. Without a follow-on government production contract, the firm could have been ruined.

Although Douglas was awarded a $6.498 million production contract for 90 B-18s, Air Corps officers had been so impressed with Boeing's entry that they convinced a dubious Secretary Woodring and the General Staff to purchase 13 experimental YB-17s for service testing (saving Boeing from bankruptcy). Air Corps bomber advocates strongly supported the B-17 over the B-18 on the spurious grounds that the

Illustration 5.5 Boeing model 299 (*Photo* Boeing Model 299, date and location unknown, VIRIN: 050406-F-1234P-028.JPG, NMUSAF, nationalmuseum.af.mil)

former was a much better coastal defense weapon. Most senior airmen remained committed to the B-17 as a strategic offensive weapon but were reluctant to say so publicly fearing a political backlash. Moreover, there was an element of inter-service rivalry in that stance. During the summer of 1937, they feared that the Navy was taking over the entire coastal defense mission through its aircraft procurement of long-range patrol bombers while the Air Corps was buying the poorly performing B-18.[105]

Westover was well aware of the contributions that the reviving commercial airline sector made to the development of the new breed of Air Corps aircraft epitomized by the YB-17. He wrote in the 1 March 1936 issue of the *Air Corps Newsletter* that "Perhaps the greatest aid given by commercial aviation directly to military aviation is the support for year-round employment of the personnel of the aircraft industry. The work of supplying the airlines with their airplanes has filled in the gaps between Army and Navy contracts…."[106] He added that "… the difference between commercial airliners and military airplanes, especially bombers and patrol boats, is so slight that the same engineering talent, mechanical personnel and, in many cases, the same tooling and shop equipment, fits into the fabrication of both."[107] Furthermore, Westover stressed that "The financial support afforded by the airlines business enables aircraft factories to maintain adequate engineering staffs, laboratories and testing equipment and the output of these is as great of an aid to the improvement of military aircraft as to the improvement of commercial aircraft."[108]

Working through its Materiel Division, NACA and industry, the Air Corps continued pushing the technology of flight in the years leading up to American military intervention in World War II. Those efforts were not limited to the more visible airframe revolution and better engines. For example, during 1935, President Roosevelt presented the 1934 Collier Trophy to Capt Albert F. Hegenberger of the Materiel Division for his achievements in blind (i.e., instrument) flying. Hegenberger was a veteran pilot and MIT-trained aeronautical engineer.[109]

In 1938, the Air Corps garnered another Collier Trophy for developing, equipping and flying an aircraft with the first successful pressurized cabin, the XC-35. That achievement had enormous implications for high altitude, long-range bombers and commercial airliners. During 1920, the Air Service's Engineering Division at McCook Field had initiated work on maintaining low-altitude temperature and pressure conditions during high-altitude flights. A tank had been constructed to fit within the fuselage of an aircraft and a wind-driven gear supercharger was used to provide pressure. Several flights were made during 1920–1922 establishing that a supercharged cabin was feasible if the valve operation, temperature control and visibility could be improved. Further tests had been abandoned at that point for unclear reasons.

Work resumed after the Air Corps developed plans in 1929 and 1934 for high-altitude airplanes. On 29 April 1935, an expenditure order was initiated at the Materiel Division to study the literature and existing development of supercharged cockpits and cabins for high-altitude airplanes. Based on that study, recommendations were developed for a project for additional study and experimentation to obtain an airplane specially designed or rebuilt for those purposes. Subsequently, the Materiel Division established such a project. It hired Dr. John E. Younger, a professor of mechanical engineering on leave from the University of California who had earlier worked as an engineer at the Materiel Division, to conduct the necessary preliminary investigations and pave the way for the procurement of a flying article. Younger began his work on 1 July 1935.[110]

Maj Carl Greene of the Materiel Division and Dr. **Younger were the guiding lights behind** the high-altitude pressurized cabin project. They and their teams, working with Lockheed engineers, heavily modified a **Model 10 *Electra*** airliner to develop a pressurized cabin and cockpit. That innovation was accomplished by using air from the engine's **turbo supercharger** channeled through a compressor outlet which was manually controlled in-flight by an onboard engineer. It provided the flight crew with a cabin altitude of about 12,000 feet even as the aircraft climbed higher, allowing them to breath without using oxygen masks.

On 5 August 1937, Army Corps pilots began flying the XC-35. Without wearing oxygen masks or flight suits of any kind, the pilots achieved a maximum speed of 350 mph at an average altitude of 20,000 feet during their 220-mile flight. On 16 September 1938, President Roosevelt personally bestowed the Collier Trophy on Maj Gen Arnold for that technological feat. Lessons learned from the XC-35 project played key roles in the designs of Boeing's B-29 *Superfortress*, the first mass-produced pressurized cabin aircraft.[111]

Notes

1. Hallion, *Strike from the Sky: The History of Battlefield Air Attack, 1911–1945*, 47.
2. Ibid.
3. *Aeronautics: Fourteenth Annual Report of the National Advisory Committee for Aeronautics [Fiscal Year] 1928* (Washington, D.C.: USGPO, 1929), 1, Hathi Trust Digital Library.
4. *Aircraft Year Book, 1929* (New York, NY: D. Van Nostrand Company, Inc. 1929), 135, AIA, aia-aerospace.org.
5. Joseph S. Ames, NACA, no title [general overview of American aviation], *New York Times*, 26 June 1927, XX15, ProQuest Historical Newspapers, Pentagon Library; "Joseph Sweetman Ames," NASA, undated, nasa.gov; John Calvin French, *A History of the University Founded by Johns Hopkins* (Baltimore, MD: Johns Hopkins Press, 1946), 407–408.
6. Crouch, *Wings: A History of Aviation from Kites to the Space Age*, 244, 245; Paper, Richard P. Hallion, "Two Hundred Years of Flight: A Bicentennial Assessment," given at the 80th Anniversary of Kitty Hawk, San Diego Aero-Space Museum, 17 December 1983, 5; Gross, *American Military Aviation: The Indispensable Arm*, 64–65.
7. Crouch, *Wings: A History of Aviation from Kites to the Space Age*, 360–361; Reynolds, *Admiral John H. Towers: The Struggle for Naval Air Supremacy*, 199; Maj Gen Oscar Westover, Chief, Air Corps, "Buying Military Aircraft," U-1169, A.C., 5/15/36 (15 May 1936), 2, RG 18, AAF Records, Entry 211, Box 13, Vol 24, National Archives; Millett and Maslowski, *For the Common Defense: A Military History of the United States of America*, 393–394; *The Aircraft Year Book for 1937* (New York, NY: Aeronautical Chamber of Commerce of America, Inc., 1937), 11, 13, 14, 15, 16, AIA, undated, aia-aerospace.org; "Our Air Supremacy Is Held Threatened," *New York Times*, 12 May 1936, 18, ProQuest Historical Newspapers, AACPL, www.aacpl.net; "Extracts from the Annual Report of the Bureau of Aeronautics submitted by Rear Admiral A.B. Cook, U.S. Navy, to the Secretary of the Navy, for fiscal year 1938," Navy Press Room, release date 25 November 1938, 5, *Current Annual Reports, Bureau of Aeronautics*, Nimitz Library.

8. Millett and Maslowski, *For the Common Defense: A Military History of the United States of America,* 361–362, 393–394; Gross, *American Military Aviation: The Indispensable Arm*, 79; Robert T. Krauskopf, "The Army and the Strategic Bombers, 1930–1939," Pt. 1, *Military Affairs*, Summer 1958, 87.
9. Fred Greene, "The Military View of American National Security Policy, 1904–1940,"*American Historical Review,* January 1961, 354–376; Rice, *The Politics of Air Power: From Confrontation to Cooperation in Army Aviation Civil–Military Relations*, 83; Millett and Maslowski, *For the Common Defense: A Military History of the United States of America*, 361–363.
10. U.S. Congress, House, Committee on Appropriations, *War Department Appropriation Bill for [Fiscal Year] 1936, Military Activities. Hearings Before the Committee on Appropriations, Part 1,* 74th Cong., 1st sess., 29 January 1935, 542, Hathi Trust Digital Library; Rae, *Climb to Greatness: The American Aircraft Industry, 1920–1960*, 58–60; Hallion, "Two Hundred Years of Flight: A Bicentennial Assessment," 5; "Boeing B-17E Flying Fortress," The National WWII Museum, undated, www.nationalww2museum.org; Gross, *American Military Aviation: The Indispensable Arm*, 64–65.
11. Rae, *Climb to Greatness: The American Aircraft Industry, 1920–1960*, 58–60; Hallion, "Two Hundred Years of Flight: A Bicentennial Assessment," 5; Freudenthal, *The Aviation Industry: From Kitty Hawk to Wall Street*, 118; "Gross, *American Military Aviation: The Indispensable Arm*, 64–65.
12. *Aeronautics: Fifteenth Annual Report of the National Advisory Committee for Aeronautics, [Fiscal Year] 1929* (Washington, D.C.: USGPO, 1930), 87, NACA Collection, UNT Libraries Government Documents Department, UNT Digital Library; *The Aircraft Year Book for 1930* (New York, NY: D. Van Nostrand Company, Inc., 1930), 1, 2, 4, 71, AIA, aia-aerospace.org; *The Aircraft Year Book for 1931* (New York, NY: D. Van Nostrand Company, Inc., 1931), 225–226, AIA, aia-aerospace.org.
13. David T. Courtwright, "The Routine Stuff: How Flying Became a Form of Mass Transportation," 209–222, in Roger D. Launius and Janet R. Daly Bednarek, eds., *Reconsidering a Century of Flight* (Chapel Hill: The University of North Carolina Press, 2003).

14. *Aircraft Yearbook 1932* (New York, NY: D. Van Nostrand Company, Inc., 1932), 282, 310, AIA, aia-aerospace.org; Rae, *Climb to Greatness: The American Aircraft Industry, 1920–1960*, 63–64.
15. Rae, *Climb to Greatness: The American Aircraft Industry, 1920–1960*, 67.
16. Ibid., 67, 68, 71; *Aircraft Yearbook 1933* (New York, NY: D. Van Nostrand Company, Inc., 1933), 248–249, AIA, aia-aerospace.org; "Douglas DC-3," SNASM, undated, airandspace.si.edu; Roger E. Bilstein, *Flight in America 1900–1983: From the Wrights to the Astronauts* (Baltimore, MD: Johns Hopkins University Press, 1984), 83; *Aircraft Yearbook 1936* (New York, NY: Aeronautical Chamber of Commerce of America, Inc., 1936), 263, 265, AIA, aia-aerospace.org; "Wright 1820 Cyclone [Engine]," NMUSAF, undated, nationalmuseum.af.mil; *Twenty-First Annual Report of the National Committee for Aeronautics, 1935* (Washington, D.C.: USGPO, 1936), 3, NACA Collection, UNT Libraries Government Documents Department, UNT Digital Library.
17. Rae, *Climb to Greatness: The American Aircraft Industry, 1920–1960*, 71.
18. U.S. Congress, House, *Department of the Navy, Appropriation Bill for [Fiscal Year] 1937, Hearing Before Subcommittee of House Committee on Appropriations*, 74th Cong., 2nd sess., 2 March 1936, 473–474, Hathi Trust Digital Library; "Bureau of Aeronautics (Navy)," NHHC, published 27 May 2021, histoy.navy.mil.
19. Stephen L. McFarland, "Higher, Faster and Further: Fueling the Aeronautical Revolution," *Innovation and the Development of Flight* (College Station, TX: Texas A&M University Press, 1999), 109, Roger D. Launius, ed.
20. Newton Copp and Andrew Zenella, *Discovery, Innovation, and Risk: Case Studies in Science and Technology* (Cambridge, MA: MIT Press, 1993), 181; "Houdry Process for Catalytic Cracking: National Historic Chemical Landmark," American Chemical Society, undated, acs.org; McFarland, "Higher, Faster and Further: Fueling the Aeronautical Revolution," 109–112.
21. "All-Metal Planes for the Navy Near: Rear Admiral Moffett in Report Tells of Better Fighting and Patrol Aircraft," *New York*

Times, 19 December 1926, 2, ProQuest Historical Newspapers, AACPL, www.aacpl.net.
22. *The Aircraft Yearbook for 1931*, 89.
23. Glenn E. Bugos, Review of *Wings of Wood, Wings of Metal: Culture and Technical Choice in American Airplane Materials, 1914–1945* (Princeton, NJ: Princeton University Press, 1999) by Eric Schatzberg, in *Technology and Culture*, April 2001, 370–372, JSTOR, Pentagon Library.
24. Gross, *American Military Aviation: The Indispensable Arm*, 65–66; "Fred E. Weick," 10 August 2015, NASA, nasa.gov; James R. Hansen, "The Cowling Story at Langley Laboratory, 1926–1936," *Aerospace Historian*. Fall, September 1985, 162–168; AP, "New Device Speeds Flight of Planes: Cowling Equipment Held Most Valuable Contribution Since War," *[Baltimore] Sun*, 10 November 1928, 9, ProQuest Historical Newspapers, AACPL, www.aacpl.net.
25. Ltr., Daniel F. Guggenheim to Herbert Hoover, Secretary of Commerce, Subj.: Establishing Fund Promoting Aeronautics In The U.S., 11 January 1926, Commerce, Box 40, Folder "Aviation – Daniel Guggenheim Fund, 1926 – undated," HHPL&M; Ltr., Daniel F. Guggenheim to Herbert Hoover, Secretary of Commerce, Subj.: Detailed Objectives of the Guggenheim Fund, 16 January 1926, HHPL&M, Commerce, Box 40, Folder "Aviation – Daniel Guggenheim Fund, 1926 – undated;" Paper, Subj.; Secretary Hoover on Daniel Guggenheim's Gift, 18 January 1926, Commerce, Box 40, Folder "Aviation – Daniel Guggenheim Fund, 1926 – undated," HHPL&M.
26. "Robert A. Millikan Facts," *The Nobel Prize*, 14 April 2023, nobelprize.org; "Dr. Theodore von Karman (1881–1963)," NASA, Jet Propulsion Laboratory, California Institute of Technology, undated, nasa.gov; Michael H. Gorn, *The Universal Man: Theodore von Karman's Life in Aeronautics* (Washington, D.C.: Smithsonian Institution Press, 1992), 38–47, 49, 52–59, 63, 68–71.
27. Gorn, *The Universal Man: Theodore von Karman's Life in Aeronautics*, 71.
28. As previously noted, three main types of fuselages evolved during the interwar period. They were: Monocoque, Semi-Monocoque and Truss Frame. Monocoque fuselages featured an exterior skin

that bore primary loads, minimizing the need for internal frameworks and benefitting aircraft with its lightweight and aerodynamically efficient design. Truss Frame fuselages used a lightweight steel alloy tube framework to ensure strength and rigidity while keeping weight low. Semi-Monocoque fuselages, merged Monocoque and Truss Frame features, providing a balanced approach that optimized strength, weight and durability.

29. *Aeronautics: Sixteenth Annual Report of the National Advisory Committee for Aeronautics [Fiscal Year] 1930* (Washington, D.C.: USGPO, 1931), 2, Hathi Trust Digital Library; Spenser, *The Airplane: How Ideas Gave Us Wings,* 182, 184; Rae, *Climb to Greatness: The American Aircraft Industry, 1920–1960*, 60–61; Futrell, *Ideas, Concepts, Doctrine: A History of Basic Thinking in the United States Air Force, 1907–1964*, 32–33; Canham, *Development and Production of Fighter Aircraft for the United States Air Force. AFLC Historical Study No. 213*, 35–36; Biddle, *Barons of the Sky: From Early Flight to Strategic Warfare, The Story of the American Aerospace Industry*, 166–167; Hallion, *Test Pilots: The Frontiersmen of Flight*, 97; *Aeronautics; Fourteenth Annual Report of the National Advisory Committee for Aeronautics*, 15–17; Phillip S. Meilinger, "The Impact of Technology and Design Choice on the Development of U.S. Fighter Aircraft," *Journal of the American Aviation Historical Society*, Spring 1991, 61, 62, 63; *Aircraft Year Book, 1928* (New York, NY: Aeronautical Chamber of Commerce of America, Inc., 1928), 143, AIA, aia-aerospace.org; *Aircraft Year Book, 1926,* 97–98, AIA, aia-aerospace.org; *Aircraft Year Book, 1927* (New York, NY: Aeronautical Chamber of Commerce of America, Inc., 1927), 111, AIA, aia-aerospace.org; Wagner, *Reuben Fleet and the Story of Consolidated Aircraft,* 34; Spenser, *The Airplane: How Ideas Gave Us Wings,* 222–223.
30. Holley, *Buying Aircraft: Materiel Procurement for the Army Air Forces. Special Studies. U.S. Army in World War II,* 21.
31. Ibid., 11.
32. Ibid., 27.
33. *Aerospace Facts and Figures 1964, 24–25*, AIA, aia-aerospace.org.
34. Ibid., 20; Gibbs-Smith, *Aviation: An Historical Survey from Its Origins to the End of the Second World War*, 190; Rae, *Climb to*

Greatness: The American Aircraft Industry, 1920–1960, 24–27, 60–61.

35. *The Aircraft Yearbook for 1938* (New York, NY: Aeronautical Chamber of Commerce of America, Inc., 1938), 459–460, AIA, aia-aerospace.org.
36. Herbert Hoover, Secretary of Commerce, 'Fifteenth Annual Report of the Secretary of Commerce [Fiscal Year] 1927," (Washington, D.C.: USGPO, 1927), 44, Hathi Trust Digital Library; Robert B. Machete, et. al., compilers, "197.4 Records of the Federal Aviation Commission 1933–35," Records of the Civil Aeronautics Board (CAB), RG 197, 1931–35, "Guide to Federal Records," National Archives, 1995, www.archives.gov; Freudenthal, *The Aviation Business: From Kitty Hawk to Wall Street,* 141, 147.
37. Millett and Maslowski, *For the Common Defense: A Military History of the United States of America,* 372–375; Millett, *Semper Fidelis: The History of the United States Marine Corps,* 333; E.H. Brainard, "Marine Corps Aviation," *Marine Corps Gazette*, March 1928, 36, Internet Archive; Johnson, *Marine Corps Aviation: The Early Years, 1912–1940*, 42, Internet Archive; Green, "The Military View of American National Security Policy, 1904–1940,"*American Historical Review*, January 1961, 361; "Annual Report of the Chief of the Bureau of Construction and Repair for the Fiscal Year Ended June 30, 1928," (Washington, D.C.: USGPO, 1928), 3, Hathi Trust Digital Library; "Annual Report of the Chief of the Bureau of Aeronautics to the Secretary of the Navy," (Washington, D.C.: USGPO, 1928), 15, 36, Hathi Trust Digital Library; "USS Lexington (CV-2)," National Museum of the US Navy, NHHC, undated, history.navy.mil; "USS Saratoga (CV-3)," National Museum of the US Navy, NHHC, published 2 September 2015, history.navy.mil; "Annual Reports of the Department of the Navy for the Fiscal Year (Including Operations to November 15, 1928) 1928," (Washington, D.C.: USGPO, 1929), 699, U.S. Congress, House of Representatives, 70th Cong., 1nd sess., House Document No. 366, ProQuest Congressional Documents, Pentagon Library; Norman Friedman, *U.S. Aircraft Carriers: An Illustrated Design History* (Annapolis, MD: Naval Institute Press, 1983), 44.

38. "Annual Reports of the Department of the Navy for the Fiscal Year (Including Operations to November 15, 1927), 1927" (Washington, D.C.: USGPO, 1928), 10, 1075, U.S. Congress, House of Representatives, 70th Cong., 1st sess., House Document No. 81, ProQuest Congressional Documents, Pentagon Library; "Admiral Asks Funds to Maintain Forces," *Washington Post*, 16 December 1928, S13, ProQuest Historical Newspapers, Pentagon Library; Millett and Maslowski, *For the Common Defense: A Military History of the United States of America*, 373–374.
39. Grossnick, *United States Naval Aviation, 1910–1995*, 64; Trimble, *Admiral William A. Moffett: Architect of Naval Aviation*, 177; "An Excellent Appointment [Assistant Navy Secretary for Aeronautics Warner]," *New York Times*, 2 July 1926, 18; ProQuest Historical Newspapers, Pentagon Library; "Annual Report of the Chief of the Bureau of Aeronautics to the Secretary of the Navy, Fiscal Year 1927," (Washington, D.C.: USGPO, 1927), 4, Hathi Trust Digital Library.
40. Blog, Roger Launius, "Edward P. Warner and the Transformation of Aeronautics in America," posted 20 June 2016, launiusr.wordpress.com.
41. Trimble, *Admiral William A. Moffett: Architect of Naval Aviation*, 167–199; "Annual Report of the Chief of the Bureau of Aeronautics to the Secretary of the Navy, Fiscal Year 1931," (Washington, D.C.: USGPO, 1931), 1–2, Hathi Trust Digital Library; "Annual Reports of the Department of the Navy for the Fiscal Year (Including Operations to September 15, 1932) 1932," (Washington, D.C.: USGPO, 1933), 486, U.S. Congress, House of Representatives, 72nd Cong., 2nd sess., House Document No. 565, ProQuest Congressional Documents, Pentagon Library; U.S. Congress, House, *Hearings Before the Subcommittee on Aeronautics Making an Investigation Into Certain Phases of the Manufacture of Aircraft and Aeronautical Accessories, and Related Subjects as the Refer to the Navy Department, House Committee on Naval Affairs.* 73rd Cong., 2nd sess., 10 April 1934, 1469, ProQuest Congressional Documents, Pentagon Library.
42. Shanahan, *Procurement of Naval Aircraft, 1907–1939*, 342; Grossnick, *United States Naval Aviation, 1910–1995*, 69; 85; "Annual Reports of the Department of the Navy for the Fiscal

Year (Including Operations to November 15, 1929) 1929," (Washington, D.C.: USGPO, 1930), 611, U.S. Congress, House of Representatives, 71st Cong., 2nd sess., House Document No. 197, ProQuest Congressional Documents, Pentagon Library; "Annual Reports of the Department of the Navy for the Fiscal Year (Including Operations to November 15, 1930) 1930," (Washington, D.C.: USGPO, 1931), 573–574, U.S. Congress, House of Representatives, 71st Cong., 3rd sess., House Document No. 540, ProQuest Congressional Documents, Pentagon Library.

43. Grossnick, *United States Naval Aviation, 1910–1995,* 80, 81, 88; "Annual Reports of the Department of the Navy for the Fiscal Year (Including Operations to September 15, 1932) 1932," (Washington, D.C.: USGPO, 1932), 480, U.S. Congress, House of Representatives, 72nd Cong., 2nd sess., House Document No. 565, ProQuest Congressional Documents, Pentagon Library; Van Fleet and Armstrong, *United States Naval Aviation, 1910–1980.NAV AIR 00-80P-1,* 80, 86; Richard Sanders Allen, *The Northrop Story, 1929–1939* (Altglen, PA: Schiffer Aviation History, 1993), 105.
44. Grossnick, *United States Naval Aviation, 1910–1995,* 78, 80; *Aircraft Yearbook 1932,* 279, AIA, aia-aerospace.org; "Boeing Model 205 Fighter," 14 February 2011, Patriots Point Naval & Air Museum, www.patriotspoint.org; Enzo Angelucci, *The American Fighter from 1917 to the Present* (New York, NY: Orion Books, 1987), 81–82; Van Fleet and Armstrong, *United States Naval Aviation, 1910–1980.NAV AIR 00-80P-1,* 78, 80.
45. Friedman, *U.S. Aircraft Carriers: An Illustrated Design History,* 57–58, 63; Norman Friedman, *Winning a Future War, War Gaming and Victory in the Pacific War* (Washington, D.C.: NHHC, 2019), source note, 98; "Report of the Bureau of Construction and Repair [To the Secretary of the Navy]," 15 October 1930, 260, Hathi Trust Digital Library; "Annual Reports of the Navy Department for the Fiscal Year 1931 (including Operations to September 15, 1931)," (Washington, D.C.: USGPO, 1932), 482, Hathi Trust Digital Library.
46. "Annual Report of the Chief of the Bureau of Construction and Repair [To the Secretary of the Navy for Fiscal Year 1932]," (Washington, D.C.: USGPO, 1931), 2, Hathi Trust Digital

Library; Trimble, *Admiral William A. Moffett: Architect of Naval Aviation*, 15, 167–199, 220–221, 225–226.

47. "Summary of the Annual Report of the Bureau of Aeronautics Submitted by Rear Admiral Ernest J. King, U.S. Navy, to the Secretary of the Navy, for Fiscal Year 1933," Navy Press Room, release date 24 December 1933, 19 (in pencil), *Current Annual Reports, Bureau of Aeronautics*, Nimitz Library; Hone and Utz, *History of the Office of the Chief of Naval Operations, 1915–2015*, 105, NHHC, history.navy.mil; Trimble, *Admiral William A. Moffett: Architect of Naval Aviation*, 229.
48. Trimble, *Admiral William A. Moffett: Architect of Naval Aviation*, 167–181, 274–275; "5-Year Navy Aircraft Plan Completed In 4 Years," *New York Times*, 29 June 1931, 3, ProQuest Historical Newspapers, Pentagon Library; "Navy Reaches Plane Quota," *Los Angeles Times*, 29 June 1931, 5, ProQuest Historical Newspapers, Pentagon Library.
49. Morris, *The Origins of American Strategic Bombing Theory*, 166.
50. "Annual Reports of the Navy Department for the Fiscal Year 1930 (Including Operations to November 15, 1930," (Washington, DC: USGPO, 1931), 569, Hathi Trust Digital Library.
51. "Annual Report of the Secretary of the Navy for the Fiscal Year 1933," (Washington, D.C.: USGPO, 1933), 10, Hathi Trust Digital Library; Thomas B. Buell, *Master of Sea Power: A Biography of Fleet Admiral Ernest J. King. Classics of Naval Literature* (Annapolis, MD: Naval Institute Press, 1995, reprint of Little Brown and Company, Boston, MA, 1980), xvii, 96–97; Trimble, *Admiral William A. Moffett: Architect of Naval Aviation*, 6, 9, 207; Turnbull and Lord, *History of United States Naval Aviation*, 283, 284; "Air Navy's Hard Tasks," *New York Times*, 30 April 1933, XX6, ProQuest Historical Newspapers, Pentagon Library; Reynolds, *Admiral John H. Towers: The Struggle for Naval Air Supremacy*, 248–250; "Fleet Admiral Ernest J. King: Ninth Chief of Naval Operations, March 26, 1942–December 15, 1945," NHHC, 15 July 2016, history.navy.mil; Grossnick, *United States Naval Aviation, 1910–1995*, 84–85, 92, 415; U.S. Congress, House, *Navy Appropriation for Fiscal Year 1935, Hearings Before a Subcommittee of the House Committee on Appropriations*. 73rd Cong., 2nd sess., 2 January 1934, 389–390, ProQuest Congressional Documents, Pentagon Library; Van Fleet and Armstrong,

United States Naval Aviation, 1910–1980.NAV AIR 00-80P-1, 83; "Asks Planes to Outfit Carriers: Admiral King Reveals That 10 Vessels Now Being Built Have No Air Armament," *New York Times*, 24 December 1933, N3, ProQuest Historical Newspapers, AACPL, www.aacpl.net; "Moffett's Body Found: Coast Guard Boat Recovers It 10 Miles From Scene of Wreck," *New York Times*, 11 April 1933, 10, ProQuest Historical Newspapers, AACPL, www.aacpl.net; U.S. Congress, House, *Department of the Navy Appropriation Bill [Fiscal Year] 1939, Hearings Before the Subcommittee of the Committee on Appropriations, House of Representatives.* 75th Cong., 3rd sess., 8 December 1937, 48, Hathi Trust Digital Library; King and Fairchild, *Fleet Admiral King: A Naval Record*, 211, 248.

52. Robert W. Love, Jr., *History of the U.S. Navy: 1775–1941, Volume One* (Harrisburg, PA: Stackpole, 1992), 591; Grossnick, *United States Naval Aviation, 1910–1995*, 87, 89; "Annual Report of the Secretary of the Navy for Fiscal Year 1935," (Washington, D.C.: USGPO, 1935), 9, Hathi Trust Digital Library.
53. U.S. Congress, House, Committee on Naval Affairs, *Hearings Before the Subcommittee on Aeronautics Making an Investigation Into Certain Phases of the Manufacture of Aircraft and Aeronautical Accessories as they Refer to the Navy Department*, 73rd Cong., 2nd sess., February 1934, 5, 6, 26, 27, 453, ProQuest Congressional Documents, Pentagon Library.
54. Ibid., 429.
55. Ibid., 430, 442.
56. Ibid., 446, 447, 486; "New Wright Motor Said to Lead World: Mystery Engine for Airplane Develops 525 Horsepower in Test at Paterson Plant. Navy Guards Its Secret. 'Cyclone' Will Be Used Only for Army and Navy Craft, Company Official Says," *New York Times*, 10 January 1927, 30, ProQuest Historical Newspapers, AACPL, www.aacpl.net; AP, "Navy Contracts For 54 Airplanes: Also Orders 48 Air-Cooled Engines....," *[Baltimore] Sun*, 21 July 1927, 2, ProQuest Historical Newspapers, AACPL, www.aacpl.net.
57. U.S. Congress, House, *Department of the Navy, Appropriation Bill for [Fiscal Year] 1937, Hearing Before Subcommittee of House Committee on Appropriations*, 74th Cong., 2nd sess., 2 March 1936, 475, Hathi Trust Digital Library.

58. "Annual Reports of the Department of the Navy for the Fiscal Year (Including Operations to November 15, 1926) 1926," (Washington, D.C.: USGPO, 1927), 638, U.S. Congress, House of Representatives, 69th Cong., 2nd sess., House Document No. 625, ProQuest Congressional Documents, Pentagon Library.
59. Weigley, *History of the United States Army*, 405, 561, 568–569; Millett and Maslowski, *For the Common Defense: A Military History of the United States of America*, 377–378.
60. Weigley, *History of the United States Army*, 406–407; Millett and Maslowski, *For the Common Defense: A Military History of the United States of America*, 376–378; D. Clayton James, *The Years of MacArthur. Vol. 1* (Boston, MA: Houghton, Mifflin, 1970), 354.
61. Paper, Roger Miller, AFCHO, Headquarters, USAF, "The U.S. Army Air Corps and the Search for Autonomy, 1926–1943," ca 10 July 1997, 3.
62. Rice, *The Politics of Air Power: From Confrontation to Cooperation in Army Aviation Civil–Military Relations*, 96.
63. Miller, "The U.S. Army Air Corps and the Search for Autonomy, 1926–1943," ca 10 July 1997, 4, 5.
64. Ibid., 6.
65. Ibid; "Infantry [and Others] Shifted Into Air Service," *New York Times*, 21 August 1929, 4, ProQuest Historical Newspapers, Pentagon Library.
66. Thomas Alexander Hughes, *Overlord: General Pete Quesada and the Triumph of Tactical Air Power in World War II* (New York, NY: The Free Press, 1995), 48.
67. Holley, *Buying Aircraft: Materiel Procurement for the Army Air Forces. Special Studies. U.S. Army in World War II*, 95; Shiner, *Foulois and the U.S. Army Air Corps, 1931–1935*, 253; Benjamin S. Kelsey, *The Dragons Teeth? The Creation of United States Air Power for World War II* (Washington, D.C.: Smithsonian Institution Press, 1982), 55; "Dayton Gives Army Air Service Large Flying Field," *U.S. Air Services*, September 1924, 23, Hathi Trust Digital Library; Paper, "Wright Field Day," ca. 31 January 1928, 1, 4, based on "Review of Air Corps Developments in 1927" by Brig Gen William E. Gillmore, Chief, Materiel Division, Wright Field, Ohio, prepared for release in public media, especially aviation publications, ASD/HO, AFSC; "Army

Dedicates New Flying Centre," *New York Times*, 13 October 1927, 3, ProQuest Historical Newspapers, Pentagon Library; "Second Annual Report of the Chief of the Materiel Division, Air Corps, Fiscal Year 1928," (Washington, D.C.: USGPO, 1928), 1, Pentagon Library.

68. Kelsey, *The Dragons Teeth? The Creation of United States Air Power for World War II*, 55; Edwin H. Rutowski, *The Politics Military Aircraft Procurement, 1926–1934. A Study in the Political Assertion of Consensual Values* (Columbus, OH: The Ohio State University Press, 1966), 75–76; Shiner, *Foulois and the U.S. Army Air Corps, 1931–1935*, 152–153; Paper, "Imitation of Development Projects and Establishment of Military Characteristics," circa January 1939, Papers of Lt Gen Frank M. Andrews, Air Corps, Personal Correspondence, Box 5, Manuscript Division, LOC.
69. "Sixth Annual Report of the Chief of the Materiel Division, U.S. Army Air Corps, Fiscal Year 1932," (Washington, D.C.: USGPO, 1932), 11, Pentagon Library; "Annual Report of the Chief of the Air Corps, Fiscal Year 1933," (Washington, D.C.: USGPO, 1933), 44, AFHRA Microfilm #167.4011; "Fifth Annual Report of the Chief of the Materiel Division, U.S. Army Air Corps, Fiscal Year 1931," (Washington, D.C.: USGPO, 1931), 10, Pentagon Library.
70. "Annual Reports of the War Department Fiscal Year Ended June 30, 1932," (Washington, D.C., USGPO, 1932), 47, Hathi Trust Digital Library.
71. Hallion, "Two Hundred Years of Flight: A Bicentennial Assessment," 6–8; "Twelfth Annual Report, Chief, Materiel Division, Air Corps, U.S. Army, Fiscal Year 1938," (Wright Field, OH: Materiel Division, 1938), 10–11, Pentagon Library.
72. Millett and Maslowski, *For the Common Defense: A Military History of the United States of America*, 383; Shiner, *Foulois and the U.S. Army Air Corps, 1931–1935*, 44–50; Crouch, *Wings: A History of Aviation from Kites to the Space Age*, 356, 357, 358; Greer, *The Development of Air Doctrine in the Army Air Arm, 1917–1941, Air Force Historical Study No. 89*, 44–75; Craven and Cate, *The Army Air Forces in World War II. Vol. 1. Plans and Early Operations. January 1939 to August 1942*, 46; "Fechet Will Head Army Air Service: Brigadier General Named to Succeed

Major Gen. Patrick When He Retires for Age," *New York Times*, 12 June 1927, 17, ProQuest Historical Newspapers, AACPL www.aacpl.net.

73. "710 Planes Were Coming to Army and Navy," *New York Times*, 7 April 1929, N1, ProQuest Historical Newspapers, AACPL, www.aacpl.net.
74. Gross, *American Military Aviation: The Indispensable Arm*, 73–74.
75. "Martin B-10," NMUSAF, undated, nationalmuseum.af.mil.
76. "Address of Maj. Gen. B.D. Foulois, Air Corps, Before the Dayton, Ohio, Chamber of Commerce," 12 November 1934, 7, Benjamin D. Foulois Papers, USAFAL, Series 7, Box 10, Folder 1; "Martin B-10," NMUSAF, published 7 April 2015, nationalmuseum.af.mil; John F. Shiner, "The Heyday of the GHQ Air Force, 1935–1939," *Winged Shield, Winged Sword: A History of the United States Air Force, Volume I, 1907–1950*, 138–139; John Pike, "Martin B-10," last modified 9 July 2018, Global Security.org; "Army Buying 38 High Speed Bombing Planes, Called Most Powerful Weapon Yet Devised," *New York Times*, 23 January 1933, 1, ProQuest Historical Newspapers, Pentagon Library; U.S. Congress, House, Committee on Appropriations, *War Department Appropriation Bill For [Fiscal Year] 1934, Military Activities, Hearings Before a Subcommittee of the House Committee on Appropriations*, Part 1. 72nd Cong., 2nd sess., 14 December 1932, 606, ProQuest Congressional Documents, Pentagon Library; Haulman, *One Hundred Years of Flight: USAF Chronology of Significant Air and Space Events, 1903–1902*, 31.
77. Bernard, Fitzsimons, ed., *The Illustrated Encyclopedia of the 20th Century Weapons and Warfare* (New York, NY: Purnell & Sons Ltd., 1969), 18–46.
78. Millett and Maslowski, *For the Common Defense: A Military History of the United States of America*, 385; Gross, *American Military Aviation: The Indispensable Arm*, 74; Robert W. Krauskopf, "The Army and the Strategic Bomber, 1930–1939," *Military Affairs*, Summer 1958, 88; Rice, *The Politics of Air Power: From Confrontation to Cooperation in Army Aviation Civil–Military Relations*, 142.
79. Lauren D. Lyman, "Army Planes Grow Deadlier," *New York Times*, 5 April 1931, 130, ProQuest Historical Newspapers,

Pentagon Library; Ross G. Hoyt, "The Curtiss Hawks," *Air Force Magazine*, October 1976, 68; "Curtiss P-6E Hawk," NMUSAF, published 7 April 2015, nationalmuseum.af.mil.

80. "Alexander de Seversky: Influential World War II Air Power Advocate," *Smithsonian National Air and Space Museum,* 9 May 2015, airandspace.si.edu; "Contract Awarded For Pursuit Airplanes," *Air Corps News Letter*, Information Division, Air Corps V-7046, A.C., 1 July 1936, 14, AFHSD, www.afhistory.af.mil/History; Joshua Stoff, *The Thunder Factory: An Illustrated History of the Republic Aviation Corporation* (Osceola, WI: Motorbooks International, 1990), 18, 19, 20; "Seversky P-35," NMUSAF, nat ionalmuseum.af.mil; William Green and Gordon Swanborough. "The End of the Beginning...The Seversky P-35", *Air Enthusiast*, July–September 1979, 8–9. 10–11; Ross G. Hoyt, "Flying the Early Birds: The P-36," *Air Force Magazine*, February 1977, 90.

81. "Annual Reports War Department Fiscal Year Ended June 30, 1934," (Washington, D.C.: USGPO, 1934), 4–5, Hathi Trust Digital Library; Newton D. Baker, Chairman, "Final Report of War Department Special Committee On Army Air Corps," 18 July 1934, 67 (Washington, D.C.: USGPO, 1934), Hathi Trust Digital Library; "Report of the Secretary of War to the President [Fiscal Year] 1934," (Washington, D.C.: USGPO, 1934), 4–5, Hathi Trust Digital Library; *Aircraft Yearbook 1934,* 95, 96, AIA, aia-aerospace.org; Paper, Clinton W. Howard, Materiel Division, "Engineering Developments of the Army Air Corps for Calendar 1933" presented at the annual meeting of the Society of Mechanical Engineers in New York City, 7 December 1933, 1, U-1076, A.C., 11/28/33, RG 18, AAF Records, Entry 211, Box 13, Vol XXII, National Archives; Haulman, *One Hundred Years of Flight: USAF Chronology of Significant Air and Space Events, 1903–1902*, 30–31; Hewes, *From Root to McNamara: Army Organization and Administration*, 389, history.army.mil; Nalty, *Winged Shield, Winged Victory: A History of the United States Air Force, Volume I, 1907–1950*, 121–122, 125–130; Rice, *The Politics of Air Power: From Confrontation to Cooperation in Army Aviation Civil–Military Relations*, 132.

82. Maurer, *Aviation in the U.S. Army, 1919–1939,* 214; Purtee, *History of the Army Air Service, 1907–1926, Air Force Logistics*

Command (AFLC) Historical Study No. 279, 67, 87–88; Purtee, *The Development of Light and Medium Bombers, AFLC Historical Study No. 196*, 19–23; Futrell, *Ideas, Concepts, Doctrine: A History of Basic Thinking in the United States Air Force, 1907–1964*, 392–393; Gordon Swanborough and Peter M. Bowers, *United States Military Aircraft Since 1909* (New York, NY: Putnam, 1964), 179; "Curtiss A-3," NMUSAF, undated, nationalmuseum.af.mil.

83. Raines, *Eyes of the Artillery: The Origins of Modern U.S. Army Aviation in World War II*, 19.
84. "Purchase of New Observation Planes," *Air Corps Newsletter*, Information Division, Office of the Chief of the Air Corps, V-7162, A.C., 1 December 1936, 3, AFHSD, www.afhistory.af.mil/History; "North American O-47B," NMUSAF, published 7 April 2015, nationalmuseum.af.mil; Holley, *Evolution of the Liaison Type Airplane, 1917–1944*, 2–23; Fahey, *U.S. Army Aircraft [Heavier-Than-Air] 1908–1946*, 31; Raines, *Eyes of the Artillery: The Origins of Modern U.S. Army Aviation in World War II*. 19; Howard K. Butler, *Army Air Corps Airplanes and Observation, 1935–1941* (St. Louis, MO: Historical Office, U.S. Army Aviation Systems Command, February 1990, Second Edition, Second Printing, 4 November 1994), 56, 60, Hathi Trust Digital Library; Ralph B. Oakley, "North American, Part 1: The Beginnings—1928–1937," *Journal of the American Aviation Historical Society*, Summer 1974, 142–143.
85. Holley, *Buying Aircraft: Materiel Procurement for the Army Air Forces*, 156.
86. Holley, *Evolution of the Liaison Type Airplane, 1917–1944. Army Air Forces Historical Studies: No. 44*, 27.
87. Henning, *Harnessing the Airplane: American and British Cavalry Responses to a New Technology, 1903–1939*, 112.
88. Ibid.
89. Raines, *Eyes of the Artillery: The Origins of Modern U.S. Army Aviation in World War II*, 19–20; Butler, *Army Air Corps Airplanes and Observation, 1935–1941*, 73–76, 140–141, Hathi Trust Digital Library.
90. The Robert J. Collier Trophy is awarded annually by the National Aeronautic Association "for the greatest achievement in aeronautics or astronautics in America, with respect to improving the

performance, efficiency, and safety of air or space vehicles, the value of which has been thoroughly demonstrated by actual use during the preceding year."

91. Raines, *Eyes of the Artillery: The Origins of Modern U.S. Army Aviation in World War II*, 20–22; George Townsend, "Autogiro. The Story of the Windmill Plane," 1985, reprinted as "The Theory of the Autogiro: All the World's Rotorcraft," undated, Internet Archive; Henning, *Harnessing the Airplane: American and British Cavalry Responses to a New Technology, 1903–1939*, 124–129; Weigley, *History of the United States Army*, 558; "Kellett K-2/K-3 Autogiro," undated, NMUSAF, nationalmuseum.af.mil; Grossnick, *United States Naval Aviation, 1910–1995*, 79, 81; "Navy Autogiro Takes Flight," posted in *History Up Close* on 23 September 2015, NHHC, history.navy.mil.
92. Raines, *Eyes of the Artillery: The Origins of Modern U.S. Army Aviation in World War II*, 22.
93. Henning, *Harnessing the Airplane: American and British Cavalry Responses to a New Technology, 1903–1939*, 130.
94. Cook, *Carl Vinson: Patriarch of the Armed Forces*, 87–91; Shanahan, *Procurement of Naval Aircraft, 1907–1939*, 360–362; U.S. Congress, House, *Navy Appropriation for [Fiscal Year] 1936, Hearings Before a Subcommittee of the House Committee on Appropriations in Charge of Navy Department Appropriations Bill for [Fiscal Year] 1936*. 74th Cong., 1st sess., 13 March 1935, 499, 507, 508, 548, ProQuest Congressional Documents, Pentagon Library; Turnbull and Lord, *History of United States Naval Aviation*, 285; Hone and Utz, *History of the Office of the Chief of Naval Operations, 1915–2015*, 106, NHHC, history.navy.mil; Millett and Maslowski, *For the Common Defense: A Military History of the United States of America*, 387, 394; Hagan, *This People's Navy: The Making of American Sea Power*, 284; Thomas Hone, "Naval Reconstitution, Surge, and Mobilization: Once and Future," *Naval War College Review*, Summer 1994, 73; Curtis Alan Utz, "Carrier Aviation Policy and Procurement in the U.S. Navy, 1936–1940," masters thesis, University of Maryland, 1989, 18–19; U.S. Congress, House, *Navy Appropriation for [Fiscal Year] 1937, Hearings Before a Subcommittee of the House Committee on Appropriations in Charge of Navy Department Appropriations Bill For [Fiscal Year] 1937*. 74th Cong.,

2nd sess., 2 March 1936, 469, 476, 477, 479, 502, ProQuest Congressional Documents, Pentagon Library; Insert, "Statement of Facilities Provided or Being Provided for Government Production of Aircraft and/or Engines: Naval Aircraft Factory," U.S. Congress, House, *Navy Appropriation for [Fiscal Year] 1941, Hearings Before a Subcommittee of the House Committee on Appropriations in Charge of Navy Department Appropriations Bill for [Fiscal Year] 1941.* 76th Cong., 3rd sess., 8 January 1940, 484, ProQuest Congressional Documents, Pentagon Library; "Extracts from the Annual Report of the Bureau of Aeronautics submitted by Rear Admiral A. B. Cook, U.S. Navy, Chief, BuAer, to the Secretary of the Navy, for Fiscal Year 1936," Navy Press Room, 15 November 1936, 4, *Current Annual Reports, Bureau of Aeronautics*, Nimitz Library; "Extracts from the Annual Report of the Bureau of Aeronautics submitted Rear Admiral A. B. Cook, U.S. Navy, Chief, BuAer, to the Secretary of the Navy, for Fiscal Year 1937," Navy Press Room, 12 November 1937, 26–27, *Current Annual Reports, Bureau of Aeronautics*, Nimitz Library; "The London Naval Conference, 1930," Office of the Historian, Foreign Service Institute, United States Department of State, undated, history@state.gov; Freudenthal, *The Aviation Business: From Kitty Hawk to Wall Street*, 237.

95. Utz, "Carrier Aviation Policy and Procurement in the U.S. Navy, 1936–1940," 5.
96. Van Fleet and Armstrong, *United States Naval Aviation, 1910–1980.NAV AIR 00-80P-1,* 85, 103; "Wasp (CV-7), 1940–1942," published 14 March 2019, NHHC, history.navy.mil; Utz, "Carrier Aviation Policy and Procurement in the U.S. Navy, 1936–1940," 5; Hone and Utz, *History of the Office of the Chief of Naval Operations, 1915–2015*, 106, NHHC, history.navy.mil.
97. Utz, "Carrier Aviation Policy and Procurement in the U.S. Navy, 1936–1940," 5.
98. U.S. Congress, House, Committee on Appropriations, *War Department Appropriation Bill for [Fiscal Year] 1933, Military Activities, Hearings Before a Subcommittee of the House Committee on Appropriations. Part 1.* 72nd Cong., 1st sess., 5 January 1932, 992–993, 996, 1069, Hathi Trust Digital Library; Vander Meulen, *The Politics of Aircraft: Building an American Military Industry*, 133; Underwood, "The Army Air Corps

Under Franklin D. Roosevelt: The Influence of Air Power on the Roosevelt Administration, 1933–1941," 1988, vii–xi; John F. Shiner, "General Foulois and the 1934 Air Mail Disaster," *Aerospace Historian,* Winter December 1978, 223; "Annual Reports War Department Fiscal Year Ended June 30, 1934," (Washington, D.C.: USGPO, 1934), 4, Hathi Trust Digital Library; Shiner, *Foulois and the U.S. Army Air Corps, 1931–1935*, 127–148.

99. Gross, *American Military Aviation: The Indispensable Arm*, 76–77; "Foulois, Air Chief, Quits Active List," *New York Times*, 9 August 1935, 19, ProQuest Historical Newspapers, Pentagon Library; Mets, *Master of Airpower, Carl A. Spaatz*, 94; Lauren D. Lyman, "Army Plane Buying Hit," *New York Times*, 13 May 1934, XX7, ProQuest Historical Newspapers, AACPL, www.aacpl.net; "Find Profits Fair on Navy Airplanes," *New York Times*, 9 March 1934, 15, ProQuest Historical Newspapers, AACPL, www.aacpl.net.

100. Rice, *The Politics of Air Power: From Confrontation to Cooperation in Army Aviation Civil-Military Relations*, 125.

101. Baker, "Final Report of War Department Special Committee on Army Air Corps," 18 July 1934, 1; Shiner, *Foulois and the U.S. Army Air Corps, 1931–1935,* 142–145; Danhoff, *Government Contracting and Technological Change*, 137–140; Foulois and Glines, *From the Wright Brothers to the Astronauts: The Memoirs of Major General Benjamin Foulois*, 233–262; U.S. Congress, House, Committee on Military Affairs, *Investigation of Profiteering in Military Aircraft Under HR 275, HR 2060,* 73rd Cong., 2nd sess., 15 June 1934, 1–3, 12–14, Pentagon Library; Eugene M. Emme, "The American Dimension," *Air Power and Warfare. Proceedings of the Eighth Military History Symposium. United States Air Force Academy.* 18–20 October 1978 (Washington, D.C.: AFCHO and the USAFA, 1979), 71; Futrell, *Ideas, Concepts, Doctrine: A History of Basic Thinking in the United States Air Force, 1907–1964*, 36–37; Robert B. Machete, et. al., compilers, "1974 Records of the Federal Aviation Commission 1933–35," Records of the Civil Aeronautics Board (CAB), RG 197, 1931–35, "Guide to Federal Records," National Archives, 1995, www.archives.gov; Rice, *The Politics of Air Power: From*

Confrontation to Cooperation in Army Aviation Civil–Military Relations, 133.

102. John F. Shiner, "The General and the Subcommittee: Congress and U.S. Army Air Corps Chief Benjamin Foulois, 1934–35," *The Journal of Military History*, October 1991, 488–489; Kelsey, *The Dragons Teeth? The Creation of United States Air Power for World War II*, 110–111; Holley, *Buying Aircraft: Materiel Procurement for the Army Air Forces. Special Studies. U.S. Army in World War II*, 128–134; "Annual Reports War Department Fiscal Year Ended June 30, 1934," (Washington, D.C.: USGPO, 1934), 30, Hathi Trust Digital Library; "Annual Report of the Army Air Corps Fiscal Year 1935," 40–41, Pentagon Library; *Aircraft Yearbook for 1934*, 126, AIA, aia-aerospace.org; Memo, Materiel Division, Army Air Corps, Subj.: "Method Used by the U.S. Army Air Corps Since January 1, 1934 for the Procurement of Airplanes and Aeronautical Accessories," circa 1934, Headquarters, AFLC, Office of History, Wright-Patterson, AFB, Ohio; Memo, Materiel Division, Army Air Corps, Subj.: "Detailed Method of Determining the Procedure for Procurement of Experimental Airplanes on a Competitive Basis," 14 June 1934, Headquarters, AFLC, Office of History, Wright-Patterson, AFB, Ohio; Ltr., George H. Dern, Secretary of War, to Representative J. J. McSwain, Chairman, Committee on Military Affairs, U.S. House of Representatives, Subj.: "War Department Competitive Aircraft Procurement," 15 August 1935, PSF, Box 14, "War Department, 1933–1941" Folder, FDRPL&M; U.S. Congress, House, Committee on Appropriations, *War Department Appropriation Bill for [Fiscal Year] 1936, Military Activities. Hearings Before the Committee on Appropriations, Part 1*, 74th Cong., 1st sess., 29 January 1935, 537, 538, Hathi Trust Digital Library.

103. John F. Guilmartin, Jr., "Air Warfare," *The Oxford Companion to American Military History*, 27; "Major General Oscar Westover," USAF, undated; Hewes, *From Root to McNamara: Army Organization and Administration, 1900–1963*, 391; U.S. Congress, House, Committee on Appropriations, *Military Establishment Appropriations Bill for [Fiscal Year] 1938, Hearings before a subcommittee of the House Committee on Appropriations.* 75th Cong., 1st sess., 25 March 1937, 516–520, ProQuest Congressional Documents, Pentagon Library.

104. U.S. Congress, House, Committee on Appropriations, *Military Establishment Appropriations Bill for [Fiscal Year] 1938, Hearings Before a Subcommittee of the House Committee on Appropriations*. 75th Cong., 1st sess., 25 March 1937, 520, 521, ProQuest Congressional Documents, Pentagon Library.
105. Ltr., Brig Gen A. W. Robbins, Chief, Materiel Division, Air Corps, to Brig Gen F. W. Andrews, Air Corps, Commanding General G.H.Q. Air Force, Subj.: Procurement Issues, 4 November 1935, Frank M. Andrews Papers, Personal Correspondence, Box 6, Manuscript Division, LOC; Ltr., Brig Gen A. W. Robbins, Chief, Materiel Division, Air Corps, to Brig Gen F. W. Andrews, Air Corps, Commanding General G.H.Q. Air Force, Subj.: Procurement Issues, 18 November 1935, Frank M. Andrews Papers, Personal Correspondence, Box 6, Manuscript Division, LOC; "Materiel Division Developments and Activities During 1935," undated, 2, U-1167, A.C., RG 18, AAF Records, Entry 211, Box 13, Vol. 24, National Archives,; "Boeing Test Bomber, Model 299," *Air Corps News Letter*, 15 March 1935, 18, V-6822, A.C., Information Division, Office of the Chief of the Air Corps, AFHSD, www.afhistory.af.mil/History; "The Crash of the Boeing Bombing Plane," *Air Corps News Letter,* Information Division, Office of the Chief of the Air Corps, V-6898, A.C., 15 November 1935, 1, AFHSD, www.afhistory.af.mil/His tory; Jean H. Dubuque, *The Development of the Heavy Bomber, 1918–1994. USAF Historical Study No. 6* (Maxwell AFB, A.L.: Historical Division, Air University, 1951), 17, 15; Mary R. Self, *History of the Development and Production USAF Heavy Bombardment Aircraft, 1917–1949. AFLC Historical Study No. 195* (Wright-Patterson AFB, OH: Historical Office, Air Force Materiel Command, 1950), 18–19; Mansfield, *Vision: A Saga of the Sky*, 117–124; Rice, *The Politics of Air Power: From Confrontation to Cooperation in Army Aviation Civil–Military Relations*, 142; Futrell, *Ideas, Concepts, Doctrine: A History of Basic Thinking in the United States Air Force, 1907–1964*, 36, 44; Roger A. Freeman, *B-17 Fortress at War* (New York, NY: Charles Scribner's Sons, 1977), 8; "Secretary of War Expresses Need for a Balanced Air Corps Program," *Air Corps News Letter*, 1 January 1938, 1, Information Division, Air Corps, V7588, A.C., AFHSD, www.afhistory.af.mil/History; "Douglas B-18 Bolo," NMUSAF,

14 April 2015, nationalmuseum.af.mil; Frederick A. Johnsen, "The Great Bolo Boondoggle," *Wings*, August 1973, 10; Tate, *The Army and Its Air Corps: Army Policy Toward Aviation 1919–1941*, 164; "Contracts for New Airplanes," *News Letter, Air Corps Training Center Number, No. 24*, Office of the Chief of the Air Corps, V-6925, A.C., 1 January 1936, 15, AFHSD, www.afhistory.af.mil/History; Ltr., Maj Gen Frank M. Andrews, to Brig Gen Delos C. Emmons, Subj.: B-17 Bomber, 1 June 1937, Frank M. Andrews Papers, Personnel Correspondence, Box 3, Manuscript Division, LOC; Ltr., Maj Gen Frank M. Andrews, to Maj Gen Charles E. Kilbourne, Subj.: Coast Defense Mission, 1 June 1937, Frank M. Andrews Papers, Personnel Correspondence, Box 5, Manuscript Division, LOC; "Model 299 Crash," NMUSAF, 25 June 2009, nationalmuseum.af.mil; Rae, *Climb to Greatness*, 96; Morris, *The Origins of American Strategic Bombing Theory*, 186.

106. Maj Gen Oscar Westover, Chief, Air Corps, "American Commercial Aviation's Contribution to National Defense," *Air Corps Newsletter*, Information Division, Office of the Chief of the Air Corps, V-6958, A.C., 1 March 1936, 15, AFHSD, afhistory.af.mil.
107. Ibid.
108. Ibid.
109. "Collier Trophy Presented to Captain Hegenberger," *Air Corps News Letter*, Information Division, Office of the Chief of the Air Corps, V-6827, A.C., 1 August 1935, 4, AFHSD, www.afhistory.af.mil/History; "Army Air Corps Wins Collier for Fourth Time," *Air Corps News Letter*, Information Division, Office of the Chief of the Air Corps, V-7844, A.C., 15 September 1938, 1, AFHSD, www.afhistory.af.mil/History; Lt. Joseph S. Edgerton, "First Solo Blind Flight A. Success," *Air Corps News Letter*, Information Division, Office of the Chief of the Air Corps, V-6271, A.C., 28 May 1932, 207, 208, AFHSD, www.afhistory.af.mil/History.
110. "Pressure Cabin Investigations Phase 1 (Aircraft Branch Report)," Air Corps Technical Report No. 4220, 1 October 1937, 1, 3, Air Corps Information Circular, Vol. VIII, No. 710, published by the Chief of the Air Corps, Washington, D.C., ProQuest Congressional Documents, Pentagon Library; "Dr. Younger Comes to Materiel Division," *Air Corps News Letter,* Information Division, Office of the Chief of the Air Corps,

V-6827, A.C., 1 August 1935, 24, AFHSD, www.afhistory.af.mil/History; "Army Again Wins Collier Air Award," *New York Times*, 16 September 1938, 44, ProQuest Historical Newspapers, Pentagon Library.

111. Maj Gen O. Westover, Subj: Air Corps Experimental Developments During 1937, Address Before the Society of Automotive Engineers, Detroit Meeting, 13 January 1938, 3, U-1215, A.C. 1/6/38, RG 18, AAF Records, Entry 211, U-Stencils, Box 14, Vol. 25, National Archives; "Collier 1930–1939 Recipients," National Aeronautic Association, undated, naa.aero/awards; "Army Air Corps Wins Collier for Fourth Time," *Air Corps News Letter*, Information Division, Office of the Chief of the Air Corps, V-7844, A.C., 15 September 1938, 1, AFHSD, www.afhistory.af.mil/History; "The Collier Trophy," *Air Corps News Letter*, Information Division, Office of the Chief of the Air Corps, V-8013, A.C., 1 April 1939, 4–5, AFHSD, www.afhistory.af.mil/History; Diana G. Cornelisse, *Splendid Vision, Unswerving Purpose; Developing Air Power for the United States Air Force During the First Century of Powered Flight* (Wright-Patterson AFB, OH: ASD/HO, USAF Publications, 2002), 128–129; Maj Gen H. H. Arnold, Subj.: "Performance and Development Trends in Military Aircraft and Accessories," delivered before the Society of Automotive Engineers, 11 January 1939, 8, U-1267, A.C., RG 18, AAF Records, Entry 211, Air Corps, U Stencils, Box 14, Vol. 26, National Archives 55–56; *The Aircraft Year Book for 1939* (New York, NY: Aeronautical Chamber of Commerce of America, Inc., 1939), 60, AIA, aia-aerospace.org; "Roosevelt Gives Collier Trophy to Army Air Corps," *Washington Post*, 17 September 1938, X2, ProQuest Historical Newspapers, Pentagon Library.

CHAPTER 6

Mobilizing for Global War, 1938–1941

During the years just before America's entry into World War II, accelerated incremental aeronautical innovation in the U.S. produced ever more capable modern aircraft. They were based upon the design features of the "airframe revolution" and increasingly powerful engines. While commercial aviation was an important factor, the armed forces and the NACA continued to be the main drivers of aviation progress.

Representative Carl Vinson, a committed navalist, continued playing a key role in convincing Congress to meet and then surpass treaty limits on the Navy's strength. The Bureau of Aeronautics' (BuAer's) leaders also promoted significant incremental advances in naval aviation technology while increasing the fleet's aircraft numbers and operational capabilities. They enjoyed more support within the battleship-oriented service than is generally recognized. Their cause was helped by the collapse of the naval armaments treaty system.

The aggressive rise of Japan, Germany and Italy provoked an international arms race that the U.S. was late to join. By the early 1930s, those developments enabled President Franklin D. Roosevelt and a reluctant Congress to launch significant rearmament programs which initially focused on strengthening America's air and naval forces.

The nation's aircraft industry, while initially smaller than some of its foreign rivals, possessed some of the finest design and manufacturing

C. J. Gross, *Technological Innovation and the Rise of Aviation, 1903–1941*, Palgrave Studies in the History of Science and Technology, https://doi.org/10.1007/978-3-031-71108-4_6

talent in the world. Largely financed by military spending, those staffs were key to the fact that Americans were flying some of the world's most innovative aircraft epitomized by two iconic ones; Boeing's B-17 *Flying Fortress* and the Douglas DC-3 airliner.[1]

Thinking Big

As author Wayne Biddle observes, when British Prime Minister Arthur Neville Chamberlain "... stepped from his personal Lockheed *Super Electra* airliner in Munich on September 15, 1938, the world turned a dark corner."[2] The threat of war had stimulated an extraordinary growth in American aviation exports. The *New York Journal and American* had reported in May 1938 that "Exports of aircraft products from the United States during the first three months of 1938 totaled $14,320, 273, a 99.3% increase over the corresponding period in 1937."[3] Such export deals carried profits of at least 18%. Meanwhile, the U.S. military began an exponential increase in spending for new aircraft. During 1938, it allocated nearly $116.8 million for that purpose, almost twice what it had allocated for that purpose three years earlier.

President Roosevelt's official position regarding the deteriorating international situation was neutrality. However, he worked around it by liberalizing policies on selling military aircraft and other arms to friendly foreign nations. Roosevelt's most significant action in that respect was his decision not to cut off the flow of arms to China after Japan invaded in 1937. Biddle notes that the President "... also approved numerous instances of letting French and British military officials inspect the latest American warplanes. The neutrality laws passed by Congress were not mere wallpaper, but Roosevelt used his discretionary powers to skirt them whenever they stood in the way of his foreign policy objectives."[4]

Foreign sales provided a significant stimulus to America's aviation industry. Between 1927 and 1937 the nation's exports had grown from 63 to 631 aircraft annually. By the latter year, they were valued at over $39 million and accounted for 26% of the industry's total sales. From 1934 to 1939, exports of aircraft and spare parts provided for 31% of the industry's sales. Prior to 1938, most foreign sales of American military aviation products had been to Asia and Latin America. During 1938, that focus shifted to threatened European democracies.[5]

Foreign sales during the last years of the 1930s provided a critical stimulus to the buildup of America's aircraft manufacturing capacity.[6] On 27

January 1939, the White House disclosed that Roosevelt had personally endorsed a plan to sell the latest U.S. military aircraft to France. The *New York Times* reported in April 1940 that Great Britain and France had negotiated tentative contracts to purchase over 1,500 warplanes in the U.S. worth at least $120 million.[7]

Convinced that fear of the German *Luftwaffe* was the main reason that Britain and France caved into Hitler's demands, Roosevelt held a secret meeting on 28 September 1938 with his top military advisors to consider the nation's aviation needs. In a dramatic reversal of his previous stance which had relied on sea power as the nation's primary defense, the President emphasized that aviation offered a solution to America's security problems. To the dismay of the officers present, Roosevelt announced that the U.S. would expand its military aircraft inventory by 10,000 planes at a cost of $500 million. He wanted to strengthen the aircraft industry so that it could produce 24,000 planes a year. At that point, not nearly enough had been done to increase the capacity of that industry to produce warplanes on that scale. The industry's production of Air Corps aircraft had been under 90 a month that year. Roosevelt, who believed that words (and aircraft) were useful in deterring Hitler's aggressive intentions, neglected to include funds for the personnel, training and repair facilities, fuel, support equipment, spare parts and supplies required to turn those new airplanes into effective air power.[8]

An internal War Department study of the American aircraft industry completed in October 1938 had concluded that assuming "... present capacity could be doubled in twelve months. We could only expect production, during the first year of war, of some 1,800 tactical planes and 1,800 training planes. However, approximately half of this production probably would have to be allocated to the Navy."[9] Although skeptical at the time about Roosevelt's military aircraft production goal (or rhetoric) in September 1938, years later then Gen of the Air Force Henry H. Arnold referred to the President's startling announcement as the Air Corps' "Magna Carta."[10]

Following Roosevelt's secret meeting, Air Corps planners and the Army's General Staff developed an ambitious plan for a "balanced force" including expansion of the service's ground units. Fearing that it was too politically controversial in a nation that remained deeply divided on the question of American military intervention in another foreign war, Roosevelt rejected it.

In early 1939, Congress passed the President Roosevelt's scaled back aviation program, allocating $300 million and authorizing the Air Corps to expand to 5,500 aircraft including 3,251 new planes. But Roosevelt's real objective remained unchanged: to obtain 10,000 aircraft over the next 10 years. The emergency expansion was rationalized as an effort to defend against the possible establishment of Nazi air bases in Latin America. To help increase the American aircraft industry's productive capacity, Roosevelt indicated that his administration was favorably disposed toward accepting French orders for new planes. The Navy, which had requested $86 million to bring its air strength up to the 3,000 planes authorized in May 1938, received only $21 million from Congress.

Arnold and senior War Department officials met with aircraft industry representatives during November 1938 and asked them to plan for an unprecedented growth of military production. Although the government could offer them neither money nor contractual guarantees at that time, those individuals worked out several informal deals. Initially, expansion plans focused on increasing the capacity of the aircraft industry rather than building government plants and adopting Detroit's mass production techniques.

The Aeronautical Chamber of Commerce concluded, after a survey of the aircraft industry's facilities and productive capacity in early 1939, that those manufacturers could more than double their military output by hiring more labor and installing machinery—all without enlarging their factories. The industry had already begun making heavy investments to enlarge its productive capacity. The Roosevelt administration was also pushing for increased exports and commercial sales to help strengthen that industry.[11]

On 4 November 1938, Assistant Secretary of War Louis Johnson requested permission from President Roosevelt to begin talks with industry on how to achieve maximum aircraft production in their commercial plants. Johnson wanted to push the industry to begin serious planning along those lines. He also forwarded plans and a memo from then Air Corps Maj Gen Arnold to Roosevelt proposing government financing of new factories each capable of producing 1,200 aircraft a year. Their estimated cost would be approximately $10 million per site if located on federal lands that included airfields. It would increase to $15 million for each facility if an airfield had to be constructed.[12]

Roosevelt had to tread carefully on rearmament because of domestic politics. Despite growing threats of international aggression, Congress and the American people remained strongly opposed to becoming involved militarily in another foreign war. Congress had passed a series of neutrality acts after 1935 to restrict governmental support and private financial assistance to any belligerent. Historians Allan R. Millett and Peter Maslowski stressed that those acts "... assumed that the United States could follow isolationist policies that would prevent entry into 'foreign wars' and still protect its national interests and physical security."[13]

In January 1939, Secretary of War Harry H. Woodring, a strict non-interventionist, had asked Congress for authority to purchase approximately 3,032 additional planes for the Air Corps as part of Roosevelt's emergency national defense proposal. That was intended to provide a total of 5,500 planes for the Army by 1941.[14] Later that month, Maj Gen Arnold testified on Capitol Hill that "The performance of airplanes depends directly upon research. For some time we have been falling behind, mainly because we have not with our research work kept abreast of the rest of the world."[15]

The War Department wasted little time initiating the Air Corps' expansion program. The day after Roosevelt signed the Army Appropriations Act on 26 April 1939, Assistant Secretary Johnson reported the award of contracts to purchase 571 aircraft and associated equipment. The price tag for the planes, engines and accessories was a then impressive $50 million. Anticipating that those funds would become available during the spring of 1939, work had already begun on the required data and contracts. Option clauses had been included in some contracts that enabled the Air Corps to acquire additional aircraft if funds became available.[16] During Fiscal Year 1939, the War Department placed aircraft, engine, spares and government-furnished equipment orders valued at $92,232,231.[17] On 10 August 1939, Johnson announced the tentative award of nearly $86 million in contracts as a result of competitive bids for additional aircraft and engines.[18] Table 6.1 shows the dramatic growth of Air Corps research and development obligations from Fiscal Year 1935 through Fiscal Year 1942.

Air Corps' procedures for implementing its accelerated modernization and buildup program had to be changed to build a much larger and more capable force quickly. For example, during late January 1939 the Air Corps sent out a circular proposal for a new twin-engine medium bomber with a maximum speed of 350 mph, 3,000 mile range and 2,000 pound bomb load to manufacturers. Based on evaluation of competing paper designs, the Air Corps approved a production contract that was signed

Table 6.1 Air Corps research and development obligations, fiscal years 1935–1942[19]

Fiscal year	*Total research and development obligations*
1935	$3,461,893
1936	$5,977,444
1937	$4,511,116
1938	$4,330,504
1939	$3,421,082
1940	$8,992,871
1941	$57,528,767
1941–1942	$33,223,006 (Supplemental)
1942	$113,327,141

with the Martin firm for 1,000 of its B-26 *Marauders* during September of that year. The aircraft made its maiden flight on 25 November 1940. Normally that entire process could have taken three years.

On 2 July 1940, Congress authorized the military to negotiate cost plus fixed-fee contracts (CPFF) with manufacturers. The legislation also allowed advance payments to a contractor not exceeding 30% of the original contract price. Military aircraft progress payments had been unknown prior to the buildup. Fixed-price contracts had been standard. Now, orders could be placed without advertising. Additional legislation granted the President authority to direct priority of deliveries for the Army and Navy over those for private accounts and exports. He could requisition items whose export had been denied as well as commandeer plants and compel acceptance of orders.

To iron out the problems with the CPFF arrangements, Assistant Secretary of War Robert P. Patterson inaugurated the "letter of intent" as a stopgap measure to speed the contracting process. It directed contractors to begin work on a project while leaving the exact details of a contract, often including the price, to be decided later. Both military departments adopted that administrative device. Congress created a new position during December 1940, the Under Secretary of War. Patterson became its first incumbent. He was responsible for all the procurement functions of the Assistant Secretary of War.

During that same period, the Air Corps and BuAer inaugurated cooperative production arrangements with the automobile industry designed to dramatically expand aircraft output. In February 1941, the Army's air arm announced that it had begun accelerating the production of new model aircraft on a limited basis by buying aircraft without requiring submission of prototypes. At that point, it limited the new policy to firms that had already produced similar satisfactory aircraft or where the designs had been proven by wind tunnel tests and similar unspecified preliminary checks. In early December 1940, officials at Wright Field reported that the first production models of new aircraft models were being put through accelerated service tests by relays of aircrews as soon as they came off production lines. In a matter of several weeks of flying around-the-clock in shifts, pilots completed service tests that could have taken operational squadrons up to a year to accomplish.[20]

Rear Adm John H. Towers, who had become BuAer's Chief on 1 June 1939, testified before Congress in January 1940 that large foreign orders were having a beneficial effect on the American aircraft industry. During the following April, Maj Gen Arnold voiced similar sentiments to the *Wall Street Journal*. Contracts for such orders contained a clause relieving the manufacturers of liability if they had to be canceled because of U.S. defense requirements. To prevent interference with American military aircraft orders, U.S. contracts also contained a penalty clause if manufacturers failed to deliver on schedule. Arnold assured Congress during November 1939 that "... in all foreign contracts there is a clause that they shall in no way interfere with production for the United States Army and Navy."[21]

Despite foreign orders, America's pre-war aircraft production capacity proved to be far short of what was eventually needed. During 1935, President Roosevelt had asked Secretary of War George H. Dern for data on U.S. aircraft production capacity. The War Department's survey concluded that, although the quality of American aircraft and accessories equaled or surpassed that of other nations, its productive capacity was well below what would be needed in a major war. It projected that, in the event of a major conflict, during the first year alone the U.S. would require 32,437 military aircraft of all types and 57,978 engines. The department estimated that the nation's aircraft industry could produce no more than 7,238 military aircraft annually. To rectify that enormous shortfall, Dern recommended that it would be necessary to convert other industries to aircraft production. That policy recommendation was finally implemented after Europe erupted in full-scale war during September 1939.

In mid-1940, T. P. Wright, the vice president for engineering of the Curtiss-Wright Corporation, estimated that the industry's total productive capacity was approximately 10,000 aircraft annually. President Roosevelt had just called for production of 50,000 American military planes a year following the rapid German conquest of Denmark, Norway, France and the low countries. Although his daring request seemed unreachable at the time, Roosevelt's habit of thinking big helped to stimulate an enormous expansion of America's aircraft industry's productive capacity. At its wartime peak during 1944, American military aircraft production reached 96,318.[22]

To place that huge wartime accomplishment in context, during 1938 Federal government appropriations for all military aviation-research and development plus production amounted to $122 million. That was a significant amount compared to the rest of the interwar period. The entire aircraft industry employed some 36,000 persons and ranked in the size of its labor force just behind the knit-hosiery makers. At that point, a contract to produce 50 aircraft was considered an enormous order. During 1938, expansion of the American aircraft industry's productivity got underway, but it was a puny 3,623 that year compared to the wartime high.[23] That was an enormous achievement, especially in light of aircraft production levels during the 1934–1941 period as shown in Table 6.2.

During 1940 the Permanent Working Committee of the Aeronautical Board issued joint procurement documents containing 60 specifications plus 23 Army-Navy standard drawings for common aircraft items. They were used by BuAer and the Air Corps to simplify aviation procurement and stocking work in both military services. Some of the more important items affected by these standardization decisions included screw threads,

Table 6.2 American aircraft production, 1934–1941[24]

Year	*Total*	*Military*	*Civil*
1934	1,615	437	1,178
1935	1,710	459	1,251
1936	3,010	1,141	1,869
1937	3,773	949	2,824
1938	3,623	1,800	1,823
1939	5,856	2,195	3,661
1940	12,813	6,028	6,785
1941	26,289	19,445	6,844

parachute silk, steels and aluminum alloys of various types, aircraft fuels and engine accessories. An especially significant accomplishment along those lines was the development of joint documents outlining general specifications for engines and model specifications for aircraft power plants plus specifications addressing the types of testing needed for them.[25]

Germany First

The Air Corps' senior leadership hadn't become especially focused on the threat posed by German military aviation until the late 1930s. Following visits to Germany by leading American aviation figures including famed flyer Charles A. Lindbergh and NACA Chairman Dr. Joseph S. Ames, the federal government began taking serious initiatives to overcome fears that the Third Reich would soon surpass the quality of U.S. aircraft because of its heavy investments in aeronautical research and development.[26]

The implications of the growing danger of German air power gradually began to sink into the minds of some Army leaders. During June 1938, Gen Malin Craig, the Army Chief of Staff, reversed his previous stance requesting that 11 B-17s and 32 B-18s be added to the already approved 1939 aircraft procurement program. When he replaced MacArthur in 1935, Craig had strongly opposed spending on research and development of new weapons including heavy bombers. Instead, he focused on buying readily available weapons for the army's combat arms. Historian Craig F. Morris writes that during August 1939 the Air Corps' Materiel Division "... warned military aviation technology in America had fallen behind that of other advanced countries, a slippage that it called 'a deplorable situation that could not be tolerated.'."[27]

Following congressional approval, the NACA established a new research facility at Moffett Field, California in December 1939 primarily because of its proximity to the growing west coast aviation industry. Because NACA's leadership had become convinced that inferior engines were retarding the development of faster American warplanes, Congress quickly approved its proposal to establish another new laboratory in June 1940. The NACA selected a site at Cleveland, Ohio, mainly because of its easy access to established engine manufacturers. It opened in 1942.[28]

The Japanese had also been making significant strides in military aviation during the 1930s. The American military establishment knew a great deal about those developments. According to historian William M. Leary, as early as the mid-1930s Army, Navy and Marine Corps intelligence officers submitted numerous "... reports on Japanese aviation [that] were timely, reasonably accurate, and generally free of racial stereotypes."[29] The Navy had kept an especially close watch on Japanese aeronautical developments. Based on numerous visits to Japanese factories and naval stations, LT Cdr Ralph A. Ofstie, an aviator and naval attaché, informed his superiors that Japanese flyers were "... well trained, had very high morale, and were capable of the very greatest determination."[30]

However, some American observers in China following the Japanese invasion of Manchuria in 1931 had not been impressed by the operational air prowess of either side. Rear Adm Harry E. Yarnell, commander of the U.S. Asiatic Fleet, concluded that Japanese "... aviation is distinctly inferior to ours. A few of our well-trained squadrons could drive their planes out of the air... in a short time."[31] But, there were dissenting opinions concerning the potency of Japanese air power. LT R. G. Lockhart, the senior naval aviator embarked on the cruiser USS *Augusta*, reported that the Japanese had developed and executed an especially impressive air operations plan which had focused on attacking Chinese air bases. Maj Gen Arnold strongly endorsed Lockhart's findings during an address at the Army War College in October 1938, warning that Japan was "... a first-rate air power which knows how to use its air strength."[32] Arnold was not alarmed by that situation though. He reasoned (incorrectly) that the Navy could take care of Asia while the Air Corps concentrated on Europe if the U.S. intervened in a global war.

In addition to congressional indifference and an isolationist public, the Air Corps' ability to analyze and deal effectively with the gathering storms overseas was hindered by some self-inflicted wounds. Its industrial war planning efforts were woefully inadequate. Furthermore, it lacked an organization to systematically collect, analyze and disseminate reliable information on significant foreign aviation developments during almost all of the interwar period. It neglected to create its own intelligence organization until November 1940 when Arnold transformed the Information Division in his office into the Intelligence Division. The Materiel Division at Wright Field was even slower to create its first technical intelligence organization. Inundated by a flood of captured foreign military equipment that needed to be evaluated, it finally established a Technical Data Library in December 1942. That temporary organization was not formalized until July 1945.[33]

The first of a series of events that brought the U.S. into military participation in World War II took place in Europe on 1 September 1939 when Germany invaded and quickly conquered much of Poland. Roughly two weeks later, the Soviet Union, then Germany's ally, gobbled up the major remaing portion of that hapless nation. In response to Hitler's aggression, Great Britain and France declared war on Germany. Things then quieted down in Europe during a period known as the "phony war." But, then during early 1940 Germany quickly conquered Denmark, Norway, the Low Countries and France. Its military style of fast moving coordinated armored and air warfare was dubbed "blitzkrieg" by western media. Those stunning German victories forced a dramatic reversal of U.S. national defense policy. America could no longer rely on the continental European balance of power to help protect it.[34]

President Roosevelt moved cautiously to prepare America for war. In the name of hemispheric defense, he sought to strengthen Great Britain against German Nazi dictator Adolph Hitler in Europe. On 19 September 1939, Treasury Secretary Henry M. Morgenthau, Jr. secured Roosevelt's backing to provide Great Britain with all the aircraft it wanted despite opposition from the U.S. armed forces. During the following May, Roosevelt ordered Secretary Woodring and Gen George C. Marshall, the Army Chief of Staff, to coordinate all new aircraft orders through Morgenthau until official administrative machinery for that purpose was established. Meanwhile, in November 1939, the President had persuaded Congress to modify the neutrality laws enabling the U.S. to sell arms to belligerents under the new "cash and carry" provisions which favored Great Britain and France because Germany could not overcome the Royal Navy and Royal Air Force (RAF) to ship U.S. arms across the Atlantic.[35]

Before World War II engulfed Europe, the Joint Army-Navy Planning Committee had begun developing a series of five color-coded "Rainbow" plans to deal with a range of potential security threats to the U.S. beyond defense of the Western Hemisphere. The planners met secretly with their British counterparts in early 1941 to map out a broad coalition strategy if the U.S. entered the war. The Americans had already concluded that Germany was the most dangerous power because of its economic and military strength. The British shared that assessment. The resulting "ABC-1 Staff Agreement" called for a "Germany First" strategy. While Roosevelt never formally approved it, American military planners based their "Rainbow" war plans developed during the summer of 1941 on that strategy.

In Asia, Roosevelt sought to deter further Japanese aggression by moving the fleet from its west coast bases to the Navy's unfinished facility at Pearl Harbor, Hawaii. He also embargoed exports of iron, steel and high octane aviation gasoline in an unsuccessful attempt to deter the Japanese.[36]

Pressured by the White House, Congress had passed the Lend Lease Act during March 1941 that authorized the President to sell, lease or lend military hardware to any country he designated as vital to American national security. Notwithstanding its repayment provisions, the legislation "... further fused the United States to the Allied war effort and gave the British a claim to American war production uninhibited by their inability to pay. In May [1941] the administration extended Lend Lease to the Chinese Nationalists with the faint hope that the Chinese would pin down the Japanese army to the mainland of Asia."[37] Ultimately, the U.S. supplied its allies—which eventually included the Soviet Union—with over 42,000 American-made aircraft during World War II.[38]

Despite German victories on the European continent and the start of the Battle of Britain in 1940, American public and congressional opinion had remained deeply divided about direct U.S. military involvement in war. Consequently, the nation's leaders were able to take only limited steps to rearm. Congress approved and Roosevelt signed the "Two Ocean Navy Bill," also known as the "Vinson-Walsh Act," that summer. It doubled the tonnage of the Navy's combat ships, authorizing 13 large new *Essex*-class fleet aircraft carriers. The USS *Essex* was commissioned on 31 December 1942. By the war's end, 20 of those giants had been launched, each with a hangar deck could accommodate up to 103 aircraft.

The 1940 legislation also increased the authorized naval aircraft strength to 4,500 "useful" airplanes (i.e., those which were not undergoing serious repairs or being used for developmental testing). That July, the authorized strength of naval aircraft was dramatically increased to 15,000 "useful" planes with a provision allowing the president to increase it further on his own authority. During January 1942, Roosevelt approved raising that authorization to 27,500 aircraft (Illustration 6.1).

The Air Corps won approval that year for its expansion plans which eventually called for up to 7,800 combat aircraft. The most controversial initiative involved a major expansion of the Army's ground forces. Over 600,000 men were drafted into the Army following passage of the "Selective Service and Training Act" in September 1940. It was the nation's first peacetime draft. During the 12 months ending in June 1941, nearly

Illustration 6.1 USS Essex underway, May 1943 (*Photo* USS ESSEX, National Archives, Catalog #: 80-G-68097, NHHC, history.navy.mil)

300,000 National Guardsmen were also called into federal active duty. Including mobilized reservists and volunteers, the U.S. Army numbered some 1.2 million personnel by the summer of 1941. But, those numbers were misleading. Historians Millett and Maslowski emphasize that "It was a force weakened, not strengthened, by manpower and materiel shortcomings."[39]

Victory Through Air Power

For Maj Gen Arnold, who became Chief of the Air Corps on 29 September 1938 after Maj. Gen. Westover was killed in an airplane crash, the airframe revolution was epitomized by the arrival of the YB-17A, the prototype of Boeing's B-17 *Flying Fortress* bomber, at Langley Field on 1

March 1937. Arnold and other senior Army airmen saw it as the technological means to turn the concept of strategic bombardment into reality and provide a practical means of attaining their goal of a separate air force.

Arnold strengthened his authority as Chief of the Air Corps during 1939. He gained control of General Headquarters (GHQ) Air Force which had been reporting directly to the Army Chief of Staff giving him jurisdiction all of the service's operational aviation assets. Moreover, all Air Corps training and equipment procurement activities were centralized directly under him in Washington, D.C. The *Washington Post* reported in September 1939 that Arnold was now effectively the Army's number two officer and one of its most powerful branch chiefs having demonstrated a strong aptitude for working with Congress and War Department officials. It could have added that the new Chief had a real flair for cultivating favorable media publicity. Arnold also had an excellent working relationship with Gen. Marshall, a strong supporter of aviation.[40]

As noted earlier, President Roosevelt had called in 1940 for the production of 50,000 military aircraft annually. That figure pressured the aircraft industry to accelerate its expansion and forced conservative-minded military officers to think big while stimulating public support for the nation's military buildup. Historian Richard J. Overy concludes that it committed the U.S. to a massive wartime military aviation program and a "general air power strategy" that played a critical role in Allied victory. Great Britain had also adopted that approach. British and American air forces fielded units that conducted massive global tactical, strategic bombing reconnaissance and airlift missions. That broad conception of air power, not just a fixation on strategic bombing, would shape postwar U.S. military aviation. In contrast, the Axis powers of Germany, Japan and Italy had focused on mounting tactical support of their ground and naval forces in comparatively limited theaters of operations during World War II.[41]

Reacting to the German invasion of the Soviet Union in June 1941, Roosevelt directed the War and Navy departments to develop an overall production estimate of the material needed to defeat the Axis powers. The Joint Army and Navy Board developed the final estimates. The military unveiled the "Victory Program" later that year. It originally provided theoretical estimates instead of exact dollar figures. A subsequent estimate placed the program's cost at nearly $150 billion. During the war, total U.S. defense production outlays totaled just over $162 billion.[42]

Marshall agreed to Arnold's request to allow his own staff to develop the air annex to the War Department's production estimates for the "Victory Program." Guided by the Army air arm's own concepts of its strategic wartime objectives, a team of four officers, all of whom had played key roles in developing the doctrine of daylight precision strategic bombing, produced the Air War Plans Division-Plan 1 (AWPD-1). The plan was completed in just nine days. AWPD-1 stressed defense of the western hemisphere followed by a strategic air offensive against Germany, while conducting defensive operations in the Pacific against Japan. Its primary goal was to build a strategic air force that could conduct weaken and topple Germany on its own if possible or support a land invasion if necessary. AWPD-1 proposed destroying the *Luftwaffe* and targeting electric power, oil and transportation systems in that order. The plan was a synthesis of the strategic air power doctrine that had evolved at the Air Corps Tactical School during the 1930s and the Rainbow Plans. AWPD-1 recommended an interim force equipped with 59,727 aircraft by 1943 or 1944. It projected an ultimate force of 63,467 planes by 1945 including a large number of long-range bombers.[43]

From 1938 through 1942, the Army's air arm grew enormously after a slow start. For example, its personnel end strength on 30 June 1938 was 21,089. By 30 June 1942, it stood at an incredible 764,416, accounting for 24.9% of the rapidly expanding Army's total manpower. The Army air arm's airplanes on hand grew from 2,402 on 31 July 1939 to 23,247 three years later.[44] To keep abreast of its explosive growth, the Air Corps was reorganized and redesignated the United States Army Air Forces (USAAF) on 20 June 1941 as part of Marshall's reorganization of the Army. On March 1942, the Army was reorganized a second time putting the USAAF on an equal footing with the Army's ground and service forces. Marshall and Arnold agreed that, in return for the Army airmen suspending their behind-the-scenes lobbying for a separate air force, the USAAF would enjoy a great deal of organizational autonomy within the Army for the war's duration.[45]

The Navy planned its portion of the "Victory Program" in isolation from the Army and the USAAF. It was essentially a repeat of the sea service's 1940 plan with additions. Importantly, it featured a new strategic rationale. The fleet would be strong enough to conduct offensive operations in both the Atlantic and the Pacific instead of remaining on the defensive in one while conducting offensive operations in the other. The Navy's plan called for a large number of new warships including 24 large aircraft carriers. It retained the already authorized force of 15,000 aircraft.[46]

The Dragon's Teeth

Ancient myth held that after a Phoenician prince tossed the teeth of a dragon he had killed into a ploughed field, a huge number of heavily armed warriors sprang up ready to do battle. Brig Gen Benjamin S. Kelsey was one of the key officers who managed the Air Corps' research and development of the pursuit aircraft that came to dominate the skies during World War II. He used that tale to illustrate his key argument that "The instantaneous response, complete with weapons, makes it a myth. The creation of U.S. air power for World War II, although almost as incredible was no myth; the time and weapons were real. Cold facts reveal numbers and time, but the seeds were planted and the tenuous growth nurtured for years before the response was needed."[47]

Regardless of their potential, there had been little support in the War Department or elsewhere in the federal government for purchasing significant numbers of long-range strategic bombers during the late 1930s. The General Staff continued to focus on the need for less expensive aircraft that would support ground forces although the Air Corps pushed back against that policy. The Navy pressed to reassert its monopoly of long-range coastal patrol activities. Congress and the public had little enthusiasm for weapons that suggested that the U.S. would adopt an offensive strategy.[48]

Despite its growing emphasis on big bombers, the Air Corps continued to allocate considerable resources to the development of more capable pursuit, attack and observation aircraft. It remained subject to the mandates of the Army's ground-oriented General Staff and the service's Chief of Staff. Most of the major pursuit and attack types employed by the Army's air arm during World War II traced their origins from the mid to late 1930s.

As noted earlier, the Seversky Company's P-35 was the Air Corps' first production model single-seat, all-metal pursuit aircraft with retractable landing gear and an enclosed cockpit. It was also equipped with a manually controlled, multi-position, constant speed propeller. The warplane had been designed by Alexander de Seversky and his chief engineer, Alexander Kartveli, a French trained immigrant of Georgian descent.

It was the forerunner of Republic Aviation's rugged P-47 *Thunderbolt* that was designed by Kartveli.

The P-47 was an outgrowth of a design series that began with the P-35. Nicknamed the "Jug," the P-47 was powered by a massive Pratt & Whitney liquid-cooled 2000 hp. R-2800 *Double Wasp* 18-cylinder radial engine. Kartveli incorporated an unusual turbo-supercharger located behind the pilot and vented its exhaust out of the plane's belly. On 6 May 1941, the XP-47B flew for the first time. Fully loaded with 2,500 hundred pounds of bombs, it weighed nearly eight tons. The initial contract for 773 aircraft was placed in September 1940. Serving primarily in a ground attack role, 15,636 P-47s were produced altogether. It was the largest and heaviest fighter in the war.[49]

The Curtiss P-40 *Warhawk*, the best available U.S. fighter-type aircraft available in quantity during December 1941, was actually a P-36 hurriedly modified with a more powerful engine. Its Pratt & Whitney R-1830 Twin Wasp 14-cylinder air-cooled radial engine was installed at the direction of Curtiss's Chief Engineer Donald Berlin in a crash effort to match the performance levels of the top German fighters of the day. The XP-40 flew for the first time during October 1938. Its initial production models were purchased the following year. During the summer of 1940, seven new P-40s were subjected to an accelerated flight testing program at Wright Field. Although agile at low and medium altitudes, the P-40 lacked power at higher altitudes. At medium and high speeds it was one of the tightest-turning early monoplane designs of the war. When production ceased in November 1944, 13,738 P-40s had been built. However, it was clearly outclassed by the top British and enemy fighters during the conflict.[50]

During February 1937, the Army's air arm, following a recommendation by the Air Corps Board, had asked the American aircraft manufacturers to submit design proposals for a high-altitude interceptor capable of a top speed of 360 mph at 20,000 feet. The aircraft had to be able to climb to that altitude from sea level within six minutes. Those advanced performance goals were authored by lieutenants Benjamin S. Kelsey and Gordon P. Saville. Many years later, Kelsey explained that he and Saville had used the word "interceptor" to bypass the inflexible Air Corps requirement that its pursuit aircraft could carry no more than 500 pounds of armament including ammunition and its limitation of single-seat aircraft to one engine. Kelsey wanted a minimum of 1,000 pounds of armament. The two officers were determined to obtain a more capable fighter, one that was superior to existing models at dog-fighting and high-altitude combat.

Any aircraft that met or exceeded those specifications could reverse the performance disparity between bombers and pursuit aircraft that had been developing since the early 1930s. Lockheed Aircraft Corporation, embarking on its first major military venture, proposed a radical twin-engine, twin-boom design that was accepted by the Air Corps that June. The firm's design team was led by chief designer Hall Hibbard and his young assistant, Clarence L. "Kelly" Johnson. The son of Swedish immigrants, Johnson went on to become one of the most storied aircraft designers in American history. He had received bachelors and masters degrees in aeronautical engineering from the University of Michigan. He was responsible for the SR-71 Blackbird, the world's fastest aircraft as well as the F-104 Starfighter and the U-2 Dragon Lady among others.

Lockheed's XP-38 prototype was armed with one 23 millimeter and four 0.50 inch machine guns, all mounted in its nose. Its proposed gross weight of 14,800 pounds was heavier than some bombers. The innovative Lockheed design incorporated tricycle undercarriage, a bubble canopy, and two 1000 hp. turbo supercharged 12-cylinder Allison V-1710 engines fitted with counter-rotating propellers to eliminate engine torque. It was the first American fighter to make extensive use of stainless steel and smooth flush-riveted aluminum skin panels. It was also the first warplane to fly faster than 400 mph in level flight.

Lockheed won the competition on 23 June 1937. The Air Corps awarded it a $163,000 contract to build a prototype; however, the firm's costs on the prototype would eventually total $761,000. Construction began in July 1938 and the prototype first flew on 27 January 1939. Although the aircraft was destroyed in a crash later that year, the Air Corps had already ordered 13 YP-38s at a price of $134,284 each for service testing. The YP-38s were developed into the P-38 *Lightning*, one of the top American fighters during World War II. The earliest production version of the plane, which entered the Air Corps' before the Japanese attack on Pearl Harbor, was capable of a top speed of 390 mph at 20,000 feet. Altogether, 10,038 of the aircraft were produced (Illustration 6.2).[51]

The Air Corps' pursuit development programs had also experienced several major setbacks during the late 1930s. The Bell Aircraft Corporation, a newcomer to the military aircraft field, was responsible for two of them. Responding to an Air Corps requirement for a long-range

Illustration 6.2 XP-38A in flight (*Photo* XP-38A inflight autographed by Lockheed's Chief Engineering Test Pilot Milo Burcham. Photo courtesy of the Stan Carpenter collection. Additional information from the NMUSAF: aircraft serial number 40-762, modified YP-38 aircraft, VIRIN 061018-F-1234P-005, flight date unknown, nationalmuseum.af.mil)

bomber destroyer aircraft, the firm proposed its XFM-1 *Airacuda* in 1936 barely beating out Lockheed in a design competition. Its radical design featured a twin-engine and carried a crew of five including two gunners. Its enclosed crew compartments were heated. In a radical departure from existing conventional designs, it was a pusher-type plane with propellers located behind its wings to give those devices greater efficiency according to Bell. Construction began on the first *Airacuda* during May 1936. The XFM-1's maiden flight was on 1 September of 1937. During May 1938, Acting Secretary of War Louis Johnson announced that Bell had been awarded a contract for a little over $3.16 million to purchase 13 YFM-1 aircraft and spares. They were used for test purposes due to the aircraft's poor performance. The staggering price for a single *Airacuda* was $219,000. During 1942, all 9 surviving YFM-1s were scrapped because of the aircraft's poor performance.[52]

While developing the *Airacuda*, Bell engineers came up with another unusual design, the P-39 *Airacobra* in response to an Air Corps pursuit requirement. During May 1937, Bell proposed a monoplane design built around a single 37-millimeter cannon located on the aircraft's centerline and protruding from its nose. The engine, in an innovative twist, was located behind the cockpit. Bell's configuration set up space for the main armament, a 37-millimeter Oldsmobile T9 cannon. Previous pursuit designs had been driven by the intended engines instead of aircraft armaments. Although devastating while it worked, the T9 cannon carried very little ammunition, had a low rate of fire and frequently jammed. The cannon would be supplemented by two nose-mounted machine guns and 500 pounds of bombs mounted externally. Unlike the *Airacuda*, the *Airacobra* was a single-seat aircraft. The Air Corps was extremely pleased with the design and issued a contract for one XP-39 prototype in October 1937. It first flew at Wright Field during April 1938. A year later additional test versions of the aircraft were ordered. That same year, the Air Corps signed its first production contract for the P-39 (temporarily designated the P-45). Deliveries began in January 1941.

But, the *Airacobra* proved to be a major disappointment in combat for the Army's air arm. Its Allison V-1710 engine was equipped with a single-stage turbocharger that limited its performance at higher altitudes. Because of its slow rate of climb and poor performance at high altitude, it could not compete with enemy fighters. However, the Soviets thought highly of the P-39s sent them through the Lend Lease program because of its ground attack capabilities. They operated 4,773 of the 9,584 *Airacobras* that were produced.[53]

North American Aviation's P-51 *Mustang*, arguably the war's finest conventionally powered fighter aircraft, was developed for the British after World War II began in Europe. Following Hitler's invasion of Poland, the British and French had intensified their efforts to purchase military aircraft in the U.S. They had settled on the P-40 which was the most advanced American pursuit aircraft at the time. However, its performance was inferior to the latest European fighters as noted previously. Air Corps officers Benjamin Kelsey and Oliver P. Echols were disturbed by the decision to increase P-40 production. They preferred to see a promising new Curtiss fighter design, the XP-46, reach production lines instead. But, Maj Gen

Arnold had refused to give them the additional four months it would take to get the XP-46 into production.[54] Kelsey recalled that "Echols made a suggestion to the British Purchasing Commission to find a manufacturer who wasn't already bogged down in high priority stuff. Curtiss-Wright and the Air Corps would make available all the data we had on the XP-46 to help them build a new fighter. This was our secret talk in the hallways to get P-46s in place of the P-40 to find some way of getting around the problem."[55]

Echols probably had the best grasp of the technical, procurement, maintenance and supply aspects of Army aviation during the conflict. He had attended Virginia Polytechnic Institute and the University of Virginia, studying aeronautical engineering, and spent most of his career in various aviation modernization assignments including stints as Chief Engineer and Assistant Chief of the Materiel Division. During World War II, he headed the Materiel Division (later Materiel Command). When Gen Marshall reorganized the Army in March 1942, Echols represented the USAAF in the critical task of coordinating all production efforts and establishing priorities for using tools, materials and manpower in the nation's aviation war production program.

Although relatively unknown outside the Army, Echols played key roles in all the major Army aircraft that were produced during World War II. Neither a West Point graduate nor holder of a significant wartime operational assignment, he retired in1946 in the rank of major general.[56] Gen Arnold paid Echols a fitting tribute in 1945 stressing that "No one knows better than you the terrific problems we have faced in the development and perfection of our equipment and in obtaining the necessary production. I know of no one who could have better carried out the Air Force's responsibilities in these fields than you have. The task could never have been done had not the Army Air Forces been represented in these matters by an officer having the full confidence of the Congress, the departments and civilian agencies of government, industry and labor....The country, the Army Air Forces and I, personally, owe you a real debt of gratitude."[57]

By the time World War II began in Europe, Sir Henry Self, who supervised the British purchasing commission's activities in the U.S., had established a strong relationship with North American Aviation which was producing training aircraft for his nation. Self approached the firm's president and general manager James H. "Dutch" Kindelberger about Echols' proposed fighter based on the Curtiss XP-46 design. Kindelberger's company specialized in training aircraft and had never designed a combat aircraft before. North American needed to construct a completely new factory and develop new processes to manufacture the proposed aircraft. Its entire future was placed at financial risk by the fighter venture.

The British attached two conditions to Kindelberger's proposal. His firm had to use the same engine as the P-40, the American Allison V-12. It had a simple, one-stage supercharger rated for low-altitude flight. Also, the company had to design and produce the first prototype in less than 120 days. Self and Kindelberger agreed to use Curtiss technical data to help develop the new warplane. In May 1940, North American signed an agreement with the Air Corps that allowed it to sell the aircraft to Britain. Echols pushed behind the scenes to include NACA's innovative new laminar wing section in the design. North American's designers were able "... to conceive of the best design possible from the newly acquired Curtiss data. In striving for the cleanest configuration possible, the engineering team led by Austrian/German engineer Edgar O. Schmued, not only incorporated the basics of the XP-46 and the laminar flow wing but departed from the rounded shapes in vogue to use innovative square cut wings and tail coupled with a low drag wing cross section."[58]

The laminar flow airfoil, developed cooperatively by North American and NACA was an outgrowth of massive aeronautical research during the 1920s and 1930s by NACA. It reduced drag at high speed by "smoothing" the turbulent airflow across the wing surfaces, thereby increasing aircraft speed and efficiency Another significant design innovation was a cooling arrangement located in the rear of the aircraft which reduced fuselage and wing drag. North American's engineers discovered that the cooling assembly could provide the "Meredith effect" in which heated air leaving the radiator generated a slight amount of jet thrust.

North American's prototype was one of the first aircraft to have its fuselage lofted[59] mathematically using conic sections which gave it smooth, low drag surfaces. To facilitate production, the aircraft's airframe was split into five main sections all of which were built entirely of aluminum.[60]

The new pursuit aircraft prototype was completed, minus an engine, in 102 days. On 26 October 1940, the prototype took its maiden flight in Southern California. After modifications, the XP-51 went into production before the end of 1940. Deliveries to the RAF began in November 1941. Meanwhile, the first XP-51 had arrived at Wright Field for performance testing during August 1941. Overcoming some skepticism about the aircraft and technical problems that had delayed production, Kelsey and other officers arranged to have the P-51 inserted into the production budget. Originally marketed as a low altitude ground support aircraft, the substitution of the superb British Merlin Rolls-Royce engine for its original power plant in October 1942 transformed the P-51 into one of the war's premier air superiority aircraft. The change was suggested by Rolls-Royce test pilot Ronald W. Harker. Adding gasoline drop tanks enabled the *Mustang* to serve as a long-range bomber escort. A new bubble canopy greatly improved pilot visibility. North American built over 15,000 P-51s before production ended in 1946.[61]

Historian Tom Crouch concludes that the P-51's reputation "... as the finest piston-engine fighter of its generation resulted entirely from good engineering." He added that "The gestation of the P-51 illuminates the complex nature of the America aeronautical community in the years immediately before the nation was swept into war. Even during an era when commercial airplanes dominated the world's airlines, military sales paid the bills (Illustration 6.3). Even in the case of export sales, American military procurement officers remained very much part of the process."[62] He also observes that the *Mustang* ".. exemplified the central role that the public–private research establishment created in the 1920s and 1930s played in the triumph of American aviation.... NACA had been a full partner in the success of the industry. Wright Field and various navy flight research facilities had had made important contributions as well...."[63]

Far less glamorous than the P-51, Army ground forces officers turned to light and relatively inexpensive civilian aircraft with reliable engines to perform observation duties for its field artillery units. Ordinarily, the fuselages of such an aircraft were built of doped canvas stretched over a welded steel-tube frame. Spruce spars and wooden ribs (aluminum ribs in Piper's planes after 1940) maintained the shape of their canvas-covered wings. Steel-tube struts, fastened to the bottom of each wing, provided additional strength. Such aircraft were powered by an innovative lightweight "opposed" engine. That meant that the engine was an internal-combustion engine with its cylinders opposite one another. While long known in Europe, this did not became a popular engine design in American aviation circles until the late 1920s. America's light aircraft industry dated from their introduction.

During August 1941, Assistant Secretary of War for Air Robert A. Lovett directed the purchase of 12 "off-the-shelf" service test stock model aircraft from the three largest light aircraft makers: Taylor, Aronca and

Illustration 6.3 XP-51 prototype (*Photo* XP-51, serial number 41-039, second aircraft built, 1941, VIRIN, 061023-F-1234P-005, courtesy of the NMUSAF, nationalmuseum.af.mil)

Piper. They were tested along with others furnished independently by Stinson, Rearwin and Interstate.

Following the large-scale Louisiana ground-air maneuvers during September 1941, the General Staff suggested that commercial light aircraft would make suitable observation aircraft. Subsequently, the Materiel Division launched negotiations with several firms to purchase 617 light aircraft. The so-called a puddle jumper conference in January 1942 established specific requirements for approximately 4,000 of those aircraft to support the ground forces in what eventually came to be known as "organic aviation" in the Army. The service's adoption of light and relatively inexpensive civilian type aircraft for its short range observation mission proved to be an effective tactical innovation during World War II despite strong opposition to purchasing them by the Air Corps and its successor organization, the USAAF.[64] The Army's air arm purchased 12,904 of those so-called liaison aircraft for the service between January 1940 and December 1945.[65]

Prompted by Hitler's invasion of Poland, the Air Corps had moved to acquire a very long-range bomber. On 15 September 1939, the Air Board recommended development of a reconnaissance-bomber with a combat radius of at least 2,000 miles. In January 1940, the Air Corps sent selected manufacturers a statement of the proposed bomber's military characteristics and asked them to submit preliminary designs and contract bids. The following April, the Air Corps requested that potential bidders include design features intended to deal with the shortcomings of American-built military aircraft identified during the latest round of European combat. Those included leak-proof gas tanks, armor, multiple gun turrets and heavier caliber machine guns.

Subsequently, a board of officers headed by Echols concluded that the designs submitted by Boeing and Lockheed were superior. Although not included in the Air Corps' specifications, Boeing's engineers decided to proceed with another innovative four-engine design because they were convinced that a two-engine plane would not assure their firm's success over its rivals. That June, Arnold and Stimson approved the board's recommendations. Boeing's engineers had drawn upon the firm's long experience producing civilian transports and the XB-15 in developing their proposed design. Boeing was awarded a contract on 6 September 1940 to build two XB-29s.

The design work that eventually resulted in the XB-29 had begun in 1934, when Boeing undertook a study for a bomber with a range of 5,000-miles, which became the XB-15. The Army and Boeing had signed a contract for the aircraft on 28 June 1934. During December 1937, Boeing delivered one to the Air Corps. Believed to be the world's largest bomber, the XB-15 had incorporated a number of features and structural details that were being used in the firm's 4-engine land and seaplanes. The process which resulted in the XB-15 had been launched during July 1933, when the Materiel Division was tasked to determine the maximum practical range for a new ultra long-range bomber with a 2,000 pound bomb load.

Although the Air Corps had originally wanted to negotiate with only one firm for the aircraft, it changed its position during the spring of 1934 and opted for two firms, Boeing and Martin, because of the project's enormous importance. It was also consistent with the War Department's policy of not encouraging monopolies, an indication that such a competition was more politically acceptable than a sole source contract to a single firm. After evaluating the two resulting proposals, the Materiel Division recommended Boeing's offering. The firm was awarded a development contract on 29 July 1935.

Responding to an Air Corps specification for a bomber with a 5,000 mile range, Boeing had designed and produced the XB-15, unofficially known as the "Super Flying Fortress." Test flights conducted during 1937 and 1938 revealed that it was too large and heavy for the engines available at that point. After only one prototype was built, the Secretary of War had canceled the XB-35 project in favor of the Douglas XB-19. The latter was a very long-range bomber that was obsolete before it was completed in May 1941 and eventually scrapped.

Meanwhile, the Army air arm's senior leadership had been enormously worried about the tremendous technical risks involved in the XB-29's development program. Boeing's proposal called for a huge aircraft with a 7,000 mile range capable of carrying a 2,000 pound bomb load at 25,000 feet with a speed of 382 mph. To drive it through the air 30% faster than the B-17, Boeing's engineers proposed using engines that only developed 83% more power than those used by the *Flying Fortress's*. Some Air Corps officers were convinced that Boeing's design could not possibly succeed. But Arnold, determined to have a super bomber that might be a war-winning weapon as well as one that made the case for a separate air force, defended the program and it forged ahead. Boeing was awarded an early cost plus fixed-fee contract for XB-29s.

The bomber's cutting-edge technological features included a pressurized fuselage designed to ease crew fatigue on lengthy missions and a computerized central fire control system. The B-29 was the first production aircraft to use the latter to remotely aim and fire its gun turrets. It was also equipped with Norden bombsights. The bomber's initial test flight didn't take place until September 1942. Problems with the engines, brakes, fire control systems and pressurized cabin delayed production. The project reached a crisis in February 1943 when a B-29 crashed after an engine caught fire, killing its entire crew. The USAAF had already taken a "$3 billion gamble" on the bomber and signed a production contract in September 1941. After a crash, the project was reorganized and the first production model B-29s starting coming off the factory assembly line during July 1943. Production and technical problems continued to delay the massive program. Finally, the first operational B-29 units departed the U.S. for China in April 1944.Boeing built a total of 3,970 *Superfortresses*. At a total cost of $3 billion for design and production, the B-29 was by far the most expensive American weapons program in World War II well exceeding the Manhattan Project's $1.9 billion price tag.[66]

A Navy Second to None

After Japan invaded China in 1937 and scrapped all treaty limitations on its fleet's size that same year, Congress reversed course and returned to the Great War-era policy to create "A Navy Second To None." In January 1938, President Roosevelt requested that Congress increase the fleet's warship tonnage by 20%. Adm William D. Leahy, the CNO, explained to the House Naval Affairs Committee, that among other things, the President wanted two new aircraft carriers and 950 new naval aircraft. Leahy's original proposal to Roosevelt had not included any additional flattops on the grounds that flying boats were more useful as scouts than carrier-based aircraft. Leahy's opinions about carrier aviation suggest a significant level of outdated thinking in the ranks of some senior naval leaders. The pioneering work of aviators with carrier planes during annual fleet exercises and the aircraft modernization programs of BuAer had rendered Leahy's views about aircraft and the primacy of battleships problematic at best.[67]

Piloted again by Representative Vinson, Congress approved the "Naval Expansion Act of 1938." Popularly known as the second Vinson Act, it authorized the sea service to build beyond treaty limits, inaugurating a period of rapid expansion and accelerated technological progress. The legislation called for the Navy to have no less than 3,000 "useful" aircraft by 1944, an increase of 47% over its previous authorization. There were no new limits on Navy procurement of types of aircraft. Although the 1938 legislation retained the existing 10% production and profit provisions, opposition to these limits was growing on Capitol Hill.

Congress easily passed the legislation and Roosevelt signed it into law on 17 May 1938. It authorized a ten-year program with a $1.1 billion price tag. Among its provisions, Congress approved the construction of two new aircraft carriers and expansion of the naval air force to 3,000 planes, nearly doubling the latter's authorized strength. Spurred by the demise of the interwar naval treaty system that had attempted to forestall a seagoing arms race, Congress increased Navy funding every year during the latter part of the decade.

Commissioned on 25 April 1940, USS *Wasp* had been designed to squeeze as many aircraft as possible into the 15,000 tons of carrier displacement allowed the U.S. under the Washington Naval Treaty after USS *Yorktown* and the USS *Enterprise* were laid down. USS *Wasp* was equipped with relatively low powered engines to save weight and space which slowed its speed. In addition, it had almost no armor and lacked torpedo protection. Although those shortcomings were recognized while ship was being built, they could not be rectified because of existing treaty limits on warship tonnage.

Congress also approved greater naval aviation appropriations throughout that period except for 1939. Air power was seen by some on Capitol Hill as a more economical method of defending the nation than building a huge surface battle fleet. In addition, isolationists wrongly regarded aviation as inherently defensive, a disingenuous idea promoted by some military aviators and industrialists, because it was politically acceptable to the American public and some members of Congress. But, those naval expansion programs failed to give the U.S. Navy superiority over the Imperial Japanese Navy which had already embarked upon its own major buildup.[68]

Like the Air Corps, BuAer continued to rely on the American aircraft industry to design and build its aircraft. The industry's biggest problem was its inadequate productive capacity for a national emergency. Costs of new naval aircraft continued to soar. There had been a 26% increase since 1936 in the average cost of new production planes and a 76% rise for experimental models during those years. The Navy, unlike the Army, bore industry's costs of developing new aircraft before they were accepted for production and was more than willing to do so to continue improving the quality of the fleet's planes.[69]

Rear Adm Towers continued to press for increased aeronautical research after taking BuAer's helm. In November 1939, his first annual report as the sea service's top aviation officer requested more money to begin "a well-considered and comprehensive experimental program" to improve the Navy's aircraft as well as "standardize materials, processes, and equipment used by BuAer and the Air Corps."[70] That month the Navy requested $34.736 million for its aviation program as part of an emergency supplemental appropriation bill. Its estimate in the bill requested funds for 518 new aircraft.[71] Writing to President Roosevelt, Jerome Hunsaker, then NACA's chairman, praised Towers ".. for being 'farsighted' in having sponsored scientific research in aerodynamics and for being constructive and helpful in urging and accelerating improved military aircraft."[72]

Towers reacted immediately to the German invasion of Poland in September 1939 going to Adm Harold R. Stark, the CNO, and asking permission to order the maximum number of Consolidated PBYs possible to strengthen the air defenses of the nation's east coast. Stark delayed, waiting for Roosevelt's response to the war in Europe. Naval historian Clark G. Reynolds stresses that, not wasting any time, Towers persuaded Consolidated's president Rueben Fleet ".. to settle on a gentlemen's agreement: Consolidated would begin work immediately on 200 PBY-5s while the navy paid to double the size of the factory. Meanwhile, Towers would call for bids; if another company got the contract, the Navy would pay Consolidated for the work already done."[73] Fleet pledged that Consolidated could sell 200 of the aircraft to the Navy for $20 million.[74]

Immediately after Roosevelt's 6 September 1939 order to begin a neutrality patrol in the Atlantic, acting Secretary of the Navy Charles Edison had approved the PBY deal. BuAer advertised bids for the aircraft but only Fleet's company responded. The contract was signed on October 25th. Historian Reynolds concludes that "... by his own initiative Towers had taken the first major step in the mobilization of the aircraft industry for a wartime navy."[75]

During January 1940, Adm Stark requested authorization to advance the Navy's warship tonnage by 25% and increase its authorized aircraft strength to 6,000. The price tag for that request was $1.3 billion.[76] After a lengthy debate, Congress reversed itself and appropriated more funds than the Navy had originally requested in its Fiscal Year 1941 budget submission. Historian Curtis A. Utz observes that "The total funds which could be obligated for new aircraft in fiscal 1941 amounted to over $146 million, a 211% increase over the funding in the 1940 appropriations act. These 1941 appropriations alone provided for the procurement of over 1,300 new aircraft.... With these funds, the BuAer could procure the large numbers of technologically advanced aircraft necessary to transform the carrier air groups."[77]

Rear Adm Towers continued to warn that the U.S. should not neglect aeronautical research and development in its rush to accelerate aircraft production. He stressed that "Important as is the necessity for accelerating production, we must not fail to realize that while we are catching up to foreign nations in productive capacity, there is a danger that lack of emphasis on development and experimentation may leave us with aircraft inferior to those of foreign countries which are able to concentrate on in this phase."[78] Towers played a key role in persuading Congress to provide major increases in naval aviation funding for Fiscal Year 1940.[79]

Reflecting a growing sense of urgency and the increasing importance of naval aviation, the 20,000-ton aircraft carrier USS *Hornet* was commissioned in October 1940, a record-breaking 25 months after its keel was laid.[80] The Navy also requested funds for a new 24,000-ton aircraft carrier in its Fiscal Year 1941 budget submission. Commissioned on 31 December 1942, USS *Essex* was the first of 20 vessels of its class that joined the fleet in World War II and served as the centerpieces of American fast carrier strike forces in the Pacific.[81] It took about three years to build one of those vessels.[82]

Responding to the inability of the U.S. Navy and the Royal Navy to put enough aircraft to sea to blunt the growing German U-boat threat, Roosevelt asked for plans in October 1940 to rapidly convert merchant vessels to small (i.e., escort) aircraft carriers. A few naval officers had advocated building such vessels as early as 1927. Towers, based on earlier discussions with the Moore-McCormack passenger steamship line, had such ship conversion plans already in hand. Two of the firm's vessels were selected. The first was fitted out by Newport News Shipbuilding with a 360-foot flight deck and commissioned as the auxiliary carrier USS *Long Island*. The second vessel became the Royal Navy's HMS *Audacity*. Work on USS *Long Island* began in March 1941 and it was commissioned on 2 June 1941. Eventually, 86 of those small carriers were built for the two navies.[83]

Writing in *Naval Aviation News*, author Scott MacDonald concludes that: "Through fulfilling a basic need of transporting large numbers of assembled aircraft to various theaters of war, the quickly conceived and executed escort carrier developed into an anti-submarine warfare weapon that defeated the German U-boat threat in the North Atlantic. They provided combat capability in the support of fleet operations in both the Atlantic and the Pacific. In short, they displayed a versatility, proved under the pressures and urgencies of a war that engulfed the world."[84]

With the international security situation continuing to deteriorate, Roosevelt became alarmed in 1941 that no new U.S. fleet aircraft carriers (i.e., *Essex*-class) were due to be completed before 1944. He proposed that some of the cruisers under construction be converted to carriers. Previous Navy studies had concluded that such vessels had too many problems to be effective. On 13 October 1941, the Navy's General Board replied that such conversions would be ineffective. Not satisfied with that response, the President directed another study. The Bureau of Ships replied on 25 October with the obvious observation that such vessels would have lesser capabilities than the fleet carriers then being built but would be available sooner. The Japanese attack on Pearl Harbor settled the issue. The *Essex*-class program was accelerated and some *Cleveland*-class light cruisers under construction were converted to aircraft carriers. During the first six months of 1942, the Navy ordered nine of those light cruisers converted to what became *Independence*-class small aircraft carriers. Equipped with approximately 30 aircraft each, the 9 vessels in that class were commissioned during 1943. While limited in capability compared to the *Essex*-class flattops, they performed yeoman service during the war.[85]

Rear Adm Towers had to deal with several other key policy issues. BuAer still relied on its established time consuming peacetime process of obtaining new aircraft.[86] To accelerate procurement, he strongly supported the use of letters of intent to contractors to start on projects while draft contracts were winding their way through the required administrative maze. BuAer's Chief fought off Air Corps efforts to dictate that all aircraft engines would be liquid-cooled. He dealt personally with giants of the aircraft industry including Fleet, Martin, Douglas and Grumman.

Towers also worked with Treasury Secretary Morgenthau to make certain BuAer did not violate the fiscal controls on aircraft procurement imposed by Congress. He resisted pressures from impatient fleet commanders to increase production of existing obsolete aircraft when vastly superior fighters were under development. Towers supported sales of military aircraft and engines to the allies as long as they did not interfere with the Navy's procurement programs. He especially valued the willingness of foreign customers to pay for plant expansion in the U.S.[87]

On 22 July 1941, Towers testified on Capitol Hill that scheduled aircraft deliveries for the first 6 months of 1941 were 20% behind the estimated 1,995 planes he had reported the previous January. He stressed that the "Failure of deliveries to materialize as scheduled is due to the inability of manufacturers to meet the high rate of production mutually established by themselves and the Navy to stimulate efforts commensurate with urgent requirements."[88] BuAer's Chief explained that it was "... not really a slow-down; it was merely an overoptimistic estimate on the part of both the Navy Department and the manufacturers. But we deliberately aimed high in order to get out of the manufacturers the maximum possible."[89]

Towers outlined several specific factors that had contributed to that situation. They were: overoptimistic initial estimates, correction of design deficiencies and incorporation of essential changes required for safe operation and military effectiveness, combat lessons learned from the European war, delayed receipt of government supplied and subcontractor-furnished component parts, the shortage of materials and processed parts (e.gs., engines, propellers and instruments), disputes between labor and management, large numbers of inexperienced workers in the expanding aircraft industry plus the dispersion of veteran civilian management and supervisory personnel caused by the rapid expansion of military production programs.

In July 1940, only nine firms were building naval planes. By the summer of 1941, that number had risen to 15. Aircraft unit costs continued to rise because the planes were getting larger and more complex. Manufactures were working to expedite production by expanding to three shifts daily. CAPT Marc A. Mitscher, Assistant BuAer Chief, reported on Capitol Hill that CPFF contracts were less expensive than straight bid awards because the latter allowed a 12% profit by law. CPFF contract profits were restricted to 7%.[90]

BuAer encouraged subcontracting. That policy saved a great deal of money that would have been spent on expanding aircraft manufacturers' existing plants to build airplanes. But, in most cases it had taken longer than anticipated to build new additions to auto body plants for aviation subcontracting work and for such firms to learn how to produce such items.[91] Towers repeatedly emphasized during a 22 July 1941 hearing on Capitol Hill that his main worry was "... the fact that most of our program, all of our program except 1,050 planes, is well down below this huge [Air Corps] bomber program."[92] That high priority program had been added to existing American aircraft production programs during May of that year.[93]

During Fiscal Year 1941, BuAer speeded up the modernization process for some aircraft by awarding production contracts on prototypes before flight testing. Towers noted that "The policy of procuring two airplanes to each promising experimental design, when economical, has been adopted in order to accelerate testing and provided insurance against accident or loss."[94] In addition, radios were installed in new aircraft by airframe makers instead of the Navy after the planes were delivered to the service.[95]

Shortly after the Japanese attack on Pearl Harbor, Towers and his BuAer staff reassessed the Navy's aircraft needs. They requested another

$500 million for 3,076 additional planes. Acting Navy Secretary James Forrestal gained congressional approval for that initiative. On December 15th Capitol Hill provided over $949 million in supplemental funding for naval aviation.[96]

To help streamline contracting procedures, the Navy had agreed in 1940 to bring top civilian legal talent with contracting experience into its materiel bureaus. Forrestal, a successful financier and the first incumbent in the new Under Secretary of the Navy post created by Congress, suggested the initiative and convinced the service's Judge Advocate General to allow those attorneys to be assigned to the Navy's bureaus instead of keeping them all under his wing. Towers supported the proposal. His BuAer became the first such organization to streamline its contracting procedures with the assistance of those individuals. Forrestal proved to be extremely effective in mobilizing industry behind the Navy's war effort. Convinced by his analysis, on 7 October 1941 Roosevelt authorized the Secretary of the Navy to negotiate contracts for the procurement of naval vessels and aircraft without advertising or competitive bidding. During the conflict, Forrestal was able to retain control of Navy procurement and logistics despite Fleet Admiral[97] Ernest J. King's determined efforts to wrest control of those critical functions from him.[98]

To accelerate deliveries of new airplanes to operational units and decrease the time naval aviators were away from their home station to a minimum, Towers had established the first Training Test and Acceptance units at San Diego and Norfolk during Fiscal Year 1941. Those organizations received new aircraft from the manufacturers, conducted thorough inspection and tests of their structures and engines, installed certain items of government-furnished material such as automatic pilot equipment and armament and calibrated radio equipment among other functions. In short, they did everything needed to make airplanes "ready-for-operation" when squadron ferry crews arrived.[99]

The major challenge facing U.S. carrier-borne air units by 1940 was the absence of new fighter aircraft comparable to those operated by potential enemies especially Japan. Then Rear Adm King, BuAer's chief, had approved the recommendations from a fighter design competition on 15 November 1935 which resulted in the development of Grumman's XF4F-1 biplane and the Brewster Aeronautical Corporation's XF2A-1 *Buffalo* monoplane. The Navy favored Brewster's monoplane design and ordered its production in 1936 but placed a backup contract with Grumman to hedge its bets. Once Grumman understood that the performance of its aircraft was inferior to Brewster's, it redesigned the XF4-1 as a mid-wing monoplane designated the XP4F-2 and equipped it with a single-stage supercharged version of Pratt & Whitney's *R-1830 "Twin Wasp"* 900 Hp. radial engine. Its construction was all metal except for control surfaces. In July 1936, BuAer's Chief approved programs for the fighters that Grumman and Brewster were developing. More powerful engines were installed in both Grumman's and Brewster's planes. A critical innovation was Leroy Grumman's folding wing (called the Sto-Wing) which enabled the Navy to embark as much as 50% more production warplanes on its aircraft carriers during the war. Although, folded wing plane designs had been developed as early as 1920, none had gone into production.

The XP4F-2 prototype became Grumman's F4F *Wildcat* that bore the brunt of aerial combat at sea for the U.S. Navy early in World War II. The first folding-wing *Wildcat* made its maiden flight on 15 April 1941 and the initial production model was delivered to the Navy six month later. The Sto-Wings added approximately $5,000 to the $26,000 cost of the original *Wildcat*. Although slower, less maneuverable and shorter-ranged than Japan's Mitsubishi*A6M Zero,* the F4F claimed a higher aerial combat victory toll because of superior tactics developed by naval aviators coupled with the *Wildcat's* rugged construction. Grumman produced 7,885 of those aircraft.

Brewster's F2A *Buffalo* was considered virtually obsolete by December 1941; it was unstable and overweight, especially compared to Japanese aircraft. Easily outperformed by Japanese fighters, the F2A suffered heavy losses and was quickly withdrawn from service.[100] The experiences learned by operating the *Wildcat* and feedback from air combat were applied to Grumman's development of its faster and more powerful F6F *Hellcat*. Those factors enabled the latter to outperform the *Zero*. Grumman had begun working on a successor to the *Wildcat* in 1938. It had signed a contract with the Navy for the prototype XF6F-1 on 30 June 1941, leading to the *Hellcat*.[101]

During early 1938, BuAer had circulated an operational requirement for another new single-seat carrier-based fighter to aircraft manufacturers. In addition to a very high top speed, BuAer specified a stalling speed not higher than 70 mph and a range of 1,000 miles. An unusual aspect of the requirement was that it allowed competitors to consider either liquid-cooled or air-cooled engines despite the established Navy policy of relying entirely on the latter. The Vought company responded with its V-166B design developed by a team led by Rex Beisel. Beisel learned his trade as an aeronautical engineer on the job at BuAer where he designed the TS-1 biplane fighter in 1921, the first Navy fighter actually built to naval specifications.

To achieve the demand for a very high performance level that matched that of land-based warplanes, Beisel's Vought team designed the smallest possible airframe to accommodate the most powerful engine then available, Pratt & Whitney's XR-2800 *Double Wasp*. A distinguishing feature of the plane was its inverted gull wing which allowed shorter and lighter landing gear to provide the necessary ground clearance for the plane's huge 14-foot diameter propeller. Vought received a contract on 30 June 1938 to build a single prototype designated the XF4U-1 by the Navy. Based on European air combat experience, what became the famed World War II *Corsair* was redesigned by adding self-sealing gas tanks and increased armament. In a departure from the all-metal aircraft common by that time, the outer wing panels were covered in fabric (Illustration 6.4).

Powered by a 2000 hp. Pratt & Whitney XR-2800-4 radial air-cooled engine, the production model aircraft conducted its first test flight on 29 May 1940. At that point, it had the biggest, most powerful engine equipped with the largest propeller of any American fighter plane. By the end of the year its top speed had reached 404 mph, faster than any other U.S. fighter then in service. The Navy ordered production of 584 of those aircraft on 30 June 1941 and deliveries began on 3 October 1942. Production aircraft were outfitted with the R-2800-8 engine and the cockpit was moved back three feet to allow more fuel to be carried in the fuselage. That change hampered the pilot's view for carrier landings. That and other features made the F4U *Corsair* difficult to operate on aircraft carriers, especially for green pilots, garnering it the dreaded nickname "Ensign Eliminator." As a result, most F4U-1s were issued to land-based Marine Corps and Navy squadrons. Because of the anticipated large need for the *Corsair*, BuAer also awarded production contracts to the Goodyear and Brewster firms. Technical fixes during the war enabled the *Corsair* to begin carrier operations in 1944.[102]

During the 1920s, the Navy had considered its torpedo-armed aircraft to be the most lethal planes to use against enemy ships. But, that role declined late in the decade. Dive bombing had become the preferred anti-ship tactic. By 1936, the sea service's only torpedo aircraft was the Great Lakes Aircraft Corporation's obsolete TG-2 underpowered biplane armed with the 1,600-pound Mark 7 torpedo. The aircraft's original version had

Illustration 6.4 F6F-3 *Hellcat* on USS *Yorktown* (*Photo* Grumman F6F-3 *Hellcat* fighter on the flight deck of USS *Yorktown* (CV-10) during Pacific operations, circa late 1943. Copyright Owner: National Archives. Catalog #: 80-G-K-16166, courtesy of the NHHC, online)

been developed before World War I. Given that situation, King had been impressed with the performance of the all-metal Douglas TBD-1 *Devastator* torpedo bomber. As noted earlier, it had originated with a contract to Douglas to develop the all-metal XTBD-1 prototype in 1934. During 1937, that aircraft became the first monoplane assigned to aircraft carriers and was armed with the Mark XIII torpedo.However, by 1940 the Navy realized that the aircraft, nicknamed "torpecker," was obsolescent and contracted for a replacement with Grumman which was designated the TBF *Avenger*. The Navy immediately withdrew the surviving TBDs from operational units after they were virtually wiped out during the Battle of Midway in June 1942.[103]

To replace the *Devastator*, BuAer had requested proposals in 1939. The bureau ordered two prototypes from Grumman in April 1940. BuAer's requirements for the new aircraft included: a top speed of 300 mph, a fully loaded range of 1,000 miles and a large internal weapons bay that could carry either a Mark XIII torpedo or a 2,000-pound bomb load. The first prototype, an all-metal low-wing monoplane was designated the XTBF-1. It made its maiden flight on 7 August 1941. The initial production model was completed in January 1942. Powered by a Wright R-2600-8, 14-cylinder double row radial engine that generated 1,700 hp. and drove a Hamilton-Standard variable pitch propeller, the TBF *Avenger* was Grumman's first torpedo bomber. With a maximum speed of 271 mph at 12,000 feet, it had a range of 1,215 statute miles. It was the first aircraft to feature the innovative new "compound angle" wing-folding mechanism created by Grumman to maximize storage space on aircraft carriers. Later *Avenger* models carried radar. The aircraft featured relatively large wings making it easier for pilots to land on pitching aircraft carrier decks and extremely rugged construction able to absorb heavy battle damage, both features Grumman trademarks. In total, 9,839 *Avengers* were built by Grumman and General Motors. Beginning with its combat debut during the Battle of Midway, the aircraft served as the Navy's standard torpedo bomber through the rest of the war.[104]

During most of World War II, the Navy's primary carrier-based scout dive bomber was the Douglas SBD *Dauntless*. It was designed by a team led by Edward H. Heinemann, a self-taught aeronautical engineer. He once quipped that his approach was simple, find the most powerful engine available and design the aircraft around it. Heinemann designed more than 20 combat planes, mostly for the Navy, during his long and distinguished career. In addition to the SDB *Dauntless*, they included the postwar A-1 *Skyraider,* A-3 *Skyhawk* and F4D *Skyray* (later redesignated the F-6 *Skyray*). As General Dynamics' Corporate Vice President of Engineering, he oversaw development of the Air Force's F-16 *Fighting Falcon* in the 1970s.

As noted earlier, the long process that led to deployment of the SBD *Dauntless* began in November 1934 when Northrop won a contract to develop its XPT-1 design. Northrop's El Segundo division was taken over by Douglas in 1937 including work on the company's evolving dive bomber design, then designated the XPT-2. So many changes had been made in the engine and aircraft's structure that it was really a new plane. It's innovative design featured perforated dive-flaps that stabilized and slowed it during very steep bombing runs. The aircraft's distinctive design also eliminated the folded wing. That feature was unusual for a World War II U.S. carrier aircraft. It was adopted to increase the plane's structural strength.

The Navy placed its first production order with Douglas for the SBD *Dauntless* on 8 April 1939. All SBD-1s were allocated to Marine Corps units. Most subsequent models were assigned to Navy squadrons, with each successive version equipped with improvements that included increased fuel capacity, illuminated gunsights, self-sealing gas tanks and armor plates to protect the two-man crew of the metal monoplane. The original production model was powered by a 1,000 hp. Wright R-1820-52 radial engine. The most produced version of the aircraft, the SBD-5, was powered by a 1,200 hp. Wright R-1820-60 power plant. Approximately 6,000 of the aircraft were built before production ceased in 1944.[105]

The Navy had continued its search for improved dive bombers during the run-up to direct U.S. military involvement in World War II. In May 1939, the sea service awarded Curtiss-Wright a contract for its XSB2C-1 scout dive bomber after a 1938 design competition that followed a similar award to Brewster for its XSB2A-1. However, serious managerial and developmental problems led the Navy to eventually scrap Brewster's program and caused long delays before Curtiss-Wright's design emerged as the SB2C *Helldiver*. The *Helldiver* was a two-place, low-wing monoplane. The sea service had developed challenging requirements for the new aircraft: it had to be able to carry a significant weight of weapons within the internally braced wing to reduce drag while incorporating specific equipment and structural features within an airframe small enough to fit two aircraft on the elevators of the new *Essex*-class aircraft carriers.

None of the SB2C's features were entirely new but its Pratt & Whitney R-2800 engine had yet to be proven on other aircraft.

The Navy ordered large-scale production on 29 November 1940 but, because of problems which required modifications, the aircraft's first production model didn't fly until June 1942. Those problems included a wingspan was eight feet wider than the *Dauntless*, with a wing area almost 25% larger to support an empty weight that was roughly 1,000 pounds heavier than the other aircraft. The *Helldiver's* tail had less area than the *Dauntless* despite needing more to control a bigger, heavier airplane. Consequently, the XSB2C-1 suffered from poor handling, directional instability and bad stall characteristics. The prototype also had structural weaknesses, while its R-2600 engine and its 3-bladed hydraulic propeller suffered their own teething problems. The Navy ordered nearly 900 internal and external changes to the design before approving it for production. Those changes, along with necessary adaptations to the production line, significantly delayed deliveries. To make matters worse, Curtiss produced the *Helldiver* at new plant in Columbus, Ohio that formally opened on 4 December 1941. That circumstance had caused its own set of delays.

The Curtiss-Wright SB2C *Helldiver* made a poor initial impression on its aircrews and maintainers, earning the pejorative nicknames "The Big-Tailed Beast" (often shortened to just "The Beast") and "Son of a Bitch, 2nd Class" (a play on the SB2C designation). Nevertheless, over 7,000 of the aircraft were built for the Navy, Marine Corps, USAAF (as the A-25A *Shrikes*) and the French Navy. The plane's problems proved too much trouble for some of its customers. The Army took delivery of about 900 A-25s before deciding it didn't need a dedicated dive bomber, and those planes never saw combat. In the Navy and Marine Corps, its wartime performance was consider poor by operational commanders in the Pacific. The *Helldiver* was the Navy's last dive bomber and the final significant combat aircraft produced by Curtiss-Wright.[106]

Flying boats remained an important component of the Navy's aviation modernization programs during the 1930s while BuAer's emphasis shifted increasingly to carrier planes. Because of funding limitations imposed during the early years of the Great Depression, the Navy relied on private firms instead of the NAF for flying boat design and production. Consolidated Aircraft's twin-engine PBY-5 *Catalina* was originally designed as a long range aircraft whose primary mission was to locate and attack enemy transport ships at sea. In World War II it did yeoman duty as a flying reconnaissance platform, most famously locating the Japanese fleet during the Battle of Midway. Many downed aircrew were also picked at sea up by those flying boats (Illustration 6.5).

During the 1930s, the Navy invested millions of dollars developing long-range flying boats for patrol work. BuAer had contracted with Consolidated, Martin and Douglas in October 1933 to build competing prototypes for a patrol flying boat. Consolidated and Douglas both delivered single prototypes of their new designs. Although the Douglas aircraft was a good design, the Navy selected Consolidated's entry because of its projected lower cost. The aircraft's development had been launched with an October 1933 contract to Consolidated. The XP3Y-1 had a parasol wing with external bracing struts, mounted on a pylon over the fuselage. Wingtip stabilizing floats were retractable in flight to form streamlined

Illustration 6.5 PBY-5 *Catalina* in flight (*Photo* Patrol bomber PBY-5 *Catalina* in flight prior to October 1938. Catalog #: NH 71941. Copyright Owner: NHHC, history.navy.mil)

wingtips. Cleaner aerodynamics gave the Model 28 better performance than earlier designs. It featured all-metal, stressed-skin with aluminum sheets except for fabric-covered ailerons and wing trailing edges. The prototype was powered by two 825 hp. Pratt & Whitney R-1830-54 Twin Wasp radial engines and could carry 2,000 pounds of bombs.

The XP3Y-1 made its maiden flight during March 1935, after which it was transferred to the Navy for service trials. BuAer requested further development to bring the aircraft into the category of *patrol bomber*. It was returned to Consolidated for modifications including installation of 900 hp. R-1830-64 engines. During October 1936, an operational unit received its first PBY *Catalina*. The Navy placed a second production order on 25 July 1936. During the next three years the aircraft was gradually improved and successive models were introduced. All told, 3,308 of the aircraft were built: 2,661 in the U.S., 620 in Canada and 27 in the Soviet Union.[107]

Jet Lag

Despite their enormous contributions to American aeronautics, there was a significant downside to American military aviation development programs during the interwar period documented by historian Holley. During August 1944, the *Luftwaffe* unveiled its ME-262 jet fighters in combat. Because of Hitler's on again, off again support for its development, the jet's operational debut has been generally dismissed as "too little, too late" to affect the war's outcome. But, would D-Day have even been possible if the allies' fighters hadn't ruled the skies over Normandy in June 1944? What would have happened to the Red Army if it had been stripped of its air support? The problem was a lack of technological insight by key allied leaders. Senior Army airmen weren't interested in the potential of jet propulsion until Maj Gen Arnold saw Frank Whittle's jet engine in Great Britain in 1941. Leading NACA and Wright Field technical experts had dismissed jet propulsion as impractical during the interwar period despite scientific developments and papers that suggested otherwise.[108]

After a meeting in Arnold's office in September 1941 following a trip to Great Britain, he decided to proceed immediately on a jet engine project. General Electric (GE) was chosen to develop the engine and Bell Aircraft was selected to build the aircraft for it. By 18 March 1942, the first American turbojet was running on a GE test stand. On 1 October

1942, the Bell XP-59A, powered by two GE Type IA turbojets, made its first test flight. During the following April, an improved turbojet known as the Type I-16 was running on a test stand. However, it was unable to give the Bell P-59A *Airacomet* an adequate speed advantage over the top existing conventionally powered fighters. Work was then launched by GE on a larger turbojet designated the Type I-40 that was later mated with Lockheed's P-80A *Shooting Star*. That aircraft was being service tested when the war ended.[109]

Historian Holley suggests that the incremental approach to aircraft and engine development fostered by the Army, despite its enormous achievements, had basically been responsible for the U.S. being surprised and outdone by the Germans in jet engine development. Focusing on the Air Corps, he observes that:

> ... the technical staff at McCook Field had been heavily committed to scientific investigation, to fundamental research, and the procurement function was largely performed in Washington. With the move to Wright field ... the situation changed drastically. The procurement staff ... who formerly had been located in the Office of [the] Air Service in Washington were moved to the newly established Materiel Division. This was done to improve coordination between those engaged in the work of contracting on the one hand and those involved in technical development on the other. While coordination was decidedly improved, the move had a number of unintended side effects. Most notably, the demands imposed by the procurement side of the house absorbed an even larger portion of the available technical manpower ... . The practical consequence of this was that purely experimental research tended to suffer.[110]

Air Corps personnel policies and organizational culture coupled with its budget priorities and congressional parsimony during the interwar period exacerbated the problem. Prior to the outbreak of war in Europe, the number of officers assigned to Wright Field never exceeded 100. As late as 1939, project officers like Kelsey had tiny staffs that consisted of one civil service engineer and a single typist. The continual rotation of Air Corps officers, most of whom lacked advanced technical degrees from top university aeronautical programs like Jimmy Doolittle, meant that continuity and corporate memory resided in the Materiel Division's civilian engineers. However, those individuals were seldom in positions of power and decision-making authority. They were also hobbled by civil service regulations that favored promotion by seniority. Consequently, many of the most talented and accomplished civilian engineers left the Air Corps for greener pastures in private industry.

The Air Corps' one-year Engineering School offered a useful introduction to technical topics but didn't begin to compare to the advanced programs offered by MIT, the University of Michigan and other top tier civilian institutions. Moreover, much of its course work had to be devoted to remedial instruction in mathematics for many of its students. On top of that, many of its graduates were assigned to duties that had nothing to do with aeronautical research and development. The immediate needs of the service came first when handing out such postings.

Moreover, assignment policies were geared to produce broad-gauged officer generalists who could occupy posts of increasing authority within the total Army. Historian Roger Miller points out that this approach had been largely successful after the tumultuous Mitchell era. For a short time early in World War II, respected airmen commanded every major overseas Army command including all of its ground and air assets thanks to Marshall. Those officers included Frank Andrews in the Caribbean and Delos Emmons at Pearl Harbor. But, the process that produced such broad gauged senior officers also shortchanged technological expertise in key Air Corps leadership positions.

Holley concludes that most senior Air Corps leaders during the interwar years, including Arnold, had a limited grasp of technological issues and were not especially receptive to increased Air Corps research cooperation with universities. Those circumstances reinforced the Army air arm's incremental approach to technological innovation. None of those senior leaders had post graduate scientific or engineering education. While the majority of them had graduated from West Point, the general engineering courses they took overlooking the Hudson River did little to prepare them for high-level planning and decision-making in aeronautical engineering.[111]

Worsening the problem, Arnold had characterized the pre-war scientific talent at Wright Field as "a mass of deadwood" according to retired Air Force Lt Gen Donald J. Putt. The latter had graduated from the Carnegie Institute of Technology with a bachelor's degree in electrical engineering in 1928. After joining the Army and earning his wings, he completed a masters degree in aeronautical engineering from the California Institute of Technology in 1938. Putt served as a test pilot and project officer for bombardment aircraft at the Materiel Division including the B-24, B-29 and B-36. He observed that Arnold's opinion was only partially true. Part of the problem was that the civilian scientific community as a whole looked down on the military as did much of its engineering counterpart that was employed outside the armed services. Putt, who commanded the Air Force's Air Research and Development Command from July 1953 to April 1954, acknowledged that World War II changed those attitudes markedly.[112]

Beyond the limited overall Air Corps aircraft modernization budgets as well as small research and development staffs as previously noted, Congress was notoriously stingy in appropriating research and development funds for the Materiel Division during most of the 1930s. Those annual appropriations never reached $3 million before Fiscal Year 1940. Aggravating that situation, at least 20% of them were allocated to overhead. Historian Holley observes that Capitol Hill "... repeatedly whittled away at what little money was available. For example, as an economy measure in the depression a 17 day furlough was imposed on all Civil Service employees. This produced a $96,000 'saving' in payroll but resulted in the loss of 114,000 man-hours not available for developmental work."[113]

A Fast Second

Following the German conquest of France in June1940 and President Roosevelt's call to produce 50,000 military aircraft annually, a reluctant Congress passed emergency procurement reform. Assistant Secretary of War Johnson had recommended a huge increase in the nation's air power program because of events in Europe. The Air Corps was to get 36,500 with the rest going to the Navy and Marine Corps, arbitrary numbers arrived at by subtracting the existing 13,500 naval aircraft program authorization from Roosevelt's annual production total. Maj Gen Arnold and others prevailed over those who advocated mass production of aircraft models currently in service like the obsolete B-18 instead of focusing on newer, more up-to-date aircraft including the B-17.

Representative Vinson continued to play a key role in strengthening America's defenses. On 14 June 1940, the day Paris fell to the Germans, the Vinson Naval Expansion Act became law. It authorized the construction of 21 combat vessels, 22 auxiliaries and 1,011 aircraft. The National Defense Expediting Act of 28 June 1940, also sponsored by Vinson, gave the Secretary of the Navy sweeping discretion to bypass competitive bidding orders in order to speed the procurement process and broaden the production base. To do so, the Secretary was authorized to use CPFF contracts as long as they didn't exceed 7% of the estimated cost. It also authorized advance payments to contractors and gave the federal government the right to furnish public facilities to private companies for defense production. The Navy's secretary was empowered to unilaterally modify a contract if warranted by national defense circumstances.

Some provisions of the legislation governed procurement by both services. Profit limits were set at 8% instead of the previous 10 and 12%. Both military services were allowed to certify to the Internal Revenue Service that the cost of any additional facilities needed by a firm to fulfill its military contracts could be written off its taxable income instead of the 5 or 10% that had been allowed under previous laws. The Second Revenue Act of 8 October 1940 suspended profit limits and authorized five-year write-offs of the costs of new production facilities.

On 1 July 1940, Roosevelt had signed the "Vinson–Walsh Act." Also known as the "Two-Ocean Navy Act," the $4 billion measure authorized construction of some 200 combatants, including 18 aircraft carriers and 20 auxiliaries. It dwarfed all previous naval construction programs authorized on Capitol Hill and made possible the Navy's tremendous wartime expansion. Some two hours after the legislation became law, the Navy let contracts for 45 ships—the largest such a single action in its history. The "War Powers Act" of 18 December 1941 granted the President even more sweeping authority. For example, it enabled him to alter contracts whenever that would help the military effort.[114]

During July 1940, the major aircraft manufacturers agreed to begin expanding their production facilities. Although they wanted to enlarge existing plants, the War Department and the Roosevelt administration pressed them to build new facilities or relocate them in America's interior because of security concerns; most of the nation's aircraft industry was located along the Atlantic and Pacific coasts. That pressure was also influenced by the administration's desire to reduce unemployment in the nation's interior and prevent blatant war profiteering by major manufacturers.[115]

The federal government ended up paying directly for most of the new aircraft plant construction instead of relying on private financing or government loans. Most commonly, it built a plant on its own dime and then leased it to a private concern. By February 1941 federal financing of new plants had reached almost ten times the sums invested by private firms. The Army and Navy divided up responsibility for financing the expansion of aircraft and engine plants so that no manufacturer would have to deal with more than one military department.[116]

Building additional aircraft industry plants alone could not meet Roosevelt's escalating aircraft production goals. Additional resources were needed. The federal government turned to the automobile industry as it had during World War I. Detroit epitomized the American practice of mass production. The Air Corps had put forward a plan during the autumn of 1938 for the automobile industry to serve as subcontractors—providing parts and subassemblies—while the aircraft makers would do all final assembly work in a national emergency.

Aircraft industry executives were reluctant to rely on Detroit for war production. They feared the resulting competition and talent drain as well as the political clout that its vastly superior financial resources could provide. However, some of the larger aircraft firms were beginning to adapt auto industry mass production methods by June 1940 including straight-line production and subcontracting of parts. Experts reported that approximately 25% of an airplane's value including engines, propellers, instruments and accessories were already being contracted out to other plants.

The automakers, benefiting in 1940 from their biggest economic boom since 1929 because of the rapid rise in motor vehicle sales, were reluctant to shift to aviation production. Some firms were eager to expand their share of aircraft engine production but that was the extent of most of their initial interest in the aviation program. The 1 February 1941 disclosure that a group of senior Army officers had met Ford Motor Company officials for two days in Detroit to make arrangements for aircraft (or aircraft parts) production at the firm's River Rouge plant signaled that the situation had changed dramatically. In August 1941, the federal government curtailed commercial automobile production to conserve critical raw materials for the military buildup. Although some motor vehicle manufacturers were already involved in aircraft and aviation engine production, Detroit didn't begin in earnest to convert its automobile plants to military production until after Pearl Harbor. The Roosevelt administration forced the automakers to convert a significant portion of their plant capacity through its control of the allocation of raw materials. Aircraft and engine plants also began operating around the clock instead of relying on single shifts.[117]

The American aircraft engine manufacturing industry had changed significantly by the eve of World War II. It was no longer dominated by Detroit as had been the case during World War I. Near the end of the 1930s, it was led by eight firms that specialized in designing and producing aircraft power plants. The Wright Aeronautical Corporation and the Pratt & Whitney Aircraft Division of the United Aircraft Corporation dominated the business. The Lycoming Division of the Aviation Manufacturing Corporation had also garnered a significant portion of engines produced for training aircraft.

On 7 March 1940, Arnold testified on Capitol Hill that Britain and France had financed major expansion of floor space at the Pratt & Whitney and Wright Aeronautical engine plants helping American engine capacity catch up with the nation's aircraft production capacity. Beyond floor space expansion, it still took firms four to six months to obtain the thousands of machine tools needed to expand their engine production, among other factors. Engine production remained a major bottleneck in the aviation expansion program. To help address that critical issue, Navy Secretary Frank Knox announced in August 1940 that his service would provide $7 million worth of machine tools to the United Aircraft Corporation to help produce a considerable portion of the 17,000 aircraft engines that its Pratt & Whitney division would be providing the armed forces during Fiscal Years 1941 and 1942. The sea service reported in early December 1940 that it had let a contract for some $14.8 million to enlarge Pratt & Whitney's engine plant. That initiative was projected to increase the plant's monthly output from 2,100 to as many as 3,000 engines by August or September 1941. Knox also explained that under a new program adopted by the military services, his department will be responsible for all Army and Navy aircraft engines produced by Pratt & Whitney. The Army would reimburse the Navy out of its own funds for the company's power plants. Under the new plan, each of the military departments would sponsor various engine manufacturers shifting agreed upon outputs from them to the other department in exchange for funds from the former.[118] Table 6.3 shows annual U.S. military aircraft engine production from 1940 through 1945.

Historian Reynolds concludes that "By early October [1940]... the big hurdles for aircraft production on a massive scale were out of the way.... Congress had appropriated the funds for American and British output. Contracts could now be let with the obstacles of excess profits and plant amortization finally removed by the passage of the excess profits law."[120] Several key metrics suggested the aviation industry's growing capacity for enormously increased production. By January 1941, American aircraft

Table 6.3 U.S. military aircraft engine production, 1940–1945[119]

Year	*Army*	*Navy*	*Total*
1940	15,346	7,321	22,667
1941	39,502	18,679	58,181
1942	103,809	34,280	138,089
1943	188,803	38,313	227,116
1944	223,861	33,050	256,911
1945	89,629	20,022	109,651
Grand total	660,950	151,665	812,615

plants had increased their factory floor space from 9,605,936 square feet the previous January to 17,943,087 square feet. Aircraft engine manufacturers had expanded plant space from 3,017,604 to 6,463,481 square feet and the size of propeller plants had grown from 3,017,604 to 6,463,401 square feet during the same 12 months. Employment in those industries had soared. During 1940, employment in the aircraft industry had grown from 65,677 to 144,583. That same year, personnel employed by engine builders had leaped from 21,257 to 42,567. Propeller makers had followed a similar trajectory with their employment rising from 2,959 to 6,743.[121]

Building the enormous air forces needed to implement a general air power strategy proved to be an enormously challenging task. But, unlike the years leading to World War I, the U.S. had built a strong aviation industry and a first-class aeronautical technology base during the interwar period, the result primarily of work and funding by the military services and the NACA in close cooperation with aircraft and engine manufacturers. Those assets were critical in helping the nation rapidly catch up and, in most respects, surpass the air power of the other major belligerents in World War II. That, coupled with the political support to mobilize for war, a vast infusion of federal money and the enormous overall industrial strength of the American economy, enabled the U.S. to become what historian Hallion calls a "fast second" in the race for air power dominance during World War II. By the time the nation declared war on the Axis powers following the attack on Pearl Harbor, the U.S. was producing more military aircraft than any other nation except Great Britain.[122]

As World War II unfolded, the performance of America aircraft equaled or surpassed those of the other powers in every major category except jet fighters. All the primary aircraft models with which the U.S. fought

World War II were either in service, being developed, under production or on the drawing boards before 7 December 1941. The advanced aeronautical base developed in the U.S. beginning in World War I coupled with the innovative work of designers and the skills of aviation managers/industrialists as well as the ability to innovate significant technological improvements in existing American aircraft during the course of the conflict made that possible.[123] Table 6.4 shows the numbers of military aircraft produced by the major powers during 1939–1945.

A similar scenario played out in the development of conventional piston aircraft engines in the U.S. The USAAF's Air Technical Service Command reported in October 1945 that the basic engines used during the conflict "... were all in operation long prior to our entry into the war and actually, those engines and other models by the same manufacturers but of different size, powered all our combat aircraft in the war. A large part of their development resulted from engine builders' initiative, export sales, commercial airline sales, as well as sponsored development by the Army and Navy."[125] Examples of that phenomenon included the B-29's Wright R-3350 engine which increased from its original 1,700 hp. in 1937 to 2,800 hp. in 1945. Pratt and Whitney's R-2800 power plant grew from 1,800 hp. in 1938 to 2,800 hp. six years later. Another example cited by the report was Allison's V-1710 which generated 1,100 hp. in 1935. The improved version delivered 1,500 hp. in 1944. Most of that lengthy and expensive development work had been accomplished by a handful of experienced aircraft engine firms: the Allison Division of General Motors, Pratt and Whitney Aircraft Corporation and the Wright Aeronautical Corporation.[126]

Table 6.4 Military aircraft production by the major powers: 1939–1945[124]

Year	*Japan*	*Germany*	*Italy*	*Great Britain*	*United States*	*U.S.S.R*
1939	4,467	8,295	1,800	7,940	2,141	10,382
1940	4,768	10,826	1,800	15,049	6,019	10,565
1941	5,088	11,776	2,400	20,094	19,433	15,735
1942	8,861	15,556	2,400	23,672	47,836	25,436
1943	16,693	25,527	1,600	26,263	85,898	34,900
1944	28,180	39,807	–	24,461	96,318	40,300
1945	11,066	7,540	–	12,070	49,761	20,900

Notes

1. Grossnick, *United States Naval Aviation, 1910–1995*, 77; Wildenberg, *All the Factors of Victory: Admiral Joseph Mason Reeves and the Origins of Carrier Airpower*, 268; Rae, *Climb to Greatness: The American Aircraft Industry, 1920–1960*, 67–71.
2. Biddle, *Barons of the Sky: From Early Flight to Strategic Warfare, The Story of the American Aerospace Industry*, 262.
3. Ibid.
4. Ibid.
5. Rae, *Climb to Greatness: The American Aircraft Industry, 1920–1960*, 78–79; Holley, *Buying Aircraft: Materiel Procurement for the Army Air Forces. Special Studies. U.S. Army in World War II*, 17–20; Oliver P. Echols, President, Air Industry Association, Subj.: "Mobilization Planning and Logistical Preparation of the Air Forces for World War II," presented to the students of the Air War College, Maxwell Field, Alabama, 21 January 1948, 10–11, AFHRA, Microfilm No. K239.716248-14.
6. Wesley Frank Craven and James Lea Cate, eds, *The Army Air Forces in World War II. Vol. 6, Men and Planes* (Chicago, IL: University of Chicago Press, 1958; Reprint, Washington, D.C: Office of Air Force History, 1983), 301.
7. "President Endorses Plan to Sell Airplanes to France," *U.S. Air Services*, February 1939, 28, Hathi Trust Digital Library; AP, "Allies Contract for 1500 Planes," *New York Times*, 19 April 1940, 8, ProQuest Historical Newspapers, Pentagon Library.
8. *Twenty-Fourth Annual Report of the National Advisory Committee for Aeronautics [Fiscal Year] 1938* (Washington, D.C.: USGPO, 1939), 1, Hathi Trust Digital Library: Arnold, *Global Mission*, 157; Millett and Maslowski *For the Common Defense: A Military History of the United States of America*, 388; Henry M. Morgenthau, Jr., "Presidential Diaries," Volume 1, January 1938 to July 1939, 53–55, FDRPL&M; Jacob A. Vander Meulen, *The Politics of Aircraft: Building an American Military Industry* (Lawrence, K.S.: University of Kansas Press, 1991), 198.
9. Memorandum for Chief of Staff from Louis Johnson, Assistant Secretary of War, Subj.: "War Requirements and War Resources,"

14 October 1938, PSF, Box 105, War Department, Louis Johnson File, FDRPL&M.

10. Arnold, *Global Mission*, 179.
11. Millett and Maslowski, *For the Common Defense: A Military History of the United States of America*, 388–389; James MacGregor Burns, *Roosevelt: The Soldier of Freedom* (New York, NY: Harcourt Brace Jovanovich, Inc., 1970), 159–167, 171–175; "Report of the Commanding General Army Air Forces to the Secretary of War," 4 January 1944, 6, Pentagon Library; "Statement of Major General Henry H. Arnold, Chief of the Air Corps Relative to the Air Corps Estimates for Fiscal Year 1941," 3 January 1940, 13, U.S. AFHRA, Microfilm #145.91–23; Daso, *Hap Arnold and the Evolution of American Airpower*, 165; Oral History, General Orval R. Cook (Ret.), interviewed by Hugh H. Ahmann and Major Richard Emmons, AFHRA, U.S. Air Force Oral History # 740, 4–5 June 1974 and 6–7 August 1974, 117–121; Holley, *Buying Aircraft: Materiel Procurement for the Army Air Forces. Special Studies. U.S. Army in World War II*, 169–170; Arnold, *Global Mission*, 172–178; Craven and Cate, *The Army Air Forces in World War II. Vol. One. Plans and Early Operations. January 1939 to August 1942*, 104; John Major, "The Navy Plans for War, 1937–1941," 243, *In Peace and War: Interpretations of American Naval History. A Second Edition. Contributions in Military History, Number 41* (Westport, CT: Greenwood Press, 1984), 243, Kenneth J. Hagan, ed., Truman R. Clark, "Flood of Victory: The American Logistics Mass in World Wars I and II," AFLC Historical Study No. 408, 1986, 69, Headquarters, AFLC, Office of History; Vander Meulen, *The Politics of Aircraft: Building an American Military Industry*, 198; *The Aircraft Year Book for 1939*, 33, AIA, aia-aerospace.org.
12. Memo, Assistant Secretary of War Johnson to President Roosevelt, Subj.: Talks With Aircraft Industry for Maximum Production, 4 November 1938, President's Secretary Files (PSF), Box 105, War Department, Louis Johnson File, FDRPL&M; Memo, Assistant Secretary of War Johnson to President Roosevelt, Subj.: New Government Financed Aircraft Plants, 4 November 1938, With Atch Memo and Plans From the Chief of the Air Corps Outlining Proposed Plans for New Government

Financed Aircraft Plants, PSF, Box 105, War Department, Louis Johnson File, FDRPL&M.

13. Millett and Maslowski, *For the Common Defense: A Military History of the United States of America*, 393–395.
14. U.S. Congress, Committee on Military Affairs, *National Defense, Hearings Before the Committee on Military Affairs on H.R. 3791.* 66th Cong., 1st sess., 17 January 1939, 5, (Washington, D.C.: USGPO, 1939), ProQuest Congressional Documents, Pentagon Library; Keith D. McFarland, *Harry H. Woodring: A Political Biography of FDR's Controversial Secretary of War* (Lawrence/ Manhattan/Wichita, KS: University Press of Kansas, 1975), passim.
15. U.S. Congress, Committee on Military Affairs, National Defense, Hearings Before the Committee on Military Affairs on H.R. 3791. 66th Cong., 1st sess., 20 January 1939, 38, Washington, D.C.: USGPO, ProQuest Congressional Documents, Pentagon Library.
16. "Award of Contracts Initiating the Air Corps Expansion Program," *Air Corps News Letter*, 1 May 1939, 11, V-8043, A.C., Information Division, Office of the Chief of the Air Corps, AFHSD, www.afhistory.af.mil/History; "Second Phase of the Program," *U.S. Air Services*, May 1939, 18, Hathi Trust Digital Library.
17. "Annual Report of the Assistant Secretary of War," 30 June 1939, 16, in the "Report of the Secretary of War to the President [Fiscal Year] 1939," Washington, D.C.: USGPO, 1939, Hathi Trust Digital Library.
18. "Results of Airplane Competitions," *Air Corps News Letter*, 1 September 1939, 13–14, V-8178, A.C., Information Division, Office of the Chief of the Air Corps, AFHSD, www.afhistory.af.mil/History.
19. "Abridged Report to the U.S. Senate Committee Investigating the National Defense Program on Comparative American and German Aircraft Development," 23 October 1945, 3 (Introduction), Records of the President's Policy Commission, Box 28, Harry S. Truman Presidential Library and Museum (HSTPL&M).
20. "Report of the Secretary of War to the President [Fiscal Year] 1941," (Washington, D.C.: USGPO, 1941), 26–29, Hathi Trust

Digital Library; Holley, *Buying Aircraft: Materiel Procurement for the Army Air Forces. Special Studies. U.S. Army in World War II,* 334–336, 474; "Wright Field Activities for Calendar Year 1940," *Air Corps News Letter*, Intelligence Division, Air Corps, V-8698, A.C., 15 January 1941, 1, SNASM Library; "Buying New Planes," *Air Corps News Letter*, Intelligence Division, Air Corps, V-8762, A.C., 15 February 1941, 16, SNASM Library; "Service Testing of Aircraft Accelerated," *Air Corps News Letter*, Intelligence Division, Air Corps, V-8666, A.C., 15 December 1940, 2, AFHSD, www.afhistory.af.mil/History; Robert H. Connery, *The Navy and the Industrial Mobilization in World War II* (Princeton, NJ: Princeton University Press, 1951), 61; Pattillo, *Pushing the Envelope: The American Aircraft Industry*, 117.

21. AP, "Heads Navy Air Bureau," *New York Times*, 2 June 1939, 8, ProQuest Historical Newspapers, Pentagon Library; U.S. Congress, House, *Navy Appropriation for [Fiscal Year] 1941, Hearings Before a Subcommittee of the House Committee on Appropriations in Charge of Navy Department Appropriations Bill for [Fiscal Year] 1941.* 76th Cong., 3rd sess., 8 January 1940, 488, ProQuest Congressional Documents, Pentagon Library; United Press (UP), "[Maj Gen Arnold] Defends Airplane Sales," *Wall Street Journal*, 29 April 1940, 1, ProQuest Historical Newspapers, Pentagon Library; U.S. Congress, *Emergency Supplemental Appropriation Bill [Fiscal Year] 1940, Hearings Before the Subcommittee of the Committee on Appropriations, House of Representatives*, 76th Cong, 3rd sess., 28 November 1939, 88, Hathi Trust Digital Library.
22. Memorandum, Secretary of War George H. Dern for the President, Subj.: U.S. Aircraft Production Capacity, 22 August 1935, PSF, Box 14, "War Department, 1933–1941" Folder, FDRPL&M; Planning Branch, Office of the Assistant Secretary of War, "A Survey of the American Aircraft Industry," 6 August 1936, 13; Craven and Cate, eds., *The Army Air Forces in World War II. Vol. 6, Men and Planes,* 350–351; Rae, *Climb to Greatness: The American Aircraft Industry, 1920–1960,* 113–114; Turnbull and Lord, *History of United States Naval Aviation,* 311.

23. Pattillo, *Pushing the Envelope: The American Aircraft Industry*, 108; Karl G. Harr, Jr. AIA president, “Embryo to Vigorous Maturity,” *Air Force/Space Digest*, September 1965, 55, AIA, aia-aerospace.org.
24. McAllister, ed., *Aerospace Facts and Figures 1964, 25*, AIA, aia-aerospace.org.
25. *The Aircraft Year Book for 1941,* 52–53, AIA, aia-aerospace.org.
26. Ltr., Maj Truman Smith, Office of the Military Attache, American Embassy, Berlin, to Col Raymond E. Lee, Military Attache, American Embassy, London, England, Subj.: German Rearmament, 7 February 1936, Truman Smith Papers, Box 2, “Germany, Military Attaché, Correspondence, 1935–1938,” USAMHI; Robert Hessen, ed., *Berlin Alert. The Memoirs and Reports of Truman Smith* (Stanford, CA.: Hoover Institution Press, 1984), xiii–xx, 83–87, 92–94; Truman Smith, “The Facts of Life. A Narrative With Documents,” unpublished autobiography, Personal Papers of Truman Smith,” USAMHI, 80–82, 84–98, 105–113; Military Attaché in Berlin, Subj.: German Air Power, “General Estimate of Nov. 1, 1937 (Extracts),” 25 October 1937, Truman Smith Papers, Box 2, “Germany, Military Attaché, Correspondence, 1935–1938,” USAMHI; Memo, Louis Johnson, Assistant Secretary of War, to President Roosevelt, Subj.: “Confidential Report on European Aircraft Production by Lawrence D. Bell,” 31 October 1938, with attached report, Bell to Chief Army Air Corps, 12 September 1938, PSF, “War Department 1934–1938,” Folder, Box 102, FDRPL&M; Memo, Col J. H. Burns, Ordnance Department Executive to Assistant Secretary of War, Subj.: “Impressions of Glenn L. Martin on European Aircraft Production,” undated, PSF, “War Department,” Folder, Box 105, Louis Johnson File, FDRPL&M; Ltr., Charles A. Lindbergh to Dr. Joseph S. Ames, NACA chairman, Subj.: Alarming European Aeronautical Developments, 28 November 1938, NASA History Office files; “12th Annual Report of the Materiel Division, Air Corps, U.S. Army, Fiscal Year 1937,” 10, Pentagon Library; *The Aircraft Year Book For 1939,* 17.
27. Weigley, *History of the United States Army*, 560; Morris, *The Origins of American Strategic Bombing Theory*, 185, 187, 191, 192–193.

28. Hessen, *Berlin Alert. The Memoirs and Reports of Truman Smith*, 94–162; Smith, "The Facts of Life. A Narrative With Documents," 93–98; *Twenty-Fifth Annual Report of the National Advisory Committee for Aeronautics [Fiscal Year] 1939* (Washington, D.C.: USGPO, 1940) 2, 4, Hathi Trust Digital Library; *Twenty-Seventh Annual Report of the National Advisory Committee for Aeronautics [Fiscal Year] 1941* (Washington, D.C.: USGPO, 1942), 1, 2, Hathi Trust Digital Library; Roland, *Model Research: The National Advisory Committee for Aeronautics, 1915–1918. Volume I*, 154–158, 160–167; George W. Gray, *Frontiers of Flight: The Story of NACA Research* (New York, NY: Alfred A. Knopf, 1945, 1948), 22–24; AP, "Our Planes are Called Interior to Germany's New Air Forces," *New York Times*, 8 April 1939, 6, ProQuest Historical Newspapers, AACPL, www.aacpl.net; Virginia P. Dawson, *Engines and Innovation: Lewis Laboratory and American Propulsion Technology* (Washington, D.C.: NASA, 1991), 2–14, Internet Archive; "Research Aircraft at NASA Glenn Research Center (1943–1958)," Anne Mills, Web Curator: Imaging Technology, NASA, last updated 17 June 2015, nasa.gov.
29. William M. Leary, "Assessing the Japanese Threat: Air Intelligence Prior to Pearl Harbor," *Aerospace Historian*, December 1987, 277.
30. Ibid, 273; Robert Jon Cox, "Short Biography of Vice Admiral Ralph Andrew Ofstie, U.S. Navy," undated, bosamar.com.
31. "Invasion of Manchuria," HSTPL&M, undated, www.trumanlibrary.gov; Leary, "Assessing The Japanese Threat: Air Intelligence Prior to Pearl Harbor," *Aerospace Historian*, December 1987, 273.
32. Leary, "Assessing the Japanese Threat: Air Intelligence Prior to Pearl Harbor," *Aerospace Historian*, December 1987, 273.
33. Arnold, *Global Mission*, 169; Roland, *Model Research: The National Advisory Committee for Aeronautics, 1915–1918. Volume I*, 147, 153–154; Craven and Cate, eds., *The Army Air Forces in World War II. Vol. 6, Men and Planes,* 480–481; Michael H. Gorn, *Vulcan's Forge: The Making of an Air Force Command for Weapons Acquisition (1950–1985). Vol. 2, Appendices* (Andrews AFB, D.C.: Office of History, Headquarters, AFSC, 1985), passim.

34. Millett and Maslowski, *For the Common Defense: A Military History of the United States of America,* 395–398.
35. Weigley, *History of the United States Army*, 580, 560; Henry Morgenthau, Jr., Secretary of the Treasury, *Presidential Diaries*, Vol. 2, 12 March 1940, 437–439, FDRPL&M; Henry Morgenthau, Jr., Secretary of the Treasury, *Presidential Diaries*, Vol. 3, 16 May to February 1941, 19 September 1940, 657, FDRPL&M; Memorandum, President Roosevelt to the Secretary of War and the Army Chief of Staff, Subj.: Coordinating Aircraft Requirements Through the Secretary of the Treasury, 24 May 1940, PSF, War Department, 1939–1941, Box 103, FDRPL&M.
36. Millett and Maslowski, *For the Common Defense: A Military History of the United States of America*, 394, 397, 398; Jean Edward Smith, *FDR* (New York, NY: Random House, 2007), 510.
37. "Historical Highlights: The Lend-Lease Act of 1941," History, Art & Archives, United States House of Representatives, undated, history.house.gov; Millett and Maslowski, *For the Common Defense: A Military History of the United States of America,* 398; "Distribution of Air Materials to the Allies: Controls, Procedures, and Policies," AAF Historical Study No. 106, Assistant Chief of Air Staff, Intelligence, Historical Division, Headquarters, USAAF, July 1944, 80, AFHRA.
38. Gross, *American Military Aviation: The Indispensable Arm*, 87.
39. Ibid, 90; Grossnick, *United States Naval Aviation, 1910–1995*, 103, 111; "USS Essex (CVS 9)," undated, NHHC, history.navy.mil; Richard Humble, *United States Fleet Carriers of World War II 'In Action,'* (Poole, Dorset, UK: Bradford Press, 1984), 36, 97; Millett and Maslowski, *For the Common Defense: A Military History of the United States of America*, 395–396.
40. Haulman, *One Hundred Years of Flight: USAF Chronology of Significant Air and Space Events, 1903–1902*, 32, 33, 34; Millett and Maslowski, *For the Common Defense: A Military History of the United States of America,* 393–394; Arnold, *Global Mission*, 145–148, 154–155; Gerald Griffin, "Army Gets New Air Corps Chief," *Baltimore Sun*, 30 September 1938, 9, ProQuest Historical Newspapers, Pentagon Library; "Centralized Control in Army Air Corps," *U.S. Air Services*, April 1939, 8, Hathi Trust Digital

Library; "Air Corps Chief Becomes Army No. 2 Man," *Washington Post*, 24 September 1939, AM7, ProQuest Historical Newspapers, Pentagon Library; Paper, Roger G, Miller, AFCHO, Headquarters, USAF, "The U.S. Army Air Corps and the Search For Autonomy, 1926–1943," 12; Emme, "The American Dimension," *Air Power and Warfare. Proceedings of the Eighth Military History Symposium. United States Air Force Academy. 18–20 October1978*, 71; Underwood, "The Army Air Corps Under Franklin D. Roosevelt: The Influence of Air Power on the Roosevelt Administration, 1933–1941," 193–199; Daso, *Hap Arnold and the Evolution of American Airpower*, 151, 161.

41. R. J. Overy, *The Air War, 1939–1945* (New York, NY: Stein and Day, 1980), 203–204; I. B. Holley, Jr., "The Management of Technological Change: Aircraft Production in the United States During World War II," *Aerospace Historian*, Fall, September 1975, 162.
42. Arnold, *Global Mission*, 245; Haywood S. Hansel, Jr., *The Air Plan That Defeated Hitler* (Atlanta, GA: Higgins-McArthur/ Longino and Porter, 1972), 61–62; Clark, "Flood, of Victory: The American Logistics Mass in World Wars I and II," AFLC Historical Study No. 408, 74–75; R. Elberton Smith, *United States Army in World War II. The War Department. The Army and Economic Mobilization* (Washington, D.C.: Office of the Chief of Military History, U.S. Army, 1959), 137; Emme, "The American Dimension," *Air Power and Warfare. Proceedings of the Eighth Military History Symposium. United States Air Force Academy. 18–20 October 1978*, 78; Futrell, *Ideas, Concepts, Doctrine: A History of Basic Thinking in the United States Air Force, 1907–1964*, 59; Joel R. Davidson, *The Unsinkable Fleet: The Politics of U.S. Naval Expansion in World War II* (Annapolis, MD: Naval Institute Press, 1996), 26.
43. Phillip S. Meilinger, "The Prescient Planners of AWDP-1," *Air Force Magazine*, 1 July 2011, www.airandspaceforces.com; Hansel, *The Air Plan That Defeated Hitler*, 5–48. 62–70, 75–76, 88, 93–94; Futrell, *Ideas, Concepts, Doctrine: A History of Basic Thinking in the United States Air Force, 1907–1964*, 59–62; Arnold, *Global Mission*, 245; Craven and Cate, *The Army Air Forces in World War II. Vol. One. Plans and Early Operations. January 1939 to August 1942*, 131–132; Emme, "The American

Dimension," *Air Power and Warfare. Proceedings of the Eighth Military History Symposium. United States Air Force Academy. 18–20 October 1978*, 78.
44. "Army Air Force Statistical Digest World War II. Second Printing," prepared by Office of Statistical Control, USAAF, December 1945, 15, 135, apps.dtic.mil/sti/citations/ADA 542518.
45. "Arnold Is Chief of the Army Air Forces," *Air Corps News Letter*, 1 July 1941, 1, Intelligence Division, U.S. Army Air Corps, AFHSD, www.afhistory.af.mil/History; Gross, *American Military Aviation: The Indispensable Arm*, 92; "Army Air Power Is Merged Under Single Command," *Chicago Daily Tribune*, 22 June 1941, 5, ProQuest Historical Newspapers, Pentagon Library.
46. Gross, *American Military Aviation: The Indispensable Arm*, 92; Davidson, *The Unsinkable Fleet: The Politics of U.S. Naval Expansion in World War II*, 26, 28.
47. Kelsey, *The Dragons Teeth? The Creation of United States Air Power for World War II*, 10.
48. Ibid, 18; Swanborough and Bowers, *United States Military Aircraft Since 1909,* 151; "Purchase of New Airplanes," *Air Corps News Letter*, Information Division, Office of the Chief of the Air Corps, V-7637, A.C., 15 February 1938, 2, AFHSD, www.afhistory.af.mil/History; Paper, Col Hugh J. Knerr, Chief of Staff, GHQ Air Force, Subj.: "Development of Large Airplanes," undated, Hugh J. Knerr Papers, Box 1, Manuscript Division, LOC; Ltr., Frank M. Andrews to Henry H. Arnold, Subj.; B-17 Purchase, 23 November 1935, Personal Correspondence, Box 1, Frank M. Andrews Papers, Manuscript Division, LOC.
49. Fact Sheet, "Seversky P-35," NMUSAF, published 20 April 2015, nationalmuseum.af.mil; Bruce Robertson, ed., *United States Army and Air Force Fighters, 1916–1961* (Fallbrook, CA.: Aero Publishers, 1961), 45, 47; Swanborough and Bowers, *United States Military Aircraft Since 1909,* 201–211, 466–467; Bilstein, *The American Aerospace Industry*, 66; Canham, *Development and Production of Fighter Aircraft for the United States Air Force. AFLC Historical Study No. 213*, 49–50; Stephen Sherman, "Alexander de Seversky: Russian Aircraft Designer & Founder of Republic Aviation," updated 16 April 2012, Acepilots.com;

George C. Larson, "Designing The P-47 Thunderbolt," *Air Force Magazine*, September 1982, 132–134, 137–138; "Contract Awarded for Pursuit Planes," *Air Corps News Letter*, Information Division, Office of the Chief of the Air Corps, V-7046, A.C., 1 July 1936, 14, AFHSD, www.afhistory.af.mil/History; Brig Gen Ross G. Hoyt, USAF (Ret.), "Metamorphosis of the Fighter," *Air Force Magazine*, October 1975, 81–82; Bernard Nalty, John F. Shiner and George M. Watson, *With Courage: The U.S. Army Air Forces in World War II* (Washington, D.C." Air Force History & Museums Program, 1994), 151; Haulman, *One Hundred Years of Flight: USAF Chronology of Significant Air and Space Events, 1903–1902*, 37; Swanborough and Bowers, *United States Military Aircraft Since 1908*, 456.

50. "Curtiss P-40 E Warhawk," NMUSAF, undated, nationalm useum.af.mil; Robertson, *United States Army and Air Force Fighters, 1916–1961*, 45–47; Swanborough and Bowers, *United States Military Aircraft Since 1909*, 201–211, 466–467; Canham, *Development and Production of Fighter Aircraft for the United States Air Force. AFLC Historical Study No. 213*, 49–50; Lindbergh, *The Wartime Journals of Charles A. Lindbergh*, 184, 188, 202–203, 215; Justin D. Murphy and Matthew A. McNiece, *Military Aircraft, 1919–1945: An Illustrated History of their Impact* (Santa Barbara, CA.: ABC-CLIO, 2009), 83; Robin Higham, *Flying American Combat Aircraft of WW II* (Manhattan, KS: Sunflower University Press, 2004), 3; Hoyt, "Metamorphosis of the Fighter," *Air Force Magazine*, October 1975, 82; Haulman, *One Hundred Years of Flight: USAF Chronology of Significant Air and Space Events, 1903–1902*, 34.
51. "Lockheed P-38L Lightening," NMUSAF, 20 April 2015, nat ionalmuseum.af.mil; Pattillo, *Pushing the Envelope: The American Aircraft Industry*, 108; Swanborough and Bowers, *United States Military Aircraft Since 1909*, 322–329; Canham, *Development and Production of Fighter Aircraft for the United States Air Force. AFLC Historical Study No. 213*, 72–73; Bernard L. Boyle, "The Search for a Long Range Escort Plane, 1919–1945," *Military Affairs*, Summer 1966, 61; Kelsey, *The Dragons Teeth? The Creation of United States Air Power for World War II*, 31, 56, 58, 113, 119; Robertson, *United States Army and Air Force Fighters,*

1916–1961, 49–51; Warren M. Bodie, *The Lockheed P-38 Lightning: The Definitive Story of Lockheed's P-38 Fighter* (Hayesville, NC: Widewing Publications, 2001), 14, 19; "Report of the Air Corps Board. Study No. 11. Development of an Interceptor Type Airplane," (Maxwell Field, AL, February 1937), 13, AFHRA; Daniel M. Tellup, "Clarence L. "Kelly" Johnson 1910–1990," *Memorial Tributes: National Academy of Engineering*, Volume 6, 1993, nae.edu.

52. Speech, Maj. Gen. O. Westover, USA, Chief of the Air Corps, Subj.: Air Corps Experimental Developments During 1937, Address Before the Society of Automotive Engineers, Detroit Meeting, 13 January 1938, 2, U-1215, A.C. 1/6/38, RG 18, Records of the Army Air Forces, Entry 211, U-Stencils, Box 14, Vol. 25, National Archives; Swanborough and Bowers, *United States Military Aircraft Since 1909*, 53–58; Robertson, *United States Army and Air Force Fighters, 1916–1961*, 48, 52, 53; Bill Norton, *U.S. Experimental & Prototype Aircraft Projects: Fighters 1939–1945* (North Branch, MN: Specialty Press, 2008), 122–125; Donald J. Norton, *Larry: A Biography of Lawrence D. Bell* (Chicago, IL: Nelson-Hall, 1981), 60, 63, 68; "The Army's New Multi-Seater Fighter," *Air Corps News Letter*, Information Division, Office of the Chief of the Air Corps, V-7421, A.C., 1 August 1937, 10, AFHSD, www.afhistory.af.mil/History; "Army's Multi-Seat Fighter Makes First Flight at Buffalo," *Air Corps News Letter*, Information Division, Office of the Chief of the Air Corps, V-7482, A.C., 15 September 1937, 6, AFHSD, www.afhistory.af.mil/History; "Army Buys Bell Fighters," *Air Corps News Letter*, Information Division, Office of the Chief of the Air Corps, V-7749, A.C., 1 June 1938, 24, AFHSD, www.afhistory.af.mil/History; *The Aircraft Yearbook for 1938*, 50–52, AIA, aia-aerospace.org; Tate, *The Army and Its Air Corps: Army Policy Toward Aviation, 1919–1941*, 162.
53. Fact Sheet, "Bell P-39Q Airacobra," NMUSAF, published 20 April 2015, nationalmuseum.af.mil; Swanborough and Bowers, *United States Military Aircraft Since 1909*, 53–58; Warren Bodie, *The Lockheed P-38 Lightning* (St. Paul, MN: Widewing Publications, 1991), 19; Ernest McDowell, *P-39 Airacobra in Action, Aircraft No. 43* (Carrollton, TX: Squadron/Signal Publications,

Inc., 1980), 10; Birch Matthews, *Cobra! Bell Aircraft Corporation 1934–1946* (Atglen, PA: Schiffer Publishing, 1996), 85, 97.

54. Jeffrey L. Ethell, "The Foaling of the Mustang," *Air Force Magazine*, June 1981, 82; Arnold, *Global Mission*, 377; Canham, *Development and Production of Fighter Aircraft for the United States Air Force. AFLC Historical Study No. 213*, 81; Robertson, *United States Army and Air Force Fighters, 1916–1961*, 62; Bernard L. Boylan, "The Search for a Long Range Escort Plane, 1919–1945," *Military Affairs*, Summer 1966, 62.
55. Ethell, "The Foaling of the Mustang," *Air Force Magazine*, June 1981, 82.
56. "Major General Oliver P. Echols," USAF, current as of 1 August 1968, af.mil; Oral History, Cook, U.S. Air Force Oral History # 740, 290–291, AFHRA; Kelsey, *The Dragons Teeth? The Creation of United States Air Power for World War II,* 142.
57. General Arnold quoted in "Major General Oliver P. Echols," USAF, current as of 1 August 1968, af.mil; Crouch, *Wings: A History of Aviation from Kites to the Space Age*, 387–388.
58. "North American P-51D Mustang," NMUSAF, published 20 April 2015, nationalmuseum.af.mil; Alan Haggerty and Roy Wood, "The P-51 Mustang: A Case Study In Defense Acquisition," *Defense Acquisition Review*, October 2020, 512–513, Defense Acquisition University, ADA529307, dtic.mil; Ethell, "The Foaling of the Mustang," *Air Force Magazine*, June 1981, 83–84; Pattillo, *Pushing the Envelope: The American Aircraft Industry*, 83.
59. Lofting was a drafting technique, sometimes using mathematical tables, where curved lines were generated for use in plans for streamlined objects such as aircraft. The lines could be drawn on wood and the wood then cut for advanced woodworking. The technique could be as simple as bending a flexible object, such as a long strip of thin wood or thin plastic so that it passed over three non-linear points, and scribing the resulting curved line; or as elaborate as plotting the line using mathematical tables.
60. Robert Jackson, *Mustang: The Operational Record* (Shrewsbury, UK: Airlife Publishing Ltd., 1992), 3, 4; Bill Yenne, *Rockwell: The Heritage of North* American (New York, NY: Crescent Books, 1989), 49; Ray Wagner, *Mustang Designer: Edgar Schmued and*

the P-51 (Herndon, VA: Smithsonian Institution Press, 2000), 61; Haggerty and Wood, "The P-51 Mustang: A Case Study in Defense Acquisition," *Defense Acquisition Review*, October 2020, 513.

61. Ethell, "The Foaling of the Mustang," *Air Force Magazine*, June 1981, 84–85; Canham, *Development and Production of Fighter Aircraft for the United States Air Force. AFLC Historical Study No. 213*, 81–83; Robertson, *United States Army and Air Force Fighters, 1916–1961*, 62–63, 65; Robert F. Door, "P-51 Mustang's Switch to Merlin Engine Made It the World-Beater of World War II," *DefenseMediaNetwork*, 26 October 2012, defensemedianetwork.com; Boylan, "The Search for a Long Range Escort Plane, 1919–1945," *Military Affairs*, Summer 1966, 62; Cook, U.S. Air Force Oral History # 740, 256–257; "North American P-51D Mustang," NMUSAF, published 20 April 2015, nationalmuseum.af.mil; "Abridged Report to the U.S. Senate Committee Investigating the National Defense Program on Comparative American and German Aircraft Development," 23 October 1945, 4 (Fighters Section), Records of the President's Air Policy Commission, Box 28, HSTPL&M.
62. "Abridged Report to the U.S. Senate Committee Investigating the National Defense Program on Comparative American and German Aircraft Development," 23 October 1945, 4 (Fighters Section), Records of the President's Air Policy Commission, Box 28, HSTPL&M.
63. Ibid.
64. Raines, *Eyes of the Artillery: The Origins of Modern U.S. Army Aviation in World War II*, 22, 28, 80, 81, 316, 317, 318, 326; Holley, *Evolution of the Liaison Type Airplane, 1917–1944*, 97, 98, 100, 101; William F. Trimble, *High Frontier: A History of Aeronautics in Pennsylvania* (Pittsburgh, PA: University of Pittsburgh Press, 1982), 218; G. Patrick Murray, "The Louisiana Maneuvers: Practice For War," *Louisiana History: The Journal of the Louisiana Historical Association*, Spring, 1972, 117–138; "Vultee L-1A Vigilant," NMUSAF, undated, nationalmuseum.af.mil; Herman S. Wolk, "Lovett," *Air & Space Forces Magazine*, 1 September 2006, airandspaceforces.com.
65. Holley, *Buying Aircraft: Materiel Procurement for the Army Air Forces. Special Studies. U.S. Army in World War II*, 551.

66. Maj Gen O. Westover, Chief of the Air Corps, Subj.: Air Corps Experimental Developments During 1937, Address Before the Society of Automotive Engineers, Detroit, Michigan Meeting, 13 January 1938, 3, U-1215, A.C. 1/6/38, RG 18, AAF Records, Entry 211, U-Stencils, Box 14, Vol. 25, National Archives; "XB-15 Airplane," *Air Corps News Letter*, Information Division, Office of the Chief of the Air Corps,V-7621, A.C., 1 February 1938, 1, AFHSD, www.afhistory.af.mil/History; Robert W. Krauskopf, "The Army and the Strategic Bomber, 1930–1939," *Military Affairs*, Summer 1958, 90, 91, 92, 93; Bill Yenne, *The Story of the Boeing Company* (Minneapolis, MN: Zenith Imprint, 2005), 43; Moy, *War Machines: Transforming Technologies in the U.S. Military, 1920–1940*, 70–72; "Super Flying Fortress At Langley Field," *Air Corps News Letter*, Information Division, Office of the Chief of the Air Corps, V-7844, A.C., 15 September 1938, 3, AFHSD, www.afhistory.af.mil/History; Self, *History of the Development and Production USAF Heavy Bombardment Aircraft, 1917–1949. AFLC Historical Study No. 195*, 40–43; Dubuque, *The Development of the Heavy Bomber, 1918–1994. USAF Historical Study No. 6*, 89–92; Mansfield, *Vision: A Saga of the Sky*, 170–178, 181–184, 203–211; Phillips Payson O'Brien, *How the War Was Won* (Cambridge, UK: Cambridge University Press, 2015), 47–48; David Willis, "Boeing B-29 and B-50 Superfortress," *International Air Power Review*, Volume 22 (Westport, CT: AIRtime Publishing, 2007), 136–169; Pattillo, *Pushing the Envelope: The American Aircraft Industry*, 109; Leo Guo, "Douglas XB-19," *Plane Encyclopedia*, 15 November 2017, plane-encyclopedia.com; "How Uncle Sam Produces Finest Wings in World [XB-19]," *Air Corps News Letter*, Intelligence Division, Air Corps, V-8762, A.C., 15 February 1941, 14, AFHSD, www.afhistory.af.mil/History; Crouch, *Wings: A History of Aviation from Kites to the Space Age*, 419–420.
67. Millett and Maslowski, *For the Common Defense: A Military History of the United States of America*, 387; "Leahy, William D: Fleet Admiral, USN (1875–1959). Chief of Naval Operations, 1937–1939; Senior Member of the World War II Joint Chiefs of Staff," NHHC, undated, history.navy.mil; Hagan, *This People's Navy: The Making of American Sea Power*, 286.

68. "Annual Report of the Secretary of the Navy for Fiscal Year 1938," (Washington, D.C.: USGPO, 1938), 10, Hathi Trust Digital Library; Utz, "Carrier Aviation Policy and Procurement in the U.S. Navy, 1936–1940," 21, 48, 49, 51–52; Grossnick, *United States Naval Aviation, 1910–1995, 93;* Tom Hone, "Naval Reconstitution, Surge, and Mobilization: Once and Future," *Naval War College Review*, Summer 1994, 73; *The Aircraft Year Book For 1939,* 69–70, 74; Love, *History of the U.S. Navy, Volume One, 1775–1941,* 608–610; Van Fleet and Armstrong, *United States Naval Aviation, 1910–1980. NAV AIR 00-80P-1,* 91; U.S. Congress, House, *Hearings Before the Committee on Military Affairs House of Representatives.* 74th Cong., 1st sess., 18 April 1935, 211, Hathi Trust Digital Library; Cook, *Carl Vinson: Patriarch of the Armed Forces*, 129–131, 137–138; "Wasp (CV-7), 1940–1942," NHHC, published 11 January 2022, his tory.navy.com; *"Ranger, Yorktown* & *Wasp* class aircraft carriers (CV 4–8). (Fiscal Years 1930, 1934, 1936 & 1939)," NHHC, 6 November 2001, history.navy.com.
69. Utz, "Carrier Aviation Policy and Procurement in the U.S. Navy, 1936–1940," 59, 60, 61, 62.
70. Reynolds, *Admiral John H. Towers the Struggle for Naval Air Supremacy*, 296.
71. U.S. Congress, *Emergency Supplemental Appropriation Bill [Fiscal Year] 1940, Hearings Before the Subcommittee of the Committee on Appropriations, House of Representatives*, 76th Cong, 3rd sess., 29 November 1939, 265, Hathi Trust Digital Library.
72. Reynolds, *Admiral John H. Towers the Struggle for Naval Air Supremacy*, 296.
73. Ibid., 298, 299.
74. Wagner, *Reuben Fleet and the Story of Consolidated Aircraft*, 213.
75. Hagan, *This People's Navy: The Making of American Sea Power*, 287; Reynolds, *Admiral John H. Towers the Struggle for Naval Air Supremacy*, 299.
76. U.S. Congress, House, *Hearing on H.R. 9848 to Authorize the Construction or Acquisition of Naval Aircraft, the Construction of Certain Public Works, and for Other Purposes, Before the House Committee on Naval Affairs.* 76th Cong., 3rd sess., 22 May 1940, 3381, 3382, ProQuest Congressional Documents, Pentagon Library; "Navy Demands 107 Ships, 6,000 Planes,"

Christian Science Monitor, 8 January 1940, 1, ProQuest Historical Newspapers, Pentagon Library.

77. U.S. Congress, House, *Hearing on H.R. 9848 to Authorize the Construction or Acquisition of Naval Aircraft, the Construction of Certain Public Works, and for Other Purposes, Before the House Committee on Naval Affairs.* 76th Cong., 3rd sess., 22 May 1940, 3394, ProQuest Congressional Documents, Pentagon Library; Utz, "Carrier Aviation Policy and Procurement in the U.S. Navy, 1936–1940," 68–74.
78. "Adm. Towers Warns on Inferior Planes," *Washington Post*, 6 December 1940, 3, ProQuest Historical Newspapers, Pentagon Library.
79. Reynolds, *Admiral John H. Towers the Struggle for Naval Air Supremacy*, 309.
80. "Annual Report of the Secretary of the Navy for Fiscal Year 1942," (Washington, D.C.:, USGPO, 1942), 28, Hathi Trust Digital Library; "Hornet (CV-8)," NHHC, published 25 September 2020, history.navy.com.
81. Utz, "Carrier Aviation Policy and Procurement in the U.S. Navy, 1936–1940," 70; "USS Essex (CV-9/CVA-9/CVS-9)," NHHC, undated, history.navy.com; Grossnick, *United States Naval Aviation, 1910–1995*, 122.
82. U.S. Congress, Senate, *Hearings Before the Committee on Naval Affairs United States Senate.* 76th Cong., 3rd sess., 24 May 1940, 34, ProQuest Congressional Documents, Pentagon Library; Reynolds, *Admiral John H. Towers the Struggle for Naval Air Supremacy*, 292–293.
83. Reynolds, *Admiral John H. Towers the Struggle for Naval Air Supremacy*, 349; Turnbull and Lord, *History of United States Naval Aviation,* 315; Grossnick, *United States Naval Aviation, 1910–1995,* 106–107; Gross, *American Military Aviation: The Indispensable Arm*, 106–107; Henry M. Dater, "Development of the Escort Aircraft Carrier," *Military Affairs*, Summer 1948, 80, 81; Van Fleet and Armstrong, *United States Naval Aviation, 1910–1980. NAV AIR 00-80P-1,* 107.
84. Scot MacDonald, "Emergence of the Escort Carriers," *Naval Aviation News*. December 1962, 53, www.history.navy.mil.
85. Norman Friedman, *U.S. Aircraft Carriers* (Annapolis, MD: United States Naval Institute, 1983), 182, 412–413; Humble,

United States Fleet Carriers of World War II 'In Action', 97; "759 Planes Urged to Equip New Navy: Air Fleet Is Required to Round Out Wilbur's Plans, Says Admiral Moffett," *New York Times*, 27 January 1928, 12, ProQuest Historical Newspapers, AACPL, www.aacpl.net; Andrew Faltum, *The Independence Light Aircraft Carriers* (Mount Pleasant, SC: Nautical & Aviation Company of America, 1 January 2002), passim.

86. "Extracts from the Annual Report of the Bureau of Aeronautics Submitted by Rear Admiral J.H. Towers, U.S. Navy, to the Secretary of the Navy, for the fiscal year 1939," Navy Press Room, release date 30 November 1939, 23–24, *Current Annual Reports, Bureau of Aeronautics*, Nimitz Library.
87. Reynolds, *Admiral John H. Towers the Struggle for Naval Air Supremacy*, 302–306; U.S. Congress, House, *Fourth Supplemental National Defense Appropriation Bill for 1941, Hearings Before the Subcommittee of the House Committee on Appropriations*. 77th Cong., 1st sess., 14 February 1941, 309, ProQuest Congressional Documents, Pentagon Library.
88. U.S. Congress, House, *Hearing on the Status of the Navy's Aircraft Procurement and Aviation Training Programs, House Committee on Naval Affairs*. 77th Cong., 1st sess., 22 July 1941, 2056, ProQuest Congressional Documents, Pentagon Library.
89. Ibid., 2064.
90. U.S. Congress, House, *Hearing on the Second Deficiency Appropriation for 1941, Hearing Before the Subcommittee of the House Committee on Appropriations*. 77th Cong., 1st sess., 6 June 1941, 280–288, ProQuest Congressional Documents, Pentagon Library; Connery, *The Navy and the Industrial Mobilization in World War II*, 299.
91. U.S. Congress, House, *Hearing on the Status of the Navy's Aircraft Procurement and Aviation Training Programs, House Committee on Naval Affairs*. 77th Cong., 1st sess., 22 July 1941, 2056–2057, 2064, 2068, ProQuest Congressional Documents, Pentagon Library.
92. Ibid., 2057, 2069, 2070.
93. Ibid., 2057.
94. "Extracts from the Annual Report of the Bureau of Aeronautics Submitted by Rear Admiral J.H. Towers, U.S. Navy, to the Secretary of the Navy, for the fiscal year 1941," Navy Press Room,

release date 14 December 1941, 23, *Current Annual Reports, Bureau of Aeronautics*, Nimitz Library.

95. Ibid., 31.
96. Reynolds, *Admiral John H. Towers the Struggle for Naval Air Supremacy*, 368.
97. President Roosevelt nominated King for that rank and the fifth star that went with it. King was sworn in to that rank on 20 December 1944.
98. Ibid., 331–332; Connery, *The Navy and the Industrial Mobilization in World War II*, 121; "James V. Forrestal: Harry S. Truman Administration," *Secretaries of Defense*, Historical Office, Office of the Secretary of Defense, undated, history.defense.gov.
99. *The Aircraft Yearbook for 1942* (New York, NY: Aeronautical Chamber of Commerce of America, Inc., 1942), 66, 67, AIA, aia-aerospace.org.
100. Utz, "Carrier Aviation Policy and Procurement in the U.S. Navy, 1936–1940," 63; Grossnick, *United States Naval Aviation, 1910–1995*, 90, 91; Swanborough and Bowers. *United States Navy Aircraft Since 1911*, 205; "F4F Wildcat (WWII)," NHHC, last posted 17 April 2014, history.navy.mil; "Brewster F2A Buffalo," *Warbird Alley*, 2016, warbirdalley.com; Van Fleet and Armstrong, *United States Naval Aviation, 1910–1980. NAV AIR 00-80P-1*, 88, 89; Thruelsen, *The Grumman Story*, 100–107, 124–125, 129; American Society of Mechanical Engineers (ASME), "Wing Folding Mechanism of the Grumman Wildcat," passim, undated, asme.org.
101. Bert Kinzey, *F6F Hellcat in Detail and Scale: Revised Edition (D&S Vol. 49)* (Carrollton, TX: Squadron/Signal Publications Inc., 1996), 6, 16; Jim Sullivan, *F6F Hellcat in Action*, (Carrollton, TX: Squadron/Signal Publications Inc., 1979), 4; Swanborough and Bowers, *United States Navy Aircraft Since 1911*, 234; Thruelsen, *The Grumman Story*, 166; Steve Ewing, *Thach Weave: The Life of Jimmie Thach* (Annapolis, MD: Naval Institute Press, 2004), 182; Steve Ewing and John B. Lundstrom, *Fateful Rendezvous: The Life of Butch O'Hare*, (Annapolis, MD: Bluejacket Books, Naval Institute Press, 2004), 155, 156.
102. "Rex Beisel Corsair Design Eng[ineer]," SNASM, undated, air andspace.si.edu; Swanborough and Bowers, *United States Navy Aircraft Since 1911*, 368, 449, 450; Barrett Tillman, *Corsair: The*

F4U in World War II and Korea (Annapolis, MD: Naval Institute Press, 1979), 5–8; "F4U Corsair," site maintained by John Pike, last modified 13 August 2012, GlobalSecurity.org.

103. Utz, "Carrier Aviation Policy and Procurement in the U.S. Navy, 1936–1940," 31–32; Thomas E. Doll, *The Douglas TBD Devastator, Aircraft in Profile Number 171* (Leatherhead, Surrey, UK: Profile Publications Ltd., 1967), 4, 32, 34; Barrett Tillman and Robert L. Lawson. *U.S. Navy Dive and Torpedo Bombers of WWII* (St. Paul, MN: MBI Publishing Company, 2001), 56, 57; Cory Graff, "Spoiler Alert: A Little Piece of Aluminum Solved the Vought Corsair's Vicious Behavior Problem," *Air & Space Smithsonian*, June/July 2021, 45, 46–47; Van Fleet and Armstrong, *United States Naval Aviation, 1910–1980. NAV AIR 00-80P-1,* 85–86; "Rex Beisel, Corsair Design Eng.[ineer]," SNASM, undated, airandspace.si.edu; Lloyd S. Jones, *U.S. Naval Fighters* (Fallbrook, CA: Aero Publishers, 1977), 14–17; Aeronautical Chamber of Commerce of America, Inc., *Aircraft Yearbook for 1935,* 86, AIA, aia-aerospace.org; "USS Ranger (CV-4)," NHHC, undated, history.navy.mil; Grossnick, *United States Naval Aviation, 1910–1995*, 81, 86.

104. Swanborough and Bowers, *United States Navy Aircraft Since 1911*, 230–233; Barrett Tillman, *TBF/TBM Avenger Units of World War 2* (Botley, UK: Osprey Publishing, 1999), 6; Stephen Sherman, "TBF/TBM Avenger: Grumman Torpedo Bomber," updated 21 January 2012, Acepilots.com; Kennedy Hickman, "World War II: Grumman TBF Avenger." *ThoughtCo,* 28 August 2020, thoughtco.com.

105. Van Fleet and Armstrong, *United States Naval Aviation, 1910–1980. NAV AIR 00-80P-1,* 86; Swanborough and Bowers, *United States Navy Aircraft Since 1911*, 183; "Douglas SBD-6 Dauntless," SNASM, undated, airandspace.si.edu; Dana T. Parker, *Building Victory: Aircraft Manufacturing in the Los Angeles Area in World War II* (Cypress, CA: Dana T. Parker Books, 2013), 25–34; Edward H. Heinemann and Rosario Rausa, *Ed Heinemann: Combat Aircraft Designer* (Annapolis, MD: Naval Institute Press, 1980), 43.

106. Grossnick, *United States Naval Aviation, 1910–1995*, 94; Van Fleet and Armstrong, *United States Naval Aviation, 1910–1980. NAV AIR 00-80P-1,* 92; Swanborough and Bowers, *United*

States Navy Aircraft Since 1911, 166–168; "Curtiss SB2C-5 Helldiver," SNASM, undated, airandspace.si.edu; Laurence M. Burke, II, "Curtiss SB2C Helldiver: The 'Big-Tailed Beast'," Aeronautics Department, SNASM, posted 17 August 2020, airandspace.si.edu; "Dive Bomber Plant Is Opened in Ohio: New Curtiss-Wright Factory Is Dedicated at Columbus," *New York Times*, 5 December 1941, 8, AACPL, www.aacpl.net.

107. Grossnick, *United States Naval Aviation, 1910–1995*, 91; Love, *History of the U.S. Navy, Volume One, 1775–1941,* 591; Len Cacutt, ed., "PBY Catalina: Ocean Patroller." *Great Aircraft of the World* (London, UK: Marshall Cavendish, 1989), 187–194; "Catalina Flying Memorial," *Catalina Flying*, undated, catalinaflying.org.au; Roscoe Creed, *PBY: The Catalina Flying Boat* (Annapolis, MD: US Naval Institute Press, 1986), 48; David Legg, *Consolidated PBY Catalina: The Peacetime Record* (Annapolis, MD: US Naval Institute Press, 2002), 285; Len Calcutt, ed., "PBY Catalina: Ocean Patroller." *Great Aircraft of the World* (London, UK: Marshall Cavendish, 1989), 187–194; Johnson, *United States Naval Aviation, 1919–1941,* 137; "PBY-5 Catalina," National Naval Aviation Museum, NHHC, undated, history.navy.mil; "Pilot's Handbook U.S. Navy PBY-5 Airplane," released by the U.S. Bureau of Aeronautics, Navy Department, 2 December 1940, 4, Consolidated Aircraft Corporation, Hathi Trust Digital Library.

108. Bernard C. Nalty, "The Defeat of Italy and Germany," *Winged Shield, Winged Sword: A History of the United States Air Force, Volume I, 1907–1950,* 312–313; Paper, I. B. Holley, Jr., "Jet Lag and the Army Air Corps," Department of History, Duke University, Durham, N.C., June 1984, 1–16, AFCHO, Headquarters, USAF; Crouch, *Wings: A History of Aviation from Kites to the Space Age*, 451; "ME-262," National Naval Aviation Museum, NHHC, undated, history.navy.mil.

109. Crouch, *Wings: A History of Aviation from Kites to the Space Age*, 451–452; "Abridged Report to the U.S. Senate Committee Investigating the National Defense Program on Comparative American and German Aircraft Development," 23 October 1945, 5 (Propulsion section), Records of the President's Air Policy Commission, Box 28, HSTPL&M; "Bell XP-59A Airacomet," undated, SNASM, airandspace.si.edu.

110. Holley, "Jet Lag and the Army Air Corps,"17.
111. Ibid., 19, 20–21, 22, 30–31; Paper, Roger Miller, AFCHO, Headquarters, USAF, "The U.S. Army Air Corps and the Search for Autonomy," 1926–1943, ca 10 July 1997, 13–14.
112. Oral History, Lt. Gen. Donald L. Putt, USAF (Ret.), interviewed by James Hasdorff, AFHRA, at Atherton, California, 1–3 April 1974, 11, 12, 15, 22–23, USAF Oral History Interview #724; "Lieutenant General Donald L. Putt," prepared by the USAF, up-to-date as of March 1957, af.mil.
113. Paper, Holley, "Jet Lag and the Army Air Corps," June 1984, 19, 20.
114. Holley, *Buying Aircraft: Materiel Procurement for the Army Air Forces. Special Studies. U.S. Army in World War II,* 230, 274–285; Danhoff, *Government Contracting and Technological Change*, 31–32; Vander Meulen, *The Politics of Aircraft: Building an American Military Industry*, 202; Memorandum for the President, Assistant Secretary of War Louis Johnson, Subj.: "Program for Creation of Air Power," 15 May 1940, 1–4, PSF, Box 105, Folder: "War Department, Louis Johnson," FDRPL&M; Lecture, Echols, "[Industrial] Mobilization Planning and Logistical Preparation of the Air Forces for World War II," 9, 10, 11; Cook, *Carl Vinson: Patriarch of the Armed Forces*, 148, 152, 153; "Naval Expansion Act, 19 July 1940," published 4 December 2017, NHHC, history.navy.mil; "Two-Ocean Navy Bill Becomes Law, 19 July 1940," *Naval History Blog*, U.S. Naval Institute, published 19 July 2010, blog.usni.org; "Report of the Secretary of War to the President [Fiscal Year] 1940," (Washington, D.C.: USGPO, 1940), 2–3, Hathi Trust Digital Library; Pattillo, *Pushing the Envelope: The American Aircraft Industry*, 120.
115. Craven and Cate, *The Army Air Forces in World War II. Vol. 6, Men and Planes*, 307–308; Reynolds, *Admiral John H. Towers the Struggle for Naval Air Supremacy*, 308; "Roosevelt Says RFC [Reconstruction Finance Corporation] Will Help Business Finance Expansions," *Wall Street Journal*, 18 May 1940, 1, ProQuest Historical Newspapers, Pentagon Library; Reynolds, *Admiral John H. Towers the Struggle for Naval Air Supremacy*, 334.

116. Craven and Cate, *The Army Air Forces in World War II. Vol. 6, Men and Planes,* 308; U.S. Congress, House, *Fourth Supplemental National Defense Appropriation Bill for 1941, Hearings Before the Subcommittee of the House Committee on Appropriations.* 77th Cong., 1st sess., 14 February 1941, 312–313, ProQuest Congressional Documents, Pentagon Library.
117. Craven and Cate, *The Army Air Forces in World War II. Vol. 6, Men and Planes,* 319–325; Hamilton Owens, "Auto Men See Snags for New Air Program," *Baltimore Sun*, 20 May 1940, 1, ProQuest Historical Newspapers, Pentagon Library; Russell B. Porter, "Aviation Men Turn to Auto Methods," *New York Times*, 3 June 1940, 5, ProQuest Historical Newspapers, Pentagon Library; "Army Men Confer With Ford Officials," *New York Times*, 2 February 1941, 33, ProQuest Historical Newspapers, AACPL, www.aacpl.net; Gross, *American Military Aviation: The Indispensable Arm*, 97, 98.
118. U.S. Congress, House, *Military Establishment Appropriation Bill for [Fiscal Year] 1941, Hearings Before the Subcommittee on Appropriations of the Committee on Appropriations, House of Representatives*, 66th Cong., 3rd sess., 7 March 1940, 480, Washington, D.C.: USGPO, 1940, Hathi Trust Digital Library; "Navy to Provide Tools for Making Plane Engines," *Christian Science Monitor*, 29 August 1940, 7, ProQuest Historical Newspapers, Pentagon Library; "Navy Moves to Buy 17,000 Air Engines," *New York Times*, 30 August 1940, 11, ProQuest Historical Newspapers, AACPL, www.aacpl.net.
119. U.S. Congress, House, *Military Establishment Appropriation Bill for [Fiscal Year] 1941, Hearings Before the Subcommittee on Appropriations of the Committee on Appropriations, House of Representatives*, 66th Cong., 3rd sess., 7 March 1940, 549, Washington, D.C.: USGPO, 1940, Hathi Trust Digital Library.
120. Reynolds, *Admiral John H. Towers the Struggle for Naval Air Supremacy*, 334.
121. *The Aircraft Yearbook for 1941* (New York, NY: Aeronautical Chamber of Commerce of America, Inc., 1941), 22, AIA, aia-aerospace.org.
122. Gross, *American Military Aviation: The Indispensable Arm*, 94, 95, 98.
123. Ibid.

124. Holley, *Buying Aircraft: Materiel Procurement for the Army Air Forces. Special Studies. U.S. Army in World War II*, 555; Overy, *The Air War, 1939–1945*, 150.
125. "Abridged Report to the U.S. Senate Committee Investigating the National Defense Program on Comparative American and German Aircraft Development," 23 October 1945, 2 (Propulsion section), Records of the President's Air Policy Commission, Box 28, HSTPL&M.
126. Ibid., 1–2, 3.

CHAPTER 7

Epilogue

The success of American air power during World War II was, to a significant degree, the product of the innovative research, development and procurement programs conducted by the nation's armed forces and the NACA in conjunction with industry during the interwar period. Perceptive American airmen like the Navy's John Towers recognized the critical role that the continuous improvement of American aircraft during the interwar years had played.[1] The global superiority of American air power which that process played a critical role in achieving continues to the present day.

American aviation achieved a global preeminence during World War II which it has arguably retained to the present day. The military air arms, working with NACA and a few select firms in the aircraft and engine industries actively promoted incremental technological innovation in aviation. They were aided immensely in that effort by foreign technologists and technologies. Their sometimes hit or miss efforts prior to 7 December 1941 created a world class technological/industrial base in aviation which, along massive federal spending and strong political support, enable the U.S. and its Allies to dominate the skies during World War II and beyond.[2]

C. J. Gross, *Technological Innovation and the Rise of Aviation, 1903–1941*, Palgrave Studies in the History of Science and Technology,
https://doi.org/10.1007/978-3-031-71108-4_7

Positive results of that process were never inevitable. Some aircraft projects, like the B-18 *Bolo* heavy bomber and the autogiro, were disappointments. Lighter-than-air vehicles proved to be very expensive mixed bags in the long run. While dirigibles performed valuable service during World War II, rigid airships like USS *Shenandoah* proved to be costly technological dead ends.

Scholars have produced many excellent in-depth studies of various aspects of the development of American military aviation up to and including World War II. This history covered a much broader horizon. It concentrated on the critical role that the U.S. armed forces played in developing aviation technology and producing high performing military aircraft from the Wright brothers to the Japanese attack on Pearl Harbor.

The rise of American air power, based largely on the development of innovative technologies and the work of determined advocates within the U.S. armed forces, the aviation industry, NACA and their political allies, was long in the making. Following the Wright brothers' initial triumph at Kill Devil Hill, a small cadre of Army and Navy officers began exploring in limited ways the potential of military aviation. Despite the skepticism of many of their seniors, they began developing relationships with civilian constituencies that would play important roles in the progress of American aviation. Officers including George Owen Squier and Washington Irving Chambers maintained close ties with scientists, engineers and inventors who formed what one historian labelled the "invisible establishment" of aeronautics. They encouraged aeronautical progress through a national program of scientific research and development and were instrumental in establishing the National Advisory Committee for Aeronautics in 1915. Subsequently, those ties matured into an extensive network of formal and informal relationships between the armed forces, the aircraft industry professional scientists and engineers which became the basis of American technological progress in aeronautics.

Prior to U.S. entry into World War I, technical progress in American aviation had proceeded very slowly. Funds remained extremely limited. Despite the political instability caused by the Mexican revolution next door and World War I overseas in Europe, the U.S. remained at peace with the world's other great powers. The American military developed neither adequate official doctrine nor the organizations needed to systematically foster the growth of aeronautics. Spurred on by the crucible of aerial combat, European aeronautical progress had left the U.S. far behind. The American aircraft industry remained a small, poorly financed handicraft operation prior to the U.S. declaration of war.

And yet, the nation embraced the incredible notion that the fastest way that the U.S. could contribute to the Allied cause and possibly win the war by itself was to build, almost from scratch, an enormous aviation program. Yankee technological ingenuity and industrial prowess would carry the day. But, in the short run, the U.S. aviation program during World War I was seen by its critics as a failure because it failed to produce clouds of American designed and produced combat planes that won the war. Its reputation had been mired in allegations of craft, corruption and mismanagement. American aviators fought with European aircraft with the exception of seaplanes and their Allied variants of seaplanes originally designed by Glenn Curtiss. The other major American technological innovation during that conflict was the "Liberty" engine. Based on a German engine design, its plans were crafted in short order by a pair of civilian experts from Detroit with years of experience building automobile engines. Despite its shortcomings, the conflict transformed American aviation into a large mass production industry based on professional science and engineering. Simultaneously, it created a nucleus of dedicated aviators in the armed services who enjoyed more support from some of their seniors than is generally recognized.

America's massive World War I aviation program had also suggested the multiple roles aviation could play in conjunction with conventional ground and naval surface forces plus the importance of technological superiority in air-to-air combat. Those lessons were not lost on a handful of U.S. military officers, including some senior non-flyers. Along with their civilian allies, they worked tirelessly to develop aviation as a vital combat arm in the armed services and commercial aviation while vigorously promoting technological innovation and a strong aviation industrial base. Relatively unknown military insiders including Mason Patrick and William A. Moffett, Sr. proved to be far more effective in promoting the development of military aviation in the U.S. during the interwar period than the controversial William "Billy" Mitchell.

Federal spending on aviation, primarily through the military services and NACA, was the main source of aeronautical innovation during the interwar period. The military air arms followed a deliberate policy of spending their development and procurement funds on a relatively small number of aircraft and engine firms which they believed had the capacity to design and produce increasingly capable aircraft.

During the 1920s, the Navy struggled to decide the best ways to take its fledgling air arm to sea with the fleet. After World War I, there were intense arguments between the advocates of seaplanes, floatplanes mounted on battleships and aircraft carriers regarding the best course to chart for naval aviation future. The Navy explored all of them. The sea service also strengthened aviation's role by establishing Bureau of Aeronautics (BuAer), placing the wartime Naval Aircraft Factory (NAF) on a permanent basis while commissioning its first aircraft carrier, the experimental USS *Langley*. It followed that up by building the twin giants USS *Lexington* and USS *Saratoga* on converted battlecruiser hulls, the world's best aircraft carriers at the time.

The postwar Army's Air Service (later Air Corps) initially focused on developing better pursuit and observation aircraft. But, gradually it shifted its emphasis to developing long-range bombers because of a combination of technological advances and the strategic bombardment doctrine emerging from the Air Corps Tactical School. However, the Air Service couldn't ignore developing aircraft that directly supported the ground forces while it remained under the control of the service's General Staff. Like their Navy counterparts, the Army's aviation arm gobbled up a disproportionate share of their service's extremely limited postwar resources.

The Army air arm's Engineering Division (later renamed the Materiel Division) and the BuAer developed policies and procedures that shaped the way the military services conducted their aviation modernization programs during the interwar years. Key legislation in the mid-1920s put American military and civil aviation on a sound legal footing while allocating significant funding increases (although inadequate according to aviation advocates) for aircraft modernization by the armed services.

Contrary to the opinions of some pioneer military flyers, the U.S. armed forces did not ignore or dramatically underfund aviation after World War I. The insatiable appetite for resources of growing military aviation programs during the early interwar period often required funding and personnel cuts to other parts of the armed forces. Those circumstances produced fierce arguments over the place of aviation within the U.S. military. Aiding aviation during that process, a number of senior non-flying Army and Navy officers were strong supporters of aviation. Many of those non-aviators viewed aviation as a valuable adjunct to established conventional ground and sea forces, not independent offensive arms of the military services. Prior to the early 1930s that was arguably a realistic viewpoint given the limited technological capabilities of current American and foreign aircraft. However, that situation began to change radically during that decade with the rise of aggressive totalitarian regimes in Germany, Italy and Japan.

Beginning in the 1920s, the U.S. fielded some of the most advanced military aircraft in the world as indicated by the numerous international performance records they held. Soon after 1918, senior U.S. military aviation leaders had decided to leave the development and production of aviation materiel (with the minor exception of a few NAF projects) to private firms. In addition to pressure from the struggling aircraft and engine industries which had barely survived after the World War I demobilization, they concluded that it would be a more cost effective method of promoting rapid aeronautical innovation. The innovative process they fostered was largely driven by technological opportunity rather than imminent threats to the nation's defense until the early or mid-1930s.

Military aviators were determined to develop aircraft that flew faster, higher and further while carrying larger armament loads in order to keep ahead of foreign rivals. Often sidestepping congressional opposition and established federal procurement laws, the services primarily relied on negotiated agreements with a few established firms that had strong records of developing the innovative aircraft and engines they promised. Production runs of successful designs were small because of funding limitations imposed by economy-minded members of Congress as the nation returned to "normalcy" after the Great War. Design contracts frequently lost money. Winning firms expected to regain their development cost overruns in those production runs.

Separate open contracts for development and production of military aircraft, mandated by federal procurement law, were largely unsuccessful. Industry and the military air arms viewed them as unfair, unproductive and inhibiting technological progress. Instead, military aviation programs relied on loopholes in federal procurement laws that enabled them to classify small production runs as experimental contracts which they were allowed to negotiate. Senior Navy and War Department civilian officials and their legal staffs usually supported that approach.

When major foreign powers began huge rearmament programs during the early 1930s, American military and naval aircraft began to lose the technological superiority that they had demonstrated earlier. Public and congressional opinion, disillusioned by the results of World War I, coupled with economy-minded presidential administrations prevented the U.S. from responding early on to the deteriorating international situation despite the increasingly alarming flow of intelligence from abroad.

The military cooperated with the NACA and industry during the interwar period to develop progressively more capable and reliable aircraft, engines and instruments in what later became known as the "airframe revolution." As a result, it replaced relatively fragile box-like, wood, wire and cloth contraptions with sleek modern streamlined aircraft equipped much more powerful and reliable conventional engines. This revolution was deeply intertwined with the evolution of modern commercial pre-jet airliners which simultaneously benefitted from and aided the technological progress of military aircraft. Outstanding individual technical innovations during that period fostered by the military included air-cooled engines, greater engine power, night and all-weather (instrument) flying, all-metal aircraft with full cantilever wings and stressed-skin monocoque fuselages, variable pitch propellers, Norden bombsights, radios, variable pitch propellers and high octane aviation gasoline.

American aviation modernization programs were heavily influenced by foreign innovations including all-metal monoplanes, variable pitch propellers, stressed-skin metal fuselages and aircraft carriers. They also benefitted enormously from the innovative talents of some highly skilled immigrants or their children. Those included Edgar O. Schmued, an Austrian/German-born engineer who led the team at North American Aviation which designed the famed P-51 *Mustang* and top notch scientists like German Max M. Munk. Historian Roger E. Bilstein writes that "The European inheritance, generously acknowledged in other areas of the history of technology, has not been similarly acknowledged in the context of American aerospace history. Having absorbed European contributions, American entrepreneurs often adapted and modified them into far more successful applications."[3]

Less well known, but equally important, was the enormous progress of conventional engine technology from World War I to the eve of direct American participation in World War II. That development was primarily stimulated in the U.S. by the military needs and funding.[4] The development of high octane aviation gasoline by the petroleum industry and the U.S. armed services also played in a key role in that process. Some of that 100 octane fuel was shared with the Royal Air Force during the Battle of Britain and arguably helped tip the scales in defeating German fighters and bombers.[5]

Representative Carl Vinson's legislative leadership during the 1930s and President Franklin D. Roosevelt's response to the 1938 Munich Crisis began the process of restoring the nation's aviation strength. Unlike World War I, this time the U.S. had significantly more time to strengthen its air power as the international security situation deteriorated in the late 1930s. As a result, every major model of American combat aircraft during World War II was either in service with operational units, being tested or on the drawing boards before the Japanese attacked Pearl Harbor on 7 December 1941. During World War II, U.S. military aircraft were able to match and in many cases outstrip the performance of the warplanes of the other major powers except German jets while far surpassing them in the overall numbers produced. The major shortcoming of the incremental American approach to aeronautical innovation was its neglect of a revolutionary new technology, jet propulsion. Nevertheless, American land and sea-based air forces, combined with those of its allies, made major contributions to the defeat the Axis powers during the war.

Enabled by their massive aircraft production programs, the U.S. and Great Britain were able to mount a general air power strategy. Unlike their Axis foes whose air forces concentrated on direct support of ground and naval forces in comparatively limited operational theaters, that strategy enabled the Allies to conduct massive tactical, strategic bombing reconnaissance and airlift campaigns around the globe. America's military aviation programs, whose technological progress had been largely financed by the Federal government during the interwar period, achieved a global preeminence which it has arguably retained to the present day.

Although most attention has focused on the key roles played by the nation's political support and massive industrial mobilization during World War II, the technological-industrial base nourished by spending on military aviation during the interwar period was a key factor in the ascendancy of American aviation during and after that global conflict. When President Roosevelt asked Congress for a declaration of war against Japan the U.S. was already producing more military aircraft than any other nation except Great Britain. Before that conflict ended in 1945, America had out-produced all of the other major powers. Over the course of that conflict, its warplanes became heavier, more sophisticated and increasingly effective (Illustration 7.1).

American aviation, built on its innovative pre-war technological foundations, played an indispensable role in America's global World War II

Illustration 7.1 B-24 assembly line (*Photo* Consolidated Aircraft Corporation's B-24 assembly line at Fort Worth, Texas illustrates the massive production of high quality combat aircraft achieved in the U.S. for World War II. U.S. Air Force photo, undated, NMUSAF, VIRIN: 060530-F-1234S-031.JPG, undated, www.nationalmuseum.af.mil)

military campaigns and those of its allies. The legacy of America's innovative interwar aeronautical research and development programs lived on after World War II.

Notes

1. Craven and Cates, *The Army Air Forces in World War II. Volume 6. Men and Planes*, 228; "Adm. Towers Warns on Inferior Planes," *Washington Post*, 6 December 1940, 3, ProQuest Historical Newspapers, Pentagon Library.
2. Ibid., 1.
3. Bilstein quoted in Gross, *American Military Aviation: The Indispensable Arm*, 297.

4. Ibid., 230–232.
5. Ibid., 113.

Index

Note: The page numbers followed by 'p' and 't' represents photos and tables.

C. J. Gross, *Technological Innovation and the Rise of Aviation, 1903–1941*, Palgrave Studies in the History of Science and Technology,
https://doi.org/10.1007/978-3-031-71108-4

The manufacturer's authorised representative in the EU is Springer Nature Customer Service Centre GmbH, Europaplatz 3, 69115 Heidelberg, Germany. If you have any concerns regarding our products, please contact ProductSafety@springernature.com

Printed and bound by CPI Group (UK) Ltd, Croydon, CR0 4YY
01/10/2026
02225967-0014